D0843233

Man-Made and Natural Radioactivity in Environmental Pollution and Radiochronology

ENVIRONMENTAL POLLUTION

VOLUME 7

Editors

Brian J. Alloway, *Department of Soil Science, The University of Reading, U.K.*
Jack T. Trevors, *Department of Environmental Biology, University of Guelph, Ontario, Canada*

Editorial Board

Man-Made and Natural Radioactivity in Environmental Pollution and Radiochronology

Edited by

Richard Tykva

Academy of Sciences of the Czech Republic,
Prague, Czech Republic

and

Dieter Berg

GSF – National Research Centre,
Munich/Neuherberg, Germany

KLUWER ACADEMIC PUBLISHERS

DORDRECHT / BOSTON / LONDON

A C.I.P. Catalogue record for this book is available from the Library of Congress.

ISBN 1-4020-1860-6

Published by Kluwer Academic Publishers,
P.O. Box 17, 3300 AA Dordrecht, The Netherlands.

Sold and distributed in North, Central and South America
by Kluwer Academic Publishers,
101 Philip Drive, Norwell, MA 02061, U.S.A.

In all other countries, sold and distributed
by Kluwer Academic Publishers,
P.O. Box 322, 3300 AH Dordrecht, The Netherlands.

Cover image:
Locations of nuclear explosions:

ALA: Alamogordo	**CHR**: Christmas Islands	**EMU**: Emu & Maralinga
JOH: Johnston Islands	**LN**: Lop Nur	**MAL**: Malden Islands
MAR: Marshall Islands	**MOR**: Mururoa & Fangataufa	**NEV**: Nevada
NZ: Novaya Zemlya	**REG**: Reganne	**SEM**: Semipalatinsk,
TRI: Trimouille Islands (Monte Bello)	**TS**: Test site	

F: France, *GB*: United Kingdom, *IND*: India, *PAK*: Pakistan, *USA*: United States of America,
USSR: Union of Soviet Socialist Republics
(Place names are noted as they were when explosions occurred.)

Printed on acid-free paper

Printed in the Netherlands.

Contents

Contributors   ix
Preface   xi

Chapter 1
BASIC TERMS OF RADIOACTIVITY   1
    Richard Tykva
1.   Development of our knowledge   1
2.   Nuclides   2
3.   Characterization of radionuclides   2
4.   Activity expressions   6
5.   Doses   7

Chapter 2
RADIONUCLIDES IN THE ENVIRONMENT   13
    Dieter Berg
1.   Naturally occurring radionuclides in the environment   14
  1.1. Primordial natural radionuclides   14
  1.2. Secondarily occurring natural radionuclides   15
  1.3. Naturally generated radionuclides   15
2.   Artificially produced radionuclides   19
  2.1. History of new artificial atomic nuclei   19
  2.2. Generation by means of a particle accelerator   19
  2.3. Production in the nuclear reactor   20
  2.4. Generation from nuclear weapon tests   24
3.   Behaviour of radioactive substances in the environment   25
  3.1. Behaviour in different ecological systems   25
  3.2. Transfer of radionuclides into food-chains   40
  3.3. Ecological models   49
  3.4. Behaviour of selected radionuclides in the environment   50

Chapter 3
RADIONUCLIDES RELEASED INTO THE ENVIRONMENT   71
    Dieter Berg
1.   Contamination of the environment by naturally occurring radionuclides 71
  1.1. Contamination without human influence   71
  1.2. Contamination by human influences excluding nuclear techniques 72
  1.3. Pollution after the introduction of nuclear techniques   77
2.   Contamination by artificially produced radionuclides   82
  2.1. Emissions during production and tests of nuclear weapons 82
  2.2. Emissions by nuclear installations at standard operation 103
  2.3. Pollution of the environment from accidents   110
  2.4. Radioactive waste   138
  2.5. Contamination by particle accelerators   142
  2.6. Contamination by medical and industrial applications of radionuclides 143
  2.7. Danger of contamination by criminal dealing in radioactive material 145

Chapter 4
**APPLICATION OF ENVIRONMENTAL RADIONUCLIDES IN
RADIOCHRONOLOGY**   147
Jan Košler, Jan Šilar, Emil Jelínek
1. General concept of the radiochronology   148
    Jan Košler
2. Radiocarbon   150
    Jan Šilar
    2.1. Origin of radiocarbon   150
    2.2. Natural radiocarbon cycle   151
    2.3. Stable isotope of Carbon ^{13}C and its behaviour   153
    2.4. Open and closed systems and dynamic equilibrium of ^{14}C   157
    2.5. The concept of the radiocarbon dating method   157
    2.6. The radiocarbon age   158
    2.7. Calculation of the radiocarbon age   161
    2.8. Precision of the radiocarbon age   162
    2.9. Limits of radiocarbon dating   162
    2.10. Basic conditions affecting radiocarbon dating   163
    2.11. Sample collection, pre-treatment and storing   164
    2.12. Factors affecting radiocarbon dating   174
    2.13. Calibration of radiocarbon age   176
    2.14. Reporting of radiocarbon ages   178
3. Tritium   179
    Jan Šilar
    3.1. Tritium in the atmosphere and hydrosphere   179
    3.2. Tritium as an environmental tracer   181
    3.3. Sampling of water for tritium analyses   181
4. Radiocarbon and tritium dating in science and technology   182
    Jan Šilar
    4.1. Different concepts of chronology: relative and absolute dating   182
    4.2. Radiocarbon in archaeology   184
    4.3. Radiocarbon in geology   188
    4.4. Radiocarbon and tritium in hydrology   193
    4.5. Radiocarbon and tritium and environmental problems   206
    4.6. Radiocarbon and tritium in technology   209
5. Other radionuclides dating methods   212
    Jan Šilar
    5.1. Silicon ^{32}Si dating   212
    5.2. Argon ^{39}Ar dating   213
    5.3. Krypton ^{85}Kr dating   214
6. K - Ar and Ar - Ar dating methods   216
    Emil Jelínek
    6.1. Chemical properties, radioactive decay and isotopic abundance   216
    6.2. Concept of the method   217
    6.3. K - Ar analytical technique   218
    6.4. Ar - Ar analytical technique   219

6.5. Step-heating and laser probe techniques   221
6.6. Applicability and limitations of K - Ar and Ar - Ar techniques   222
6.7. Representative examples   223
7.   Rb - Sr dating method   224
Emil Jelínek
7.1. Chemical properties, radioactive decay and isotopic abundance   224
7.2. Concept of the method   225
7.3. Rb - Sr analytical techniques   227
7.4. Whole-rock and mineral dating of magmatic and metamorphic events   227
7.5. Sr isotopic composition of sediments and ocean water   229
7.6. Evolution of Sr isotope in time and Sr model ages   231
7.7. Representative examples   233
8.   Sm - Nd method   237
Jan Košler
8.1. Chemical properties, radioactive decay and isotopic abundance   237
8.2. Concept of the method   238
8.3. Sm - Nd analytical techniques   240
8.4. Nd in rock-forming minerals   240
8.5. Whole-rock and mineral dating of magmatic and metamorphic events   241
8.6. Evolution of Nd isotopes in time   241
8.7. Age calculation based on Nd model   243
8.8. Nd isotopes: key to the petrogenesis of igneous rocks   245
8.9. Representative examples   248
9.   U - Th - Pb dating methods   249
Jan Košler
9.1. Chemical properties, radioactive decay and isotopic abundance   249
9.2. Concept of the method   250
9.3. U(Th) - Pb analytical techniques   253
9.4. U-Pb dating-concordia diagrams, models of lead loss and intercept age 255
9.5. Single-Zircon evaporation dating, $^{207}Pb/^{206}Pb$ apparent age   258
9.6. Common lead method of dating, age data from Pb model   259
9.7. Representative examples   262
10. Re - Os dating method   263
Emil Jelínek, Jan Košler
10.1. Chemical properties, radioactive decay and isotopic abundance   263
10.2. Concept of the method   263
10.3. Re - Os analytical techniques   264
10.4. Applicability and limitations of the Re - Os method   265
10.5. Representative examples   266
11. Lu - Hf dating method   267
Emil Jelínek
11.1. Chemical properties, radioactive decay and isotopic abundance   267
11.2. Concept of the method   268
11.3. Lu - Hf analytical techniques   268
11.4. Evolution of Hf isotopes in time   269
11.5. Applicability and limitations of the Lu - Hf method   270
11.6. Representative examples   270

Chapter 5
RADIONUCLIDE ANALYSES   273
Richard Tykva, Jan Košler
1. Activity measurements   273
    Richard Tykva
    1.1. Basic terms   273
    1.2. Detection efficiency and background   274
    1.3. Counting and spectrometry   282
    1.4. Sample treatment   285
    1.5. Categories of detectors   288
    1.6. Special detection arrangements   303
    1.7. Accuracy of activity measurements   311
2. Analytical techniques in radiogenic dating   322
    Jan Košler
    2.1. Sample dissolution and chemical separation   323
    2.2. Mass spectrometry with thermal ionisation source   325
    2.3. Accelerator mass spectrometry   332
    2.4. Isotope Dilution   334

References   337

Abbreviations   391

Index   395

Contributors

Dieter Berg
GSF – National Research Centre, Institute of Radiobiology
D – 85764 Neuherberg, Germany

Emil Jelínek
Charles University, Faculty of Science, Department of Geochemistry,
Mineralogy and Mineral Resources,
CZ – 128 43 Prague 2, Czech Republic

Jan Košler
Charles University, Faculty of Science, Department of Geochemistry,
Mineralogy and Mineral Resources
CZ – 128 43 Prague 2, Czech Rebuplic

Jan Šilar
Charles University, Department of Hydrology, Engineering Geology and
Applied Geophysics
CZ – 128 43 Prague 2, Czech Republic

Richard Tykva
Academy of Sciences of the Czech Republic, Institute of Organic Chemistry
and Biochemistry, Head, Department of Radioisotopes
CZ – 166 10 Prague 6, Czech Republic

Preface

Radioactivity can be detected at different levels in almost all objects all over the world, including the human body. This omnipresence of naturally occurring radioactivity is of immediate and crucial concern to people who work in the nuclear industry, to state and local authorities responsible for environmental protection and control of nuclear weapons, and to researchers in scientific and technological disciplines, such as physics (e.g., interaction of radiation with matter), chemistry (e.g., management of radioactive wastes), biology (e.g., radiation bioeffects and risks), ecology (e.g., remediation of environmental pollution), electronics (e.g., measurement instruments), etc.

Unlike other environmental pollutants, such as heavy metals and pesticides, some other scientific disciplines, for example, archaeology, hydrology and geology, profit by the environmental radionuclides, using methods based on their application in radiochronology.

The basic goal of this book is to examine the complex state of radioactivity in the environment, including its sources and applications. In principle, there are two sources of environmental radioactivity, namely man-made and natural. The authors of this book set out to analyze mainly empirical aspects of the activities of both groups. On one hand, a detailed analysis of the sources releasing radionuclides into the environment by human activities should, while describing environmental pollution and its dangers, contribute to its decrease in the future. On the other hand, the analyses of natural radionuclides, as well as their influences and use in different fields, serve to complete an evaluation of the present state of environmental radioactivity. All auxiliary parts (e.g., principles of radionuclide analyses) are included to the extent necessary for understanding the basic themes.

The many recent examples contained in the book will be useful in studying various problems of radioactivity in the present environment, and can help, not only in preparing, carrying out and evaluating outdoor and laboratory experiments, but also in protection of the environment and human health through analyses of possible sources of radioactive pollution.

January 2003

Richard Tykva   and   Dieter Berg

Chapter 1

BASIC TERMS OF RADIOACTIVITY

Richard Tykva
Academy of Sciences of the Czech Republic, Institute of Organic Chemistry and Biochemistry, Department of Radioisotopes. CZ – 16610 Prague 6, Czech Republic

1.   DEVELOPMENT OF OUR KNOWLEDGE

Radioactivity was first described in 1896 by Antoine Henri Becquerel (1852-1908) shortly after the discovery of X-rays by Konrad Wilhelm Röntgen (1845-1923) in 1895. Radioactivity was accidentally discovered by the exposure-producing effect on a photographic plate by a mineral containing uranium - pitchblende, when wrapped in a black paper and kept in the dark. Soon after the fundamental discoveries, Pierre Curie (1859-1906) together with his wife Maria Curie-Sklodowska (1867-1934) extracted two new sources of radioactivity from pitchblende – radium and polonium. In 1898, Ernest Rutherford (1871-1937) found that there are at least two components of the radiation emitted by these elements. In 1899, the same physicist distinguished alpha and beta particles. The following year, Paul Villard (1860-1934) discovered and described the gamma rays emitted by radium.

In 1919 E. Rutherford carried out the first artificial transmutation of one element to another. In his experiments, he was able to produce stable oxygen ^{17}O and ^{1}H by bombarding stable nitrogen ^{14}N with alpha particles. At the beginning of 1930s, other important discoveries contributed considerably to our knowledge of radioactivity. The neutron was discovered in 1932 by James Chadwick (1891-1974), the positron in the same year by Carl David Anderson (1905-1991), and deuterium (^{2}H) in 1933 by Harold C. Urey (1893-1981).

A further significant step in the study of radioactivity consisted in the discovery of artificial radionuclides by Frédéric Joliot (1900-1958) and Irène Curie (1897-1956) in 1934. They bombarded boron and aluminium with alpha particles from polonium and obtained radioactive nitrogen (^{13}N) and phosphorus (^{30}P), respectively. These two reactions may be represented simply as ^{10}B (α,n) ^{13}N and ^{27}Al (α,n) ^{30}P. Very soon after the discovery of

1

R. Tykva and D. Berg (eds.), Man-Made and Natural Radioactivity
in Environmental Pollution and Radiochronology, 1-11.
© 2004 *Kluwer Academic Publishers. Printed in the Netherlands.*

artificial radioactivity, scientists began to bombard practically every element of the periodic system with accelerated protons, deuterons and alpha particles – using early Cockcroft-Walton accelerators, Van de Graaff accelerators and cyclotrons. In this way, they were able to produce and identify hundreds of new radionuclides. The use of the electron linear accelerator provides the additional possibility of producing new radionuclides.

At present, we have identified something like 2600 nuclides: 260 stable nuclides, 25 very long-lived naturally occurring radionuclides, and more than 2300 man-made radionuclides. The recent advances in particle accelerators, nuclear instrumentation and experimental techniques have led to an increased ability to prepare new nuclides.

2.   NUCLIDES

A nuclide can be defined as any species of atom having a certain number of protons and neutrons in its nucleus. The sum of the numbers of protons and neutrons represents the mass number of the nuclide.

Nuclides can be either stable or unstable. The unstable nuclides – radionuclides – decay to stable or unstable products with lower atomic mass, the difference in masses being emitted in the form of energy radiation.

Radioactivity represents the spontaneous emission of sub-atomic particles and/or high-frequency electromagnetic radiation by radionuclides. Radioactive material may consist of one or more radionuclides.

In this chapter, the characterisations of radioactivity and radionuclides are presented in the scope required for this book. All the other relevant data can be found in detail in an extensive handbook that was recently published [Shleien et al. (1998)].

3.   CHARACTERIZATION OF RADIONUCLIDES

The main quantity used for quantification of radioactive sources is the activity, which is defined as the quotient of the number of radioactive transformations in a radionuclide dN, and the time interval dt in which these transformations (sometimes called disintegrations) occurred, i.e.,

$$A = \frac{dN}{dt} \qquad (1)$$

In environmental systems, the term activity is sometimes used in quite different contexts (e.g., biological activity). Therefore, in relation to quantification of nuclear transformations, the term radioactivity (correctly describing a natural effect) is often used instead of activity. A nucleus can undergo different nuclear transformations, the main ones being represented by α (alpha) decay - emission of helium nuclei, β (beta) decay – either β (electrons) or β⁺ (positrons), respectively, and emission of γ (gamma) rays – emission of electromagnetic quanta. In principle, there can be other nuclear transformations, although they are rare, represented by orbital electron capture, internal conversion, isomeric transmission, spontaneous fission, neutron emission and other processes [Shleien et al. (1998)].

All these processes are characterized by statistical fluctuations, which usually follow a simple Poisson distribution (Chapt. 5, Eq. 7). Although the actual numbers of repeated observations of nuclear transformations differ, the average number of such observations exhibits decreased fluctuation depending on the measuring time. As the number of observations increases, e.g., number of readings or measurements, the average value approaches the expectation value of the quantity. In our case, this quantity is the activity, which actually corresponds to the mean value of the number of spontaneous nuclear transformations dN divided by the time interval dt, during which they occurred.

The activity of a radionuclide is numerically equal to the decay rate of this radionuclide and is given by the fundamental law of radioactivity

$$\frac{dN}{dt} = -\lambda \cdot N \qquad (2)$$

where N is the number of radioactive (unstable) nuclei and λ is the decay constant.

After integration a very useful equation is obtained

$$N(t) = N(0) \cdot e^{-\lambda \cdot t} \qquad (3)$$

where $N(0)$ is the number of radioactive nuclei present at a reference time $(t_0 = 0)$, $N(t)$ is their number present at a time t later, and e is the base of the natural logarithm.

This exponential decay relationship is of fundamental importance in working with radioactive materials. Taking into account Eq. 1 and Eq. 2, equation (3) can be rewritten in the form

$$A(t) = A(0) \cdot e^{-\lambda \cdot t} \qquad (4)$$

where $A(0)$ and $A(t)$ represent activity in time $t = 0$ and t, respectively.

In practice, radionuclides are characterized by their half-lives rather than decay constants. The half-life is defined as the time period required for the activity to decrease to just one half of its initial value. It is actually the time interval over which the chance of survival of a particular radioactive atom is exactly one-half. From Eq. 4 we obtain

$$\frac{1}{2} = e^{-\lambda \cdot T_{1/2}} \qquad (5)$$

where $T_{1/2}$ is the half-life, for which we get

$$T_{1/2} = \frac{ln\,2}{\lambda} = \frac{0.693}{\lambda} \qquad (6)$$

Sometimes, also a mean life may be used; it can be introduced on the basis of the probability $p(t) \cdot dt$ that a radioactive nucleus survives up to time t and decays in the interval between t and $t+dt$, i.e.,

$$p(t)dt = e^{-\lambda \cdot t} \cdot dt \qquad (7)$$

from where we can calculate the mean life T_{ml}:

$$T_{ml} = \frac{\displaystyle\int_{0}^{\infty} t \cdot p(t)dt}{\displaystyle\int_{0}^{\infty} p(t)dt} = \frac{1}{\lambda} \qquad (8)$$

The mean life T_{ml} and the half-life (also called half-period) are bound together as:

$$T_{ml} = \frac{T_{1/2}}{0.693} \qquad (9)$$

The relations between the mean-life T_{ml}, the half-life $T_{1/2}$, the decay constant and the initial number of radioactive atoms N_0 is graphically illustrated in Fig. 1.

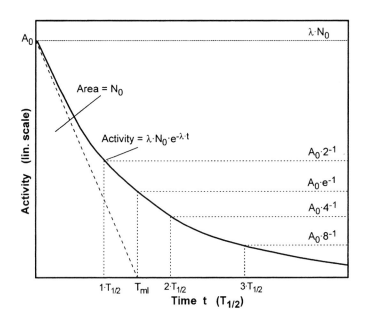

Figure 1:   Relationships of characteristics in radioactive decay.

It is quite common in the decay of a radioactive atom that the resulting atom (daughter or decay product) is not stable and can again undergo a nuclear transformation. The process may continue in a series until it comes to an end stable product. In general, radioactive nuclide A decays to nuclide B, which is also radioactive. Then nuclide B decays to a radioactive nuclide C, and so on. For example, ^{232}Th decays to a series of ten successive radioactive nuclides.

Let us consider the initial part of a radioactive series consisting of three radioactive nuclides A, B, C and stable D with decay constants λ_A, λ_B and λ_C, i.e.,

$$A \xrightarrow{\lambda_A} B \xrightarrow{\lambda_B} C \xrightarrow{\lambda_C} D \tag{10}$$

If, at the beginning, at time $t = 0$, N_{A0} atoms of type A are present, the numbers N_A, N_B and N_C of atoms of types A, B and C which will be present at a later time t, are given by the equations

$$N_A = N_{A_0} e^{-\lambda_A t} \tag{11}$$

$$N_B = N_{A_0} \frac{\lambda_A}{\lambda_B - \lambda_A} \left(e^{-\lambda_A t} - e^{-\lambda_B t} \right) = N_A \frac{\lambda_A}{\lambda_A - \lambda_B} \left(e^{(\lambda_A - \lambda_B) \cdot t} - 1 \right) \tag{12}$$

$$N_C = N_{A_0} \left(\frac{\lambda_A}{\lambda_C - \lambda_A} \frac{\lambda_B}{\lambda_B - \lambda_A} e^{-\lambda_A t} + \frac{\lambda_A}{\lambda_A - \lambda_B} \frac{\lambda_B}{\lambda_C - \lambda_B} e^{-\lambda_B t} + \right.$$
$$\left. + \frac{\lambda_A}{\lambda_A - \lambda_C} \frac{\lambda_B}{\lambda_B - \lambda_C} e^{-\lambda_C t} \right) \tag{13}$$

where the activities of individual radionuclides A, B and C are given as:

$$A_A = \lambda_A \cdot N_A, \quad A_B = \lambda_B \cdot N_B, \quad A_C = \lambda_C \cdot N_C \tag{14}$$

As long as a radioactive parent is initially a pure source whose activity at $t = 0$ is A_{A0}, the ratio of the activities of the parent A and the daughter B will change with time according to the equation:

$$\frac{A_B(t)}{A_A(t)} = \frac{\lambda_B}{\lambda_B - \lambda_A} (1 - e^{-(\lambda_B - \lambda_A) t}) \tag{15}$$

If the half-life of radionuclide B is greater than that of A, namely $\lambda_A > \lambda_B$, then this ratio will continuously increase with time. On the other hand, if the half-life of the parent A is greater than the half-life of its daughter B, i.e., $\lambda_B > \lambda_A$, as t increases, the ratio $A_B(t)/A_A(t)$ also increases, but as t becomes large, this ratio becomes a constant greater than one. Consequently, for sufficiently large values of t, the daughter will decay with the half-life of the parent, but its activity will be greater than the parent activity by the factor $\lambda_B/(\lambda_B - \lambda_A)$. Such a situation is referred to as a transient equilibrium.

The other type of equilibrium occurs when the parent is very much longer-lived than the daughter so that $\lambda_B >> \lambda_A$. In this case, taking into account Eq. 11 and Eq. 12, for the ratio of the activities A_B/A_A one obtains

$$\frac{A_A(t)}{A_B(t)} = 1 - e^{-\lambda_B \cdot t} \qquad (16)$$

The situation represented by the above equation where, after a sufficiently long time, the daughter activity will reach the parent activity, is called secular equilibrium. An example of such secular equilibrium is the decay of ^{226}Ra ($T_{1/2} = 1600\ y$) to ^{222}Rn ($T_{1/2} = 3.82\ d$).

In addition to the previous two cases, characterized by a certain degree of equilibrium, there is a third case of interest where $\lambda_A > \lambda_B$ (a short-lived parent and a long-lived daughter) and no equilibrium occurs.

4.   ACTIVITY EXPRESSIONS

The amount of a radionuclide or a mixture of radionuclides is expressed as its activity. The unit of activity is "becquerel" with the symbol Bq. 1 Bq corresponds to one nuclear transformation per second i.e., $1\ Bq = 1\ s^{-1}$ (decay or disintegration rate – dps or dpm, resp.) The becquerel unit is actually a special name for the reciprocal second.

Some 70 years ago, when radium was the most important radioactive source, the amount of radioactivity was given in mass, usually in terms of mg or g of radium. The former unit of activity, curie (Ci), was related to the disintegration rate of 1 gram of ^{226}Ra, for which early experiments gave a value of $3.7 \cdot 10^{10}$ per second i.e., $3.7 \cdot 10^{10}$ Bq.

Formerly, when the unit Ci was in use, the small amount of radionuclides was expressed by means of sub-multiples formed by the normalized prefixes such as femto- (10^{-15}, f), pico- (10^{-12}; p), nano- (10^{-9}, n), micro- (10^{-6}, μ) or milli- (10^{-3}, m). This was because 1 Ci was too big a unit, not only for low-level radioactivities, but also for most other applications. With the unit Bq, the situation is quite the opposite (the terms kilo- (10^3, k), mega- (10^6; M); giga- (10^9; G), tera- (10^{12}; T); peta- (10^{15}; P), exa- (10^{18}, E)): even for low-level counting, it is sometimes considered to be too small, so that its multiples are often used.

Another quantification of activity is used very occasionally, e.g., in the case of tritium as T.U. (Chapt. 4, 3.1).

A radioactive source can sometimes be considered as a point source, whose dimensions are small enough so that they may be neglected. Quite often, however, sources have different sizes, shapes and forms, so other derived quantities can be more useful. The first two such quantities are usually referred to as the mass a_m or volume a_v activity (concentration) and are defined as:

$$a_m = \frac{A}{m} \qquad (17)$$

$$a_v = \frac{A}{V} \qquad (18)$$

where A is activity of the material having mass m or volume V, respectively.

The basic units of mass activity and volume activity are $Bq \cdot kg^{-1}$ (sometimes also $Bq \cdot mol^{-1}$) and $Bq \cdot m^{-3}$ (sometimes also $Bq \cdot L^{-1}$), respectively. The activity concentration per unit volume is used mainly for radioactive gases, where other parameters, such as temperature and pressure, also have to be stated. It may sometimes be useful to give the activity of a gas in terms of its value per mole as a unit of substance.

The mass activity concentration of a pure (carrier-free, unmixed with any other nuclear species) radioactive source, also called the specific activity (a_{ms}), can be calculated using the formula

$$a_{ms} = \frac{\lambda \cdot N}{(N \cdot M)/n_A} \qquad (19)$$

where N is the number of radioactive atoms, M is the molar mass of the source and n_A is the Avogadro number.

When a radionuclide is spread on a surface, e.g., in the case of surface contamination, the appropriate quantity would be area activity given by the quotient of the activity A and the relevant area or surface S, i.e.,

$$a_s = \frac{A}{S} \qquad (20)$$

with the unit $Bq \cdot m^{-2}$.

5.   DOSES

Ionizing radiation emitted by radionuclides in low chronic exposures is considered to produce damage (stochastic effects), e.g., precancerous lesions, genetic effects etc., and, in high exposures, also nonstochastic effects, such as, e.g., cataracts or skin changes, respectively. A special risk for human health is represented by internal contamination, where radon plays an important role amongst environmental radionuclides. The biokinetic data for various radionuclides are rather varied [Turner (1986)] (see Chapt. 2, 3.). Considering this relation together with physical disintegration data, the same activity of different radionuclides does not necessarily result in the same radiation hazard [Shapiro (1981)], [Shleien et al. (1998)]. Recent detailed data on this aspect, including a descriptive commentary, can be found either on the Internet for ingestion, inhalation and external exposure, (http://www.epa.gov/radiation/ionize2.htm) or in the International Commission on Radiological Protection Recommendations [ICRP 81 (2000)]. An extensive description of the biological responses of molecules, cells, tissues, organs, and organisms to low doses of radiation can be found in the framework of the Low Dose Radiation Research Program of the U.S. Department of Energy, Office of Biological and Environmental Research.

For general quantification of radiation exposures, the quantity "effective dose equivalent" is used as the most important radiation protection quantity. It is based on the absorbed dose and the dose equivalent, which have been defined as follows. The absorbed dose of radiation is the energy imparted per unit mass of the irradiated material. The basic unit of this quantity is "gray" (Gy), which is a special name for SI unit $J \cdot kg^{-1}$. On the other hand, the dose equivalent H represents the product of the absorbed dose D in the tissue and quality factor Q, which depends on the ionization ability of the radiation in question, i.e.,

$$H = D \cdot Q \tag{21}$$

The SI unit of any of the quantities expressed as dose equivalent is called sievert (Sv). The dose equivalent in sieverts is equal to the absorbed dose in grays multiplied by the quality factor.

The effective dose equivalent H_E is introduced by the definition:

$$H_E = \sum_T w_T \cdot H_T \tag{22}$$

In other words, it is the sum, over the selected tissues, of the product of the dose equivalent H_T in tissue T and the weighting factor w_T, representing the ratio of the total stochastic effects resulting from irradiation of this tissue, to the total risk when the whole body is exposed uniformly [Xu and Reece (1996)].

In order to assess the exposure due to the intake of various radionuclides, the International Commission on Radiological Protection (ICRP) introduced such quantities as the committed dose equivalent and committed effective dose equivalent, which will soon be replaced by the recently adopted committed effective dose.

The committed dose equivalent to a given organ or tissue T from a single intake of radioactive material into the body was introduced by the expression

$$H_T(50) = \int_{t_0}^{t_0+50} \dot{H}_T(t) \cdot dt \tag{23}$$

where $\dot{H}_T(t)$ is the relevant dose equivalent rate at time t and t_0 is the time of intake.

The period of integration was then taken as 50 years. The committed effective dose equivalent $H_E(50)$ was defined as

$$H_E(50) = \sum_T H_T(50) \qquad (24)$$

In the ICRP Recommendations from 1991 [ICRP 60 (1991)] other radiation protection quantities have been suggested, namely the equivalent dose and the effective dose, and, especially, the committed effective dose which represents the quantity for the assessment of irradiation of the human body from incorporated radionuclides.

The equivalent dose is derived from the absorbed dose averaged over a tissue or organ T by the expression:

$$H_T = \sum_R w_R \cdot D_{T,R} \qquad (25)$$

where w_R is the radiation weighting factor, depending on the type and energy of radiation, and $D_{T,R}$ is the average absorbed dose from radiation R in tissue T.

It has been found that the relationship between the probability of stochastic effects and equivalent dose depends on the organ or tissue irradiated. This is why a further quantity, based on the equivalent dose, had to be introduced to indicate the combination of different doses applied to several different organs or tissues in such a way that it would be likely to correlate well with the total stochastic effects. Following this concept, the effective dose E is defined as the sum of the weighted equivalent doses in all the tissues and organs of the body, i.e.,

$$E = \sum_T w_T \cdot H_T \qquad (26)$$

where H_T is the equivalent dose in tissue or organ T and w_T is the weighting factor for this tissue or organ.

Radionuclides can irradiate the human body both externally and internally. While the externally applied radiation results in the simultaneous deposition of energy in irradiated tissues or organs, their irradiation due to the incorporated radionuclides is spread out in time. The time distribution of energy deposition will vary with the physicochemical form of the radionuclide and its subsequent metabolism. In order to take into account this distribution, the ICRP recommended the use of the committed equivalent dose $H_T(\tau)$ (for a single intake of activity at time t_0) defined as

$$H_T(\tau) = \int_{t_0}^{t_0+\tau} \dot{H}_T(t) \cdot dt \tag{27}$$

where $\dot{H}_T(t)$ is the relevant equivalent dose rate in an organ or tissue T at time t and τ is the time period over which the integration is performed.

This period is now usually taken as 50 years for adults and 70 years for children and infants.

If the committed tissue or organ equivalent doses resulting from an intake are multiplied by the appropriate weighting factors w_T, and then summed, the result will be the committed effective dose

$$E(\tau) = \sum_T w_T \cdot H_T(\tau) \tag{28}$$

When referring to an equivalent or effective dose accumulated in a given period of time, it is implicit that any committed doses from intakes occurring in that same period are included [ICRP 60 (1991)].

Much attention has been devoted to the radon problem, as the radiation dose from its inhaled decay products is the dominant component of natural radiation exposure of the general population. For radon decay products, the collective quantity that is most commonly used is the equilibrium-equivalent decay-product concentration $a_{Veq}(Rn)$, which is also called the equilibrium-equivalent radon concentration, although it is used as a measure of the decay-product concentration. The equilibrium-equivalent radon concentrations for ^{222}Rn and ^{220}Rn are given in terms of their individual decay-product concentrations a_{Veq} as

$$a_{Veq}\left(^{222}Rn\right) = 0.106 \cdot a_v\left(^{218}Po\right) + 0.513 \cdot a_v\left(^{218}Po\right) + 0.381 \cdot a_v\left(^{214}Bi\right) \tag{29}$$

and

$$a_{Veq}\left(^{220}Rn\right) = 0.913 \cdot a_v\left(^{212}Pb\right) + 0.087 \cdot a_v\left(^{212}Bi\right) \tag{30}$$

In both the above equations these concentrations are given in $Bq \cdot m^{-3}$.

The extent to which, under real conditions, the concentrations of short-lived decay products are lower than the values of their concentrations corresponding to equilibrium with radon as a parent radionuclide can be characterized by the equilibrium factor (F) defined as the ratio of the equilibrium-equivalent radon concentration to the actual radon concentration at the place of interest, i.e.:

$$F = \frac{a_{Veq}(Rn)}{a_V(Rn)} \tag{31}$$

For the assessment of radiological hazards due to exposure to radon (and especially to its decay products), it is very important to have information about the potential alpha-energy concentration (PAEC). This quantity represents the energy that would eventually be released in a specified volume of undisturbed air by the short-lived decay products of radon through the emission of alpha particles. The PAEC is expressed in units of $J \cdot m^{-3}$ or sometimes in a special unit, widely used in radon monitoring, the so-called working level (*WL*). The working level is defined as any combination of short-lived radon products in one litre of air that will result in the ultimate emission of potential alpha particle energy equal to $1.3 \cdot 10^5$ MeV. At secular equilibrium (the case where the half-life of a decay product is much shorter than that of the parent), 1 WL corresponds to a ^{222}Rn concentration of 3.7 kBq·m^{-3} (originally this was related to a radon concentration equal to 100 pCi·L^{-1}).

The concentrations of individual radon decay products can be expressed in *WL* units as:

$$WL = \frac{13.69 \cdot n_v \left(^{218}Po\right) + 7.69 \cdot \left\{ n_v \left(^{214}Pb\right) + n_v \left(^{214}Bi\right) \right\}}{1.3 \cdot 10^5} \qquad (32)$$

where $n_v(^{218}Po)$, $n_v(^{214}Pb)$ and $n_v(^{214}Bi)$ are concentrations of the relevant decay products in numbers of their atoms per litre.

Chapter 2

RADIONUCLIDES IN THE ENVIRONMENT

Dieter Berg

GSF - National Research Centre, Institute of Radiobiology, D – 85764 Neuherberg, Germany

The onset of nuclear techniques can be closely connected with the discovery of X-rays at the end of the nineteenth century and since then, vigorous progress has been made. As in the case of each new technological achievement, not only benefits but also dangers to humanity arise. By the use of ionizing radiation and radioactive isotopes in engineering, medicine and science completely new techniques have been established. However, in strong contrast to these benefits, a lot of dangers and risks never conceivable before have been evoked. Acceptance by the public is controversial, not least due to the horror scenarios of nuclear weapons [Fabrikant (1983a)].

The greenhouse effect caused by the increase of CO_2-concentration in the atmosphere more and more forces mankind to give up techniques of power generating that set free lots of CO_2 by combustion of fossil fuels like coal, oil or gas. The energy produced by nuclear techniques is connected only with low emissions of CO_2 and at the same time largely enables conservation of fossil fuels. In medical disciplines such as common diagnostics and oncology, ionizing radiation and radioactive isotopes are beneficially applied to patients. The involuntary, accidental, or criminal release of radioactive material into the biosphere, even the possibility of this happening is in its catastrophic consequences a fact which has to be evaluated.

Environmental contamination by radioactive substances did not start with the development of the nuclear technologies. It resulted already earlier from human activities such as mining or combustion, which unknowingly augmented naturally occurring radioisotopes in the biosphere.

A great deal of radioactive material has entered the environment as a result of nuclear tests. For military purposes about 200 t of plutonium have been produced [Koch (1995)], about 10 t of this plutonium and a large number of other radionuclides were released during the production process and at the tests of atomic weapons.

R. Tykva and D. Berg (eds.), Man-Made and Natural Radioactivity
in Environmental Pollution and Radiochronology, 13-69.
© 2004 *Kluwer Academic Publishers. Printed in the Netherlands.*

1. NATURALLY OCCURRING RADIONUCLIDES IN THE ENVIRONMENT

1.1. Primordial Natural Radionuclides

The unstable nuclei present at the creation of the earth are called "primordial" natural radionuclides [Jammet et al. (1982)]. Today only those are still present in the environment which possess very long half-life periods of more than 10^8 years. These nuclides are listed in Table 1.

The most frequently occurring radionuclides ^{40}K, ^{238}U, ^{235}U, and ^{232}Th together with their instable progenies contribute to about one half of the natural radiation exposure of humans [UNSCEAR (1988)]. ^{238}U, ^{235}U, and ^{232}Th

Table 1: Primordial natural radionuclides, half-lives, typical range of concentrations in the earth, isotopic abundance and the type of decay (α, β^-, γ, electron capture EC), based on [Chang (1999)], [a][Lederer et al. (1967)], [b][UNSCEAR (1988)], [c][Bunzl (1991)].

Nuclide	Half-life (y)	Abundance (%)	Decay	Concentration (Bq/kg)	
^{40}K	$1.27 \cdot 10^9$	0.012	EC,β^-, γ	[b]100 [c]0.2	- 700 - 1200
^{238}U	$4.47 \cdot 10^9$	99.275	α	[b]10 [c]8	- 50 - 110
^{232}Th	$1.41 \cdot 10^{10}$	100.000	α	[b]7 [c]4	- 50 - 78
^{235}U	$7.04 \cdot 10^8$	0.720	α, γ		
^{176}Lu	$3.78 \cdot 10^{10}$	2.590	EC, β^-, γ		
^{187}Re	$4.35 \cdot 10^{10}$	62.600	(α), β^-		
^{87}Rb	$4.75 \cdot 10^{10}$	27.835	β^-	[c]20	- 560
^{147}Sm	$1.06 \cdot 10^{11}$	15.000	α		
^{138}La	$1.05 \cdot 10^{11}$	0.090	EC, β^-, γ		
^{190}Pt	$6.50 \cdot 10^{11}$	0.010	α		
^{115}In	$4.41 \cdot 10^{14}$	95.710	β^-		
^{180}W	[a]$1.10 \cdot 10^{15}$	[a]0.135	α		
^{144}Nd	$2.29 \cdot 10^{15}$	23.800	α		
^{50}V	$1.40 \cdot 10^{17}$	0.250	EC,β^-,γ		
^{142}Ce	$>5.00 \cdot 10^{16}$	11.080	α		
^{152}Gd	$1.08 \cdot 10^{14}$	0.200	α		
^{152}Gd	[a]$>1.00 \cdot 10^{14}$	[a]0.146	α		
^{209}Bi	[a]$>2.00 \cdot 10^{18}$	[a]100.000	α		

can be found in minor but still well-detectable traces in nearly all soil and rock samples of different sources. All other radionuclides, with the exception of ^{40}K (see Chap. 4, 6.) occur only in very low amounts. The concentrations in rock samples are extremely variable and generally higher in primeval rock than in sediments. Some sedimentary rocks, for instance slate and phosphate containing rocks, are exceptions, their radioactivity can be considerably high.

1.2. Secondarily Occurring Natural Radionuclides

A part of the primarily occurring radioisotopes generate during their decay further radioactive atomic nuclei. These secondarily generated radionuclides continuously produced from their parent substances are called secondarily occurring radionuclides. Since a lot of these radionuclides form radioactive progenies themselves this results in complete decay series which finally come to an end with a stable isotope.

In Fig.1 the naturally occurring decay series of ^{238}U, ^{235}U and of ^{232}Th are presented. The secondarily formed radionuclides with longer half-life periods are ^{234}U, ^{230}Th, ^{231}Pa and ^{226}Ra. All other nuclides are short lived.

In soil as well as in the air, the radioisotopes from all three natural decay series are present. For human radiation exposure ^{222}Rn from the ^{238}U-series and ^{220}Rn from the ^{232}Th-series are of high importance together with their progenies which are also radioactive (see 3.4.6). In contrast to their solid precursors these radioisotopes are volatile and diffuse from ground and building material into the air [SSK (1992ab)]. About one half of natural radiation exposure can be attributed solely to this phenomenon.

1.3. Naturally Generated Radionuclides

In addition to the radioactive material dating back to the genesis of the earth and resulting decay products, which are partly radioactive themselves, further radionuclides are continuously produced by nuclear processes in the atmosphere and soil. These processes are initiated by natural radiation, for instance by cosmic rays.

1.3.1. Cosmogenic Radionuclides

The main contribution to natural generation of radionuclides is of cosmogenic. These nuclides are produced in the stratosphere, as well as in the upper troposphere, by interaction with neutrons, protons, α-particles, pions, mesons, and other particles with oxygen, nitrogen, or argon from the atmosphere. Some of the radionuclides generated by cosmic radiation are compiled in Table 2. Because of their importance for human exposure ^{3}H, ^{14}C, ^{7}Be and ^{22}Na are of special interest.

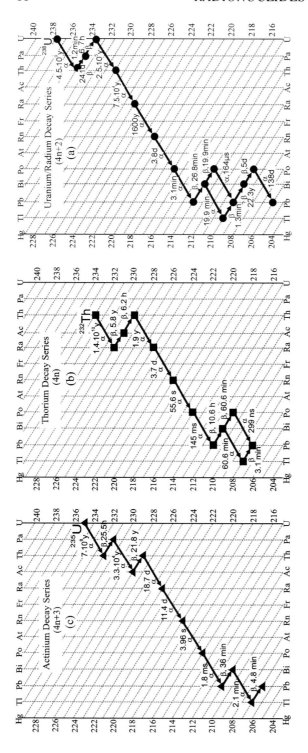

Figure 1: The three natural decay series: (a) Uranium/Radium-series, (b) Thorium-series, (c) Actinium-series.

Table 2: Natural radionuclides produced by cosmic radiation, based on [Edwards (1962)], [Jammet et al. (1982)], [UNSCEAR (1982)].

Radionuclide	Half-Life		Radionuclide	Half-Life	
^{3}H	12.3	y	^{32}P	14.3	d
^{7}Be	53.3	d	^{33}P	25.3	d
^{10}Be	$1.6 \cdot 10^6$	y	^{35}S	87.5	d
^{14}C	5730	y	^{38}S	2.8	h
^{22}Na	2.6	y	^{34m}Cl	32	min
^{24}Na	15	h	^{36}Cl	$3 \cdot 10^5$	y
^{28}Mg	20.9	h	^{38}Cl	37.2	min
^{26}Al	$7.4 \cdot 10^5$	y	^{39}Cl	55.6	min
^{31}Si	2.6	h	^{39}Ar	269	y
^{32}Si	172	y	^{81}Kr	$2.3 \cdot 10^5$	y
^{129}I	$1.6 \cdot 10^7$	y	^{85}Kr	10.7	y

Table 3: Production rate of cosmogenic radionuclides in earth's atmosphere, based on ^{a}[NCRP 81 (1985)], ^{b}[NCRP 62 (1977)], [UNSCEAR (1982)].

	Formation Rate (Number of Atoms $\cdot$ m^{-2}s^{-1})			
	^{14}C	^{3}H	^{7}Be	^{22}Na
Total	16000 –	2500	810	0.86
Troposphere	11000	840	270	0.24
Activity: (Bq/y)	$^a 1.4 \cdot 10^{15}$	$^b 1.5 \cdot 10^{17}$		

Natural ^{3}H (see 3.4.2 and Chap. 4, 3.) is generated by the interaction of neutrons produced by cosmic radiation with nitrogen ^{14}N (n,^{3}H) ^{12}C, neutron energy greater than 4.4 MeV, or by the reaction ^{6}Li (n,α) ^{3}H. Natural ^{14}C (see 3.4.3, Chap. 4, 1.) is produced by an (n,p) reaction with the nitrogen from the air. In the 19th century, before the ^{14}C-content of the atmosphere was diluted by the rapidly increased release of CO_2 from fossil fuels, the specific activity of the air was 230 Bq/kg carbon [Suess (1965)] corresponding to about 63 Bq/kg CO_2.

The formation rates of ^{14}C, ^{3}H, ^{7}Be, ^{22}Na are listed in Table 3 and Chapter 4, Table 2. Per second and per square meter of the surface of the earth, between 16000 and 25000 ^{14}C-atoms are produced, 11000 of these in the troposphere [Lal et al. (1967)], [UNSCEAR (1977)]. The ^{3}H-formation rate is smaller by about a factor of ten that of ^{7}Be about 20, and that of ^{22}Na even 20000 times smaller. The global reserves of ^{14}C, ^{3}H, ^{7}Be and ^{22}Na corresponding to the natural equilibrium amounts to about 8.5 EBq, 1.3 EBq,

40 PBq and 400 TBq respectively [UNSCEAR (1982), (1988)], [Lal et al. (1967)]. Before the first nuclear test explosion the natural ^{3}H-concentration in terrestrial surface water amounted 200 to 900 Bq/m^3 [Kaufman et al. (1954)], [Buttlar et al. (1955)] and in sea water 100 Bq/m^3 [Jammet et al. (1982)]. The concentrations of activity of ^{7}Be and ^{22}Na in surface air are stated to be 3000 and 0.3 μBq/m^3 respectively [Kolb (1974)], [UNSCEAR (1982)]. In Table 4 the percentage distributions in the environment are listed.

Table 4: Distribution of cosmogenic radionuclides in different strata, based on [Lal et al. (1967)], [UNSCEAR (1977), (1982)], ^{a}[NCRP 81 (1985), ^{b}[NCRP 62 (1979)]

	Distribution		(% of global inventory)	
	^{14}C	^{3}H	^{7}Be	^{22}Na
Stratosphere	0.3	6.8	60	25
Troposphere	1.6	0.4	11	1.7
Land Surface	4	27	8	21
Mixed Oceanic Layers	2.2	35	20	44
Deep Ocean	92	30	0.2	8
Ocean Sediments	0.4			
Global Inventory (Bq)	$8.5 \cdot 10^{18}$ $^a 1.4 \cdot 10^{17}$	$1.3 \cdot 10^{18}$ $^b 2.6 \cdot 10^{18}$	$3.7 \cdot 10^{16}$	$4 \cdot 10^{14}$

1.3.2. Fission Generated Radionuclides

In uranium ores, spontaneous nuclear fission occurs rarely but is a perceptible phenomenon. Fission products then result from these spontaneous reactions and react further with neutrons to produce the transuranic elements which can be detected as traces in the ore. The content of ^{90}Sr in the earth's crust generated in this way is estimated to be 50 PBq at a concentration of about 2 μBq/kg [SSK (1992a)], [Shukoljukow et al. (1970)].

In 1972, during the search for uranium in the west African republic of Gabon, the fossil remains of a natural reactor were detected. This surprising discovery was called the "Oklo" phenomenon. Nowadays 0.72 % of the isotope ^{2358} is present in natural uranium. But $2 \cdot 10^9$ years ago, due to the short half-life of ^{235}U compared to ^{238}U, its content was large enough to enable significant fission reactions. The criticality lasted for more than 10^9 years with a total energy release of 15000 MW-years and a loss by fission of about six tons of ^{235}U. The remains of this fission reaction as well as activation products can be found in close proximity [Eisenbud (1987)], [IAEA (1975)], [Cowan (1976)].

2. ARTIFICIALLY PRODUCED RADIONUCLIDES

2.1. History of New Artificial Atomic Nuclei

The prerequisites for the production of new artificial atomic nuclei were detected by the knowledge that nuclear transmutations may result from bombarding of atomic cores with various particles.

In 1919 the English physicist E. Rutherford [Rutherford (1919ab)] succeeded in performing the first artificial nuclear transmutation. He irradiated nitrogen with α-particles of a thorium-C' (^{212}Po) preparation, producing a stable oxygen isotope

$$^{14}N \ (\alpha,p) \ ^{17}O \qquad (1)$$

In 1934, I. Curie and F. Joliot [Curie et al. (1934)] first described a nuclear transmutation in the course of which an artificial radionuclide was created

$$^{10}B \ (\alpha,n) \ ^{13}N \qquad (2)$$

By bombarding boron atoms with α-particles the hitherto unknown ^{13}N was generated. The new nuclide disintegrates by β^+ decay which disintegrates to ^{13}C with a half-life of ten minutes.

2.2. Generation by means of a Particle Accelerator

In accelerators charged particles, such as electrons, protons, deuterons, α-particles, or heavy ions, are focused by means of electromagnetic fields to a beam and accelerated to defined energy levels. Depending on the design, such accelerators are used in research on elementary particles or especially for the production of radionuclides. Accelerators, which can be built very compactly in the form of cyclotrons, are widespread and have gained great importance for a large number of medical applications, especially the positron emission tomography (PET).

Table 5 shows those radionuclides which are relevant for medical applications with short half-lives. For environmental hazards they at are most of local importance (see Chap. 3, 2.5). Within construction materials, however, nuclides can be activated with longer half-lives, but then activities can be mostly neglected. In the case of high energy accelerators of some GeV, radionuclides with half-life periods of more than 15 days are produced close within a neighbourhood of the beam, by aberration of the beam or by insufficient shielding from secondary particles.

Table 5: Radionuclide production for medical use by particle accelerators, based on [Stöcklin et al. (1995)].

Radio-nuclide	Half-Life		Decay	Formation	Particle Energy (MeV)
^{201}Tl	72.9	h	EC,γ	^{203}Tl (p,3n) ^{201}Pb $\rightarrow$ ^{201}Tl	28 - 20
^{123}I	13.3	h	EC,γ	^{124}Te (p,2n) ^{123}I	15.5- 10
				^{127}I (p,5n) ^{123}Xe $\rightarrow$ ^{123}I	65 - 45
^{81}Rb $\downarrow$ ^{81m}Kr	4.6 13.1	h s	EC,β$^+$,γ	^{82}Kr (d,3n) ^{81}Rb	
^{67}Ga	78.3	h	EC,γ	^{68}Zn (p,2n) ^{67}Ga	26 - 18
^{68}Ge $\downarrow$ ^{68}Ga	270.8 67.6	d min	EC,β$^+$,γ	RbBr (p,spall)	800, 500
^{18}F	109.8	min	EC,β$^+$,γ	^{18}O (p,n) ^{18}F	16 - 3
^{11}C	20.4	min	EC,β$^+$,γ	^{14}N (p,α) ^{11}C	13 - 3
^{13}N	10	min	EC,β$^+$	^{16}O (p,α) ^{13}N	16 - 7
^{15}O	122.2	s	EC,β$^+$	^{15}N (p,n) ^{15}O	8 - 0

2.3.   Production in the Nuclear Reactor

The production of artificial radionuclides in a nuclear pile results from two completely different physical reactions, from nuclear fission or from activation by charged particles, neutrons or γ-quanta.

In the case of the nuclear fission of ^{235}U, for instance, the bombardment with thermal neutrons results in the extremely unstable, intermediate nucleus ^{236}U that bursts into two, or occasionally also three parts:

$$^{235}_{92}U + n \rightarrow\ ^{236}_{92}U \rightarrow\ ^a_b X +\ ^c_d Y + \kappa \cdot n + 195\,MeV \qquad (3)$$

with κ=number of neutrons (n), $a + c + \kappa = 236$ and $b + d = 92$.

Observing a larger number of fissions, it can be shown that the fission products $^a X$ and $^c Y$ are not always identical, but merely have a certain probability of formation. The rate of this probability ranges over some orders of magnitude (see Fig. 2). Besides ^{235}U, there are further isotopes that have appropriate effective cross sections for thermal neutrons for nuclear fission, for instance ^{233}U, ^{239}Pu and ^{241}Pu.

Numerous fission products primarily generated in a nuclear reactor have only a short half-life and most of them finally disintegrate to stable nuclei by passing through a series of decay. Some examples of such decay chains are shown in Fig. 3. From this it becomes evident that these chains mostly consist of short lived nuclides but there are also long-lived radioactive radionuclides such as ^{90}Sr and ^{137}Cs. The activity inventory in the reactor core at a certain moment depends on the type of reactor, on the chronological course of the generated thermal efficiency, as well as on the burn-up of the fuel rods. Table 6, for example, presents the stock of fission products and the radionuclides resulting from their decay chains for a pressurized water moderated and cooled reactor (PWR) after a prolonged running period.

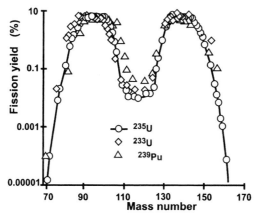

Figure 2:   Probability for the formation of fission products after bombardment of ^{235}U, ^{233}U and ^{239}Pu with thermal neutrons.

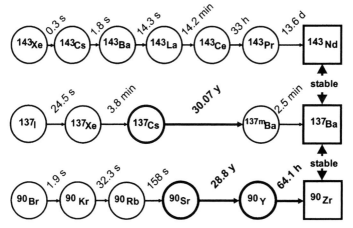

Figure 3:   α-, β- radioactive decay series for fission products, based on [Lederer et al. (1967)], [Jammet et al. (1982)].

Table 6: Inventory of some fission products and their progenies of a PWR with an electric power of 1300 MW_{el} 0, 24 and 120 hours after shutdown of the reactor, based on [SSK (1989)].

	Activity	(Bq/MW)			Activity	(Bq/MW)	
	0 h	24 h	120 h		0 h	24 h	120 h
^{85}Kr	$7.53 \cdot 10^{12}$	$7.53 \cdot 10^{12}$	$7.53 \cdot 10^{12}$	^{127}Sb	$8.55 \cdot 10^{13}$	$7.23 \cdot 10^{13}$	$3.51 \cdot 10^{13}$
^{85m}Kr	$2.79 \cdot 10^{14}$	$6.91 \cdot 10^{12}$	$2.46 \cdot 10^{6}$	^{129}Sb	$3.05 \cdot 10^{14}$	$6.59 \cdot 10^{12}$	$1.34 \cdot 10^{6}$
^{87}Kr	$5.22 \cdot 10^{14}$	$1.10 \cdot 10^{9}$	0.00	^{127}Te	$8.38 \cdot 10^{13}$	$7.77 \cdot 10^{13}$	$4.39 \cdot 10^{13}$
^{88}Kr	$7.42 \cdot 10^{14}$	$2.13 \cdot 10^{12}$	$1.40 \cdot 10^{2}$	^{127m}Te	$1.04 \cdot 10^{13}$	$1.04 \cdot 10^{13}$	$1.03 \cdot 10^{13}$
^{133}Xe	$2.03 \cdot 10^{15}$	$1.96 \cdot 10^{15}$	$1.26 \cdot 10^{15}$	^{129}Te	$3.03 \cdot 10^{14}$	$3.70 \cdot 10^{13}$	$2.67 \cdot 10^{13}$
^{135}Xe	$4.42 \cdot 10^{14}$	$4.74 \cdot 10^{14}$	$5.68 \cdot 10^{11}$	^{129m}Te	$4.53 \cdot 10^{13}$	$4.47 \cdot 10^{13}$	$4.10 \cdot 10^{13}$
^{131}I	$9.72 \cdot 10^{14}$	$9.03 \cdot 10^{14}$	$6.48 \cdot 10^{14}$	^{131m}Te	$1.42 \cdot 10^{14}$	$8.20 \cdot 10^{13}$	$8.92 \cdot 10^{12}$
^{132}I	$1.43 \cdot 10^{15}$	$1.17 \cdot 10^{15}$	$4.98 \cdot 10^{14}$	^{132}Te	$1.40 \cdot 10^{15}$	$1.14 \cdot 10^{15}$	$4.85 \cdot 10^{14}$
^{133}I	$2.03 \cdot 10^{15}$	$9.35 \cdot 10^{14}$	$3.80 \cdot 10^{13}$	^{134}Cs	$9.40 \cdot 10^{13}$	$9.38 \cdot 10^{13}$	$9.35 \cdot 10^{13}$
^{134}I	$2.20 \cdot 10^{15}$	$4.90 \cdot 10^{7}$	0.00	^{136}Cs	$3.59 \cdot 10^{13}$	$3.40 \cdot 10^{13}$	$2.76 \cdot 10^{13}$
^{135}I	$1.89 \cdot 10^{15}$	$1.53 \cdot 10^{14}$	$6.51 \cdot 10^{9}$	^{137}Cs	$8.01 \cdot 10^{13}$	$8.01 \cdot 10^{13}$	$8.01 \cdot 10^{13}$
^{89}Sr	$1.03 \cdot 10^{15}$	$1.02 \cdot 10^{15}$	$9.64 \cdot 10^{14}$	^{140}Ba	$1.79 \cdot 10^{15}$	$1.69 \cdot 10^{15}$	$1.36 \cdot 10^{15}$
^{90}Sr	$5.95 \cdot 10^{13}$	$5.95 \cdot 10^{13}$	$5.95 \cdot 10^{13}$	^{140}La	$1.83 \cdot 10^{15}$	$1.80 \cdot 10^{15}$	$1.54 \cdot 10^{15}$
^{91}Sr	$1.27 \cdot 10^{15}$	$2.21 \cdot 10^{14}$	$2.00 \cdot 10^{11}$	^{141}Ce	$1.69 \cdot 10^{15}$	$1.66 \cdot 10^{15}$	$1.52 \cdot 10^{15}$
^{90}Y	$6.21 \cdot 10^{13}$	$6.16 \cdot 10^{13}$	$6.03 \cdot 10^{13}$	^{143}Ce	$1.57 \cdot 10^{15}$	$9.56 \cdot 10^{14}$	$1.27 \cdot 10^{15}$
^{91}Y	$1.32 \cdot 10^{15}$	$1.31 \cdot 10^{15}$	$1.25 \cdot 10^{15}$	^{144}Ce	$1.09 \cdot 10^{15}$	$1.08 \cdot 10^{15}$	$1.07 \cdot 10^{15}$
^{95}Zr	$1.72 \cdot 10^{15}$	$1.70 \cdot 10^{15}$	$1.63 \cdot 10^{15}$	^{143}Pr	$1.55 \cdot 10^{15}$	$1.53 \cdot 10^{15}$	$1.33 \cdot 10^{15}$
^{97}Zr	$1.71 \cdot 10^{15}$	$6.40 \cdot 10^{14}$	$1.22 \cdot 10^{13}$	^{239}Np	$1.93 \cdot 10^{16}$	$1.45 \cdot 10^{16}$	$4.45 \cdot 10^{15}$
^{95}Nb	$1.71 \cdot 10^{15}$	$1.71 \cdot 10^{15}$	$1.70 \cdot 10^{15}$	^{238}Pu	$1.20 \cdot 10^{12}$	$1.20 \cdot 10^{12}$	$1.21 \cdot 10^{12}$
^{99}Mo	$1.84 \cdot 10^{15}$	$1.43 \cdot 10^{15}$	$5.22 \cdot 10^{14}$	^{239}Pu	$3.16 \cdot 10^{11}$	$3.16 \cdot 10^{11}$	$3.16 \cdot 10^{11}$
^{99m}Tc	$1.61 \cdot 10^{15}$	$1.37 \cdot 10^{15}$	$5.04 \cdot 10^{14}$	^{240}Pu	$3.67 \cdot 10^{11}$	$3.67 \cdot 10^{11}$	$3.67 \cdot 10^{11}$
^{103}Ru	$1.51 \cdot 10^{15}$	$1.48 \cdot 10^{15}$	$1.38 \cdot 10^{15}$	^{241}Pu	$8.65 \cdot 10^{13}$	$8.65 \cdot 10^{13}$	$8.65 \cdot 10^{13}$
^{105}Ru	$9.62 \cdot 10^{14}$	$2.34 \cdot 10^{13}$	$7.23 \cdot 10^{6}$	^{241}Am	$7.37 \cdot 10^{10}$	$7.39 \cdot 10^{10}$	$7.55 \cdot 10^{10}$
^{106}Ru	$3.70 \cdot 10^{14}$	$3.70 \cdot 10^{14}$	$3.70 \cdot 10^{14}$	^{242}Cm	$2.08 \cdot 10^{13}$	$2.08 \cdot 10^{13}$	$2.08 \cdot 10^{13}$
^{105}Rh	$9.05 \cdot 10^{14}$	$6.51 \cdot 10^{14}$	$9.99 \cdot 10^{13}$	^{244}Cm	$8.28 \cdot 10^{11}$	$8.30 \cdot 10^{11}$	$8.30 \cdot 10^{11}$

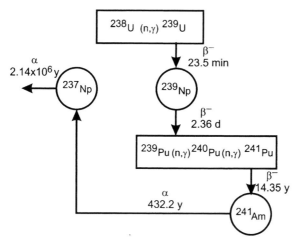

Figure 4: Activation in the nuclear fuel, formation of plutonium and americium, based on [Jammet et al. (1982)], [Chang (1999)].

Apart from the nuclear fission in the reactor, radionuclides are formed by the activation. At first in the fuel itself (see Fig. 4) ^{239}U is built from ^{238}U by neutron capture, which itself decays via its radioactive daughter ^{239}Np to ^{239}Pu. After further neutron activation ^{241}Pu is generated from ^{239}Pu.

Contaminants of the fuel, coolant and structural elements of the reactor cause further neutron activation products (see Table 7). The sort and amount of radionuclides varies considerably with the type of the reactor.

For instance, in PWRs ^{3}H is produced by neutrons of lithium and boron [Jammet et al. (1982)] or in the fuel and in the cooling water by the reaction from ^{1}H or ^{2}H [Bonka (1982)]. ^{14}C is produced in light-water cooled graphite moderated reactors (LWR) or heavy-water moderated and cooled reactors (HWR), formed by the reaction from oxygen ^{17}O (n,α) ^{14}C, or from nitrogen ^{14}N (n,p) ^{14}C in the fuel, or from the moderator or the coolant [Smith (1982)]. During normal operation only a minor percentage of the radioactive stock in the fuel rods diffuses into the coolant, so that the main part reaches the reprocessing plant later. Besides these radionuclides mentioned above, further isotopes are activated by the nuclear reactions in the structural parts of the reactor (see Table 7). The pollution of the biosphere from the different fission and activation products generated during normal function of the reactor, or in the case of an accident, will be discussed later in Chapter 3.

In practice the most important nuclear reaction for the production of artificial radioisotopes for medicine, science or for the detection and determination of trace elements in different samples is the (n,γ)-reaction. Reactors are also good neutron sources, since at nuclear fission more neutrons are produced than necessary for maintaining of the chain. For that reason numerous radionuclides can be produced by irradiating appropriate targets in the beam of the neutron source. An example for the production of ^{131}I is the irradiation of a tellurium-target. The short-lived intermediate ^{131}Te disintegrates to ^{131}I, which can be separated chemically from the irradiated target:

$$^{130}Te(n,\gamma)\,^{131}Te \xrightarrow{\ \beta^-,25min\ } \,^{131}I \qquad (4)$$

Table 7: Some activation products with their progenies formed in nuclear reactors (LWR's), based on 3.5 GW$_{th}$ [a][Bonka (1982)], 3.7 GW$_{th}$ [b][SSK (1989)], 3.2 GW$_{th}$ [c][Eisenbud (1987)], 1.4 GW$_{el}$ [d][Izqierdo et al. (1993)].

	Half-life		Inventory (Bq)	Comments
^{3}H	12.3	y	[a] $7.4 \cdot 10^{14}$	fuel, coolant
^{14}C	5730	y	[a] $3.7 \cdot 10^{11}$	fuel, coolant
^{51}Cr	27.7	d	[a] $3.7 \cdot 10^{17}$	structure material, coolant
^{54}Mn	312.3	d	[d]	structure material, coolant
^{58}Co	70.9	d	[a] $7.4 \cdot 10^{16}$, [c] $2.9 \cdot 10^{16}$	structure material, coolant
^{60}Co	5.3	y	[a] $7.4 \cdot 10^{16}$, [c] $1.1 \cdot 10^{16}$	structure material, coolant
^{55}Fe	2.7	d	[a] $9.3 \cdot 10^{16}$	structure material, coolant
^{59}Fe	44.5	d	[d]	structure material, coolant
^{65}Zn	244.3	d		structure material, coolant
^{95}Nb	35.0	d	[b] $6.4 \cdot 10^{18}$, [c] $5.6 \cdot 10^{18}$	fuel, struct. mat., coolant
^{95}Zr	64.0	d	[a] $5.2 \cdot 10^{18}$, [b] $6.4 \cdot 10^{18}$, [c] $5.6 \cdot 10^{18}$	fuel, structure material
^{110m}Ag	249.8	d	[d]	structure material, coolant
^{124}Sb	60.2	d	[d]	structure material, coolant
^{238}Pu	87.7	y	[a] $5.5 \cdot 10^{15}$, [b] $4.5 \cdot 10^{15}$, [c] $2.1 \cdot 10^{16}$	fuel
^{239}Pu	$2.4 \cdot 10^{4}$	y	[a] $7.4 \cdot 10^{11}$, [b] $1.2 \cdot 10^{15}$, [c] $7.8 \cdot 10^{14}$	fuel
^{240}Pu	$6.6 \cdot 10^{3}$	y	[b] $1.4 \cdot 10^{15}$, [c] $7.8 \cdot 10^{14}$	fuel
^{241}Pu	14.4	y	[a] $3.0 \cdot 10^{17}$, [b] $3.2 \cdot 10^{17}$, [c] $1.3 \cdot 10^{17}$	fuel
^{239}Np	2.4	d	[b] $7.2 \cdot 10^{19}$, [c] $6.1 \cdot 10^{19}$	fuel
^{241}Am	$4.3 \cdot 10^{2}$	y	[a] $1.1 \cdot 10^{14}$, [b] $2.8 \cdot 10^{14}$, [c] $6.3 \cdot 10^{14}$	fuel
^{242}Cm	$1.6 \cdot 10^{2}$	d	[a] $7.4 \cdot 10^{16}$, [b] $7.8 \cdot 10^{16}$, [c] $1.6 \cdot 10^{16}$	fuel
^{244}Cm	18.1	y	[a] $5.5 \cdot 10^{15}$, [b] $3.1 \cdot 10^{15}$, [c] $8.5 \cdot 10^{14}$	fuel

2.4.   Generation from Nuclear Weapon Tests

After dropping the uranium bomb onto Hiroshima on August 6th, 1945 and a plutonium bomb three days later onto Nagasaki, numerous tests of nuclear weapons were undertaken in the atmosphere (see Chap. 3, 2.1). The majority of those weapon tests were performed before 1963. The amount of radioactive substances in the biosphere increased rapidly, especially in the northern hemisphere and including the northern polar region. For the most important nuclear states this was the reason to agree on August 5th, 1963 in Moscow on a stop of all nuclear weapon tests in the atmosphere. Only China and France carried out further tests in the atmosphere. France declared the end of tests in the atmosphere in 1974 and underground in 1996. It is estimated [UNSCEAR (1982)] that by 1980, during 423 nuclear tests in the atmosphere, 217 Mt fission energy (545 Mt TNT-equivalents) was released. The spectrum of the most important fission products (see Table 8) generated by nuclear explosions and of the radioisotopes formed by activation does not, with respect to its relevance for the environment, essentially differ from that present in a reactor.

Table 8:   Activity in Bq of some fission and activation products including transuranium elements created by nuclear weapon tests, based on [a][Jammet et al. (1982)], [UNSCEAR (1993)].

Fission Products				Activation Products			
	Half-life		Activity (Bq)		Half-life		Activity (Bq)
^{89}Sr	50.5	d	$9.1 \cdot 10^{19}$	^{3}H	12.3y		$2.4 \cdot 10^{20}$
^{90}Sr	28.6	y	$6.0 \cdot 10^{17}$	^{14}C	5730	y	$2.2 \cdot 10^{17}$
^{95}Zr	64.0	d	$1.4 \cdot 10^{20}$	^{54}Mn	312.3d		$5.2 \cdot 10^{18}$
^{103}Ru	39.3	d	$2.4 \cdot 10^{20}$	^{55}Fe	2.7y		$2.0 \cdot 10^{18}$
^{106}Ru	373.4	d	$1.2 \cdot 10^{19}$	^{238}Pu	87.7y		$3.3 \cdot 10^{14}$
^{131}I	8.0	d	$6.5 \cdot 10^{20}$	^{239}Pu	24110	y	$6.5 \cdot 10^{15}$
^{136}Cs	13.2	d	[a]$7.0 \cdot 10^{18}$	^{240}Pu	6564	y	$4.3 \cdot 10^{15}$
^{137}Cs	30.1	y	$0.9 \cdot 10^{18}$	^{241}Pu	14.4y		$1.4 \cdot 10^{17}$
^{140}Ba	12.8	d	$7.3 \cdot 10^{20}$	^{242}Pu	$3.7 \cdot 10^{5}$ y		[a]$1.6 \cdot 10^{13}$
^{141}Ce	32.5	d	$2.5 \cdot 10^{20}$	^{242m}Am	141.0y		[a]$3.7 \cdot 10^{11}$
^{144}Ce	284.9	d	$3.0 \cdot 10^{19}$	^{244}Cm	18.1y		[a]$2.6 \cdot 10^{11}$

3.   BEHAVIOUR OF RADIOACTIVE SUBSTANCES IN THE ENVIRONMENT

3.1.   Behaviour in Different Ecological Systems

Ecology concern the interaction between living organisms and their habitat surroundings. Ecological investigations therefore concentrate on the biosphere where all life on earth takes place. The biosphere consists of a part of the atmosphere, of a part of the crust of the earth, of the lithosphere and of nearly the whole hydrosphere. Sections of the biosphere are called ecosystems, as for instance woodlands, rivers, lakes, deserts, oceans, or the deep sea. In these systems, by exchange processes close correlations exist between the living creatures and their habitat. Radionuclides entering the spheres of existence are involved in this exchange. Radioisotopes from the habitat can be taken up by the animate one, exchanged within the organisms of an ecosystem or passed through to others. This is the reason why they are distributed in a manner typical for the ecosystem. Plants living autotrophically by photosynthesis are at the beginning of the food-chain. In contrast, animals are dependent on heterotrophical organic nourishment. Besides the incorporation of radioactive substances from the inorganic compounds, an exchange or passage respectively between the organisms has to be taken into account. To finish the circuit, particularly adapted organisms like moulds or bacteria exist which are nourished saprophytically by decomposition of organic compounds.

In contrast to most chemical substances for which toxicity is commonly related to specific molecular structures, radiotoxicity of the isotopes depends on the radiological characteristics as well as on the molecular structure of the radioactive substance. The damage itself normally is an effect of the radiation emitted during disintegration. An alteration of the chemical structure of the molecule comprising a radionuclide does not remove the radionuclide from the environment but merely changes its kinetic behaviour in the biosphere or can lead to augmented deposition in different organs. The pathways through the environment are extraordinarily complex.

At reactor accidents and nuclear tests (see Chap. 3, 2.3 and 2.1) a great number of different radionuclides are released. The chemical structures into which the radionuclides are emitted and further conversions are scarcely or only incompletely known. In the following, the very short-lived radionuclides are taken into consideration only on a small scale since on transportation through the food-chain their concentrations decline very rapidly by physical decay and finally do not contribute significantly to the total radiation dose. Besides those naturally occurring, the following artificially produced radionuclides and radioelements are of relevance for the food-chain: ^{3}H, ^{14}C, ^{85}Kr, ^{90}Sr, ^{129,131}I, ^{106}Ru, ^{134,137}Cs, ^{139,141,143,144}Ce, Pu, Am, Np, and to a lesser extent also ^{132}Te, ^{99}Tc, ^{60}Co, and ^{54}Mn.

The main source of artificial radioactivity entering the environment globally results from nuclear weapon tests. Radionuclides released into the environment not accidentally, but by regular operation of a nuclear fuel circuit (see Chap. 3, 2.2) are more relevant for the population residing immediately in the neighbourhood of the place of release. Exceptions are radionuclides that on their passage through the environment are long-lived and mobile enough to be widely distributed. Some of these nuclides are ^{3}H, ^{14}C, ^{85}Kr, and [129]I. After accidents (see Chap. 3, 2.3) numerous additional radionuclides ^{131}I, [132]Te, [106]Ru, [134,137]Cs and ^{90}Sr can be present. This became obvious after the Chernobyl reactor accident when numerous European countries were contaminated with many of radionuclides.

Important parameters for the spreading and radiological relevance of a radionuclide depend on the mediums of emission, i.e., air, water, or soil. Furthermore the temporal contribution of the emission, singularly or permanently, and the kind of ecosystem concerned are of importance.

Fig. 5 shows clearly that the release of fallout into the air (e_a) or into water (e_w) starts a complicated migration pattern through the biosphere. Fallout brought to the air is deposited onto the soil, the vegetation and on water surfaces. Transportation by water contaminates rivers, lakes and the sea, many of which are situated far away from the release point. From soil and water, the radionuclides enter the air by resuspension (r), for example by wind and vaporization. Erosion of the soil (t_e) transports radioactive substances from the ground to rivers, lakes and the oceans or reversibly by flooding and artificial irrigation (t_i) again onto the soil or the surface of the vegetation. From the aqueous media the fallout can be deposited as sediment or be remobilized.

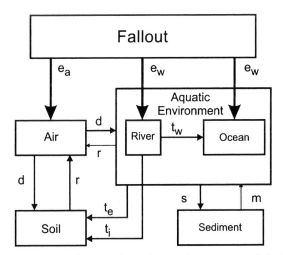

Figure 5: Release of radionuclides in air or in aquatic systems and their main pathways through the different ecological systems.

d: deposition by setting, rain-out/wash-out (dry or wet), e_a: emission into air, e_w: emission into aquatic environment (river, lake, ocean), **m:** mobilization from sediment, **r:** resuspension by wind or by evaporation, **s:** settle or store into the sediment, t_e: transport by erosion, t_i: transport by artificial irrigation or by floods (inundation), t_w: transport by water

The mobility in an ecosystem is not only specific for an element but depends also on its chemical characteristics or the chemical compound originally released. Interactions with other substances or organisms of the ecosystem can exert a substantial influence on the biokinetics of animals. Besides climatic influences determined by the meteorological dynamics, spreading and of elements depends not only on the physicochemical influences of inanimate nature, but also on "migration" through organisms, according to the motto "eating or be eaten", that is to say the food-chain.

3.1.1.   Behaviour in the Air

Depending on the meteorological situation, radioactive material emitted into the atmosphere is widely spread [UNSCEAR (1962)]. Turbulent air streams provide for mixing and winds for transportation over far distances. Since the transport conditions in turbulent streams are extraordinarily complicated there exist no exact mathematical procedures to describe the spatial and temporal behaviour of radioactive material immediately after release into the atmosphere. However, some methods based on statistical procedures have been developed. They, at least to a certain extent, permit predictions of how pollutants will be distributed by defined meteorological conditions [NCRP 123 (1996)].

Until the 1940s pollutants had been released exclusively into "the friction layer" reaching about 300 m above the ground level. In the second half of the twentieth century the contaminants directly entered the upper air layers up to 30 km and beyond it by nuclear weapon tests, new techniques of energy production, high-flying airplanes, and by activities in space.

Within the atmosphere, gases or aerosols are diluted by molecular and turbulent diffusion. The contribution of molecular diffusion to dilution [Gifford (1968)] amounts to a diffusion coefficient of about 0.2 cm^2/s which is many orders of magnitude smaller compared to the stormy turbulent stream ($\approx 10^{11}$ cm^2/s). Although the atmosphere contains a lot of gases, the three elements nitrogen, oxygen, and argon contribute 99 % of the weight. The atmosphere has a total weight of approximately $5 \cdot 10^{21}$ g and a total volume of $3.96 \cdot 10^{24}$ cm^3 at standard conditions [NCRP 44 (1975)]. An additional component is $1.2 \cdot 10^{17}$ kg of vapour [Eisenbud (1987)] which varies extraordinarily with meteorological conditions.

Air pressure and density (dry) decrease exponentially with altitude. The pressure under standard condition, (i.e., mean sea level) is 1013.25 hPa at a density of 1.3 kg/m^3. Both, pressure and density decrease by a factor of 10^{-3} at an altitude of 50 km. The ideal temperature change in the altitude is demonstrated in Fig. 6. In the troposphere temperature normally decreases with altitude, by about 6.5 °C per 1000 meters. This temperature gradient is called "lapse rate of temperature". Above the troposphere, in the stratosphere in the so-called tropopause, the temperature gradient changes abruptly and a nearly isothermal zone begins. Above the stratosphere and within the mesosphere, temperature increases again. The troposphere ranges to 11 km of altitude and contains 75 % of the atmospheric mass and nearly all of the water content. The border to the stratosphere, the tropopause, varies with latitude and season (see Fig. 9). The central point of thermodynamic events influencing the

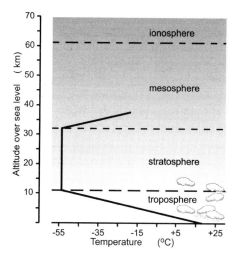

Figure 6: Idealized profile of the temperature.

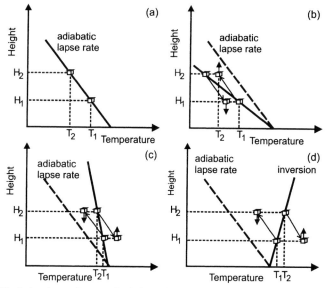

Figure 7: The behaviour of a parcel of air under different temperature profiles: (a) idealized dry adiabatic lapse rate, (b) the lapse rate is lower or, (c) higher than adiabatic lapse rate, (d) increasing temperature with the idealized profile of the temperature.

weather is situated at this level. The stratosphere containing only low humidity ranges up to 32 km of altitude. Compared to the troposphere it is a place of relatively stable conditions. At a height of 61 km, above an intermediate zone, the mesozone, begins the ionosphere which is filled with ions.

The intermixture of the compounds of the atmosphere is vertically influenced essentially by the temperature gradient and horizontally by the wind. In Fig. 7 the influence of the thermal distribution onto vertical air movements is demonstrated. The dry adiabatic lapse rate commonly amounts to -6.5 °C/km. In the case of an element of the air volume ascending from an altitude H_1 to the altitude H_2 this air package will cool down corresponding to the adiabatic lapse rate. When the prevailing temperature distribution corresponds to the adiabatic temperature gradient (see Fig. 7a) the element of the air and the surrounding atmosphere have identical temperatures and densities and therefore will neither ascend nor descend. If the vertical temperature distribution (see Fig. 7b) is super-adiabatic, which means that the temperature decreases with height faster than in case of the "adiabatic standard lapse rate", the air package ascending from height H_1 and temperature T_1 to height H_2 will be warmer than the surrounding air. Because of its minor density it will further ascend. Inversely when air is descending from height H_2 to height H_1 it will, because of its minor temperature and larger density, continue descending. At super-adiabatic conditions vertical movements are accelerated. This is called an unstable state of the atmosphere. When the temperature decreases slower with height than under standard conditions (see Fig. 7c) a volume element from height H_1 will be colder in the height H_2 than the surroundings and cease ascending but will descend back to its original height. Inversely the air package from H_2 will be warmer in H_1 and ascend again. In this case the atmosphere is in a stable state. In particular stable conditions in the case of inversion are found when temperature even increases with height (see Fig. 7d).

Temperature variation in the atmosphere, especially near the surface of the earth, is influenced by the day/night rhythm. At intense solar radiation during day-time, the surface air near the earth surface is warmed up faster than the upper air layers. A super-adiabatic state of the temperature gradient is created and the vertical air streams become very unstable. After sunset the surface of the earth cools faster than the surrounding air. A positive temperature gradient is formed, called inversion, resulting in more stable arranging of the air layers. Hence the vertical temperature gradient is dependent on day-time, solar irradiation, clouding, the seasons, and from warm and cold fronts which induce shifting of the temperature profile. Changes in the temperature profile thus influence the spreading of radioactive gases and aerosols in exhaust air trails.

The expansion of radioactive substances in plumes correlating to the vertical temperature distribution is schematically demonstrated in Fig. 8. When the temperature profile is super-adiabatic (see Fig. 8a) and hence highly unstable conditions in the atmosphere predominate, the plume shows a bizarre formation of loops accompanied by highly ascending and widely descending clouds of exhausted gas which even can touch the ground. When the decrease (see Fig. 8b) of the temperature with height is weaker than adequate to the dry adiabatic lapse rate, then the exhausted air trail tends to form cones. The highly stable conditions of an inversion causes to be the air is hardly transported vertically. In this case the plume shows thread formation (see Fig. 8c). In reality such simple circumstances are scarcely found and temperature distributions are substantially more complicated. In Fig. 8d inversion in funnel height and above, a profile of decreasing temperature was supposed. Because of the stable conditions, air from the plume does not sink down but may rise. In Fig. 8e, however, it is supposed that inversion begins at the height of the funnel. In this case the plume cannot spread to the top but descends and even touches the soil.

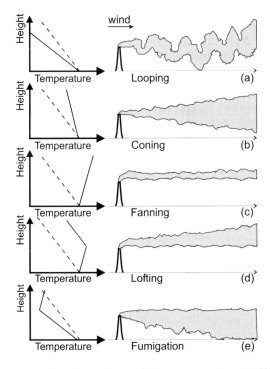

Figure 8: The behaviour of a plume during different temperature profiles, after [Eisenbud (1987)]:

——— real lapse rate,   - - - - - - - dry adiabatic lapse rate.

At contact with the plume the radionuclides transported in this way are precipitated as suspended solids or as aerosols on the surface of the earth (dry deposition). At higher altitudes the particles can serve as condensation seed and drop with the rain (rain-out) onto the soil or are washed out (wash-out) by the rainfall and thus be deposited (wet deposition) on the surface of the ground or waters (see Fig. 5).

While nuclear material emitted at standard operation of nuclear piles has only a rather local meaning for the radiation exposure, the fallout after acci-dental release, as well as after nuclear weapon tests are of more than local importance. As a result of the high temperatures in the centre of nuclear explosions, radioactive substances reach high altitudes. The ascending partially gaseous products chill down, condense and form aerosols. The resulting particle sizes are of a broad divergence and in the further process influence the distribution of the enclosed and adherent radioactive sub-stances. The altitude such particles can reach depends on the force of the ex-plosion. While larger particles are precipitated relatively fast by the force of gravity and are most likely deposited nearby the test site, small particles less than 0.4μm diameter reach larger altitudes but do also stay considerably longer in the atmosphere. From this the probability is augmented that these debris are mixed thoroughly, transported and distributed over an essentially larger distance. Since radioactive particles precipitate as local fallout empiri-cally within 100 km, smaller particles are injected into the troposphere or at adequate explosive force also into the stratosphere. Radionuclides present in the troposphere, the tropospheric fallout, which could not pass the tro-popause are transported depending on the actual wind conditions (trajecto-ries) and scarcely are well mixed. From the troposphere the radioactive de-bris are eliminated relatively quickly by "rain-out", which means formation of drops within the clouds, by "wash-out", insertion into falling drops, depo-sition onto the soil or plant surfaces and by precipitation and catching by surfs of the oceans. The medium residence time within the troposphere amounts to about 30 days [UNSCEAR (1982)]. Because of the short residence most radionuclides have no chance to pass from one hemisphere to another. Exceptions from this rule are gaseous radioactive material as for instance ^{85}Kr, which may be exchanged within 1.5 y [UNSCEAR (1977)] between the hemispheres.

Radioactive fragments passed by explosive forces up to the stratosphere are called global or "stratospheric" fallout (see Fig. 9). This leads to a world-wide contamination although the largest part stays in the hemisphere, into which the injection originally took place. The mixing processes in the strato-sphere and the air movements there are, despite numerous investigations with radionuclides, not completely understood. In general the mixing proc-esses in the stratosphere proceed slower than in the troposphere. [Stewart

(1957)] developed a model for the exchange between the stratosphere and the troposphere which is to a great extent consistent with the observations of the behaviour of the fallout from nuclear weapon tests. Especially near the equator the tropospheric air passing over to the stratosphere can ascend to an altitude of up to 30 km. The western jet streams are with their transport velocity of 100 to 300 km/h responsible for a fast transmission. Simultaneously a strong vertical mixing takes place. The tropopause depends on the latitude (see Fig. 9), at the equator it is situated at about 17 km of altitude and in the polar regions at about 9 km. The latitude where the nuclear event took place is of decisive importance for the entrance of radioactive substances into the stratosphere. While in the regions of the upper stratosphere and above the elimination of particles is essentially determined by gravity, in the lower regions adjacent to the tropopause also intermixture results by eddy diffusion with the troposphere (see Fig. 9). The circular air streams in the troposphere at lower latitude are determined by Hardley´s cell circulation [UNSCEAR (1982)], [Newell (1971)]. These cells seasonally grow, decrease, or shift. The medium residence times of the aerosols in the polar regions range from 3 to 12 months, and from 8 to 24 months in the equatorial areas respectively. In the months of springtime more of the debris or the aerosols is eliminated than in the following months. The half-life period for changing from the upper stratospheric layers to deeper levels is 6 to 9 months and from heights of more than 50 km it is about 24 months [Eisenbud (1987)].

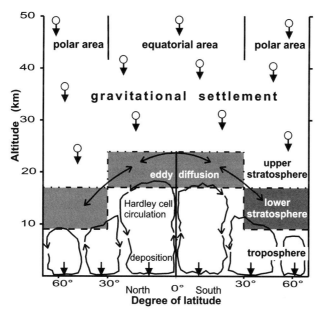

Figure 9: Atmospheric transport and deposition processes of radionuclides, after [UNSCEAR (1982)].

3.1.2.   Behaviour in Aquatic Systems

Fig. 5 demonstrates that the input of pollutants into waters can occur directly or indirectly. Besides the direct release of radioactive substances into rimlets, rivers, lakes or the ocean, the deposition of pollutants, originally discharged into the air, plays a non-negligible role, since the water surface of the oceans of $3.6 \cdot 10^8$ km^2 cover a large part of the earth. The oceans represent with a medium depth of 3800 meters [Sverdrup et al.(1963)], [Kinne (1970)] and a volume of $1.37 \cdot 10^9$ km^3 [Revelle et al. (1957)] a considerably potential distribution area. The behaviour of radioactive substances in water is, in contrast to the atmosphere, far more complicated. Besides diffusion processes, turbulent as well as not turbulent flows, resembling those in the atmosphere, the physical and chemical interactions of the pollutants with the water itself or with therein dissolved or suspended substances, the further fate of the pollutants is essentially influenced by the organisms of the water (see 3.2). Furthermore the type of bottom, the nature of soil adjacent to the shore line, the depth of waters as well as horizontal and vertical circulation, influence the transport, mixing and distribution of radionuclides. All waters like rivulet, river, stream, lake, river delta, bay or the open sea have individual properties that influence intermixture processes. These are dependent not only on the locality but also on the seasons. That is the reason why model calculations are applicable only under special assumptions.

A pollutant present as solid matter suspended in water can be precipitated to the ground by gravitational forces, filtered by microorganisms, or stuck to plant surfaces. Dissolved pollutants can be bound to organic or inorganic materials. They may enter the metabolism of animals, may be absorbed and stored in the tissues of organisms or can terminate as excretion product. The organisms serve alive or after fading out as organic substrate for other symbiosis systems, or form sediments. Sediments often are permanent or temporary sinks of radioactive substances. They may be resuspended and by this way reenter the available biosphere by turbulent currents, by the tides, by change of flow conditions or change of the salt content [Lentsch et al. (1972)]. The elements Sr, Sb, and Cr tend to stay in solution whereas Cs, Mn, Fe, Co and the actinides are prone to be suspended particles.

In the ocean (see Fig. 10), near the surface at a depth of 200 m, there is a region which is pushed by the winds, mechanically well mixed. The thickness of this layer varies geographically and by the predominating meteorological conditions. In this layer the vertical changes of temperature, salt content, and density are extraordinarily small. However, 75 % of the volume of the oceans consists of cold depth water with temperatures of 1° to 4° C and a salt content of 3.47 % [Eisenbud (1987)]. Between the well-mixed superficial water and the depth water there exists a intermediate region in which the temperature rapidly decreases, the salt content and the density, however, rapidly increase. The zone with the largest temperature variation is called the "thermocline"; the area where the density gradient is greatest, is called the "pycnocline." This region is situated at a depth of about 1000 m, or, quite often less than this. Since lighter water lies over denser water, vertical movements are reduced. The exchange between shallow and deep water, i.e., the vertical transport between these regions is slight.

The characteristic currents of the ocean surface are correlated with the wind directions and are responsible for the horizontal transport. The motion of the shallow water in the Florida current (Gulf stream) amounts to 144 km/d and in the Kuroshio current in the Western Pacific to 66 km/d [Nas-NRC (1957a)].

From measurements of the vertical ^{226}Ra- and ^{230}Th-distributions, [Koczy (1960)] developed a model for the vertical diffusion of substances from the bottom of the oceans to the surface (see Fig. 10). Substances having been dissolved from the floor of the ocean, diffuse slowly through the "layer" of 20 to 50 m of thickness. Above this friction layer a quicker intimate mixture occurs, the diffusion constants increase up to values of 30 cm^2/s and then decrease with rising level above the sea bottom. At about 1000 m of depth, in the thermo- or pycnocline zone a second minimum of 10^{-2} cm^2/s is reached. Above this layer in the thoroughly mixed region the diffusion coefficients increase rapidly to 50 to 500 cm^2/s. In water depths of 750 to 1750 m. [Koczy (1960)] estimated the vertical migration rate to be 0.5 to 2 m per year for the Atlantic Ocean. This would mean, applying similar migration velocities at a depth of 3000 m, that dissolved radioactive material would need more than 1000 years to enter the surface water. [Prichard et al. (1971)] estimated a vertical migration velocity of 6.6 m per year from a vertical ^{14}C-concentration profile in the North-East Pacific at a depth of about 1000 m.

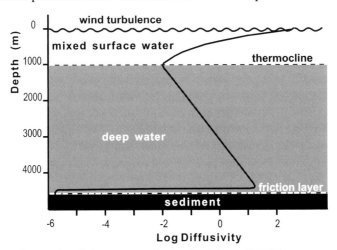

Figure 10: Diffusivity in relation to ocean depth, after [Koczy (1960)].

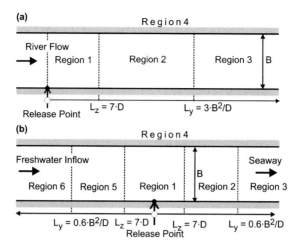

Figure 11: River model based on [NCRP 123 (1996)]: B = river width, D = river depth.
(a): River model   (b): Estuaries model

The [NCRP 123 (1996)] report describes transportation models which estimate the concentrations of radionuclides released into rivers, estuaries, coastal waters, or lakes. These models are suitable only for long-term effluents starting from a point source (stable conditions) and are not applicable in case of released contaminants caused by accidental or temporary purging. In the Fig. 11 the ideas, that the modelling is based, on are sketched.

For a release into a river (see Fig. 11a) at depth D and width B, four different regions are defined. Region 1 is the zone where a vertical intermixture of the radionuclides takes place. L_z is the distance from the release point which is necessary to guarantee a complete vertical mixing. This distance depends essentially on the depth of the river. Within this region an incomplete vertical intermixture is considered, the lateral expansion may be neglected in this zone. Region 2 is the zone where, after vertical mixing, the lateral intermixture over the whole river profile takes place. L_y is the distance from the releasing point to the section where vertical and lateral mixture is completed. The broader and the flatter the river, the larger is this distance. Region 3 is the section of the river behind which the mixing processes are mostly terminated. In this region the concentration of radionuclides is considered to be constant. Region 4 is the opposite river side, the "far" bank. In this section the concentration of radionuclides should be less than or equal to the concentration attained after complete mixing.

An estuary is the mouth of a river where the flow direction of water changes, caused by the tides. The model differs from a pure river model solely by an augmented number of regions which takes into account the temporal inversion of the water flux. In region 1 (see Fig. 11b) no complete intermixture is attained by the up or down flow. In regions 2 and 5 it is supposed that the vertical mixing in the up and down flow is finished and that in these zones the lateral mixture by the up and down flow is completed. Regions 3 and 6 are zones where the radionuclides in the estuary or upcurrent profile are mixed thoroughly both vertically and laterally.

The spreading of radionuclides in lakes depends on the size and the depth of the lake. While in small lakes the distribution tends to uniformity this is not the case for large lakes where, similarly to the oceans, the intermixture decreases rapidly with increasing depth.

3.1.3.   Behaviour in Soil

A large portion of the airborne radionuclides is deposited on the surface of the ground (see Fig. 5). Soil as a complicated physicochemical system contains numerous components that can influence the mobility of radionuclides. It consists of a mixture [Eisenbud (1987)] of organic and mineral material, water and air. In the vertical profile (see Fig. 12), four layers can be distinguished, they are called horizons. Depending on structural conditions, these horizons may be split into further layers.

Horizon A, about 30 to 60 cm of thickness, is the upper horizon and is called surface soil. In this region most of the biological processes take place. In the uppermost layer A_{00} loose foliage and organic fragments are found which mostly are not yet decomposed. The following layer A_0 contains partially decomposed organic material. In the next layers A_1, A_2, and A_3 the organic substances are mixed with mineral material. The share of organic substance diminishes with increasing depth.

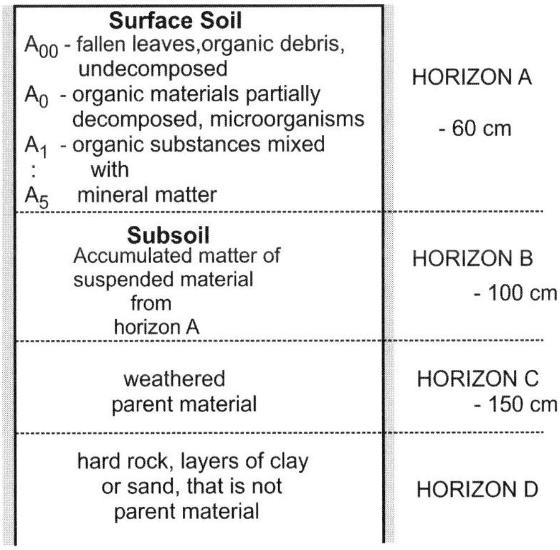

Figure 12: Horizons in a hypothetical soil, based on [Eisenbud (1987)].

Horizon B is also called subsoil and reaches to about one meter below the soil surface. In this zone suspended material from horizon A accumulates.

Horizon C reaching to a depth of 1.5 meters, consists of layers of loose stones, partially broken into small pieces, and of weathered rocks, which are the parent materials of the soils.

In horizon D, layers follow that were not parent material for the soil, but exert a decisive influence on the overlying ground.

Several mechanisms exist for the transportation of radionuclides within the soil, e.g., by diffusion or by convection. Radionuclides may reenter the air by resuspension processes from the uppermost layers of the soil (<1 cm), for example by wind, whirling up, or volatilization. The vertical migration rate in soil [Bunzl (1991)] is not a constant rate but depends on the depth of the ground. It determines the residence time of radioactive substances in diverse layers, for holding nearby plant roots and available for incorporation into plants. Furthermore, the mobility of a radionuclide is a measure for the period in which the radionuclides may enter the ground water and the drinking water. The migration rate, therefore, influences considerably the transfer to the human food-chain. The particular distribution in soil, the depth profile, has a remarkable influence on the external as well as on the internal radiation exposure of human.

Natural radionuclides are present in soil (see Table 1, Fig. 1 and Chap. 3, 1.2), and the content of those may changed by anthropogenic activities like the use of fertilizers, or by emissions from combustion of coal, oil or gas. A large number of artificial radionuclides produced by nuclear weapon tests or by nuclear facilities has been added in the last decades (see Chap. 3, 2.).

After the deposition onto the earth surface, the vertical transport results from the following phenomena. The diffusion of radionuclides plays a dominant role only in the case of very dry soils. This mostly is a very slow process with diffusion coefficients of about $1 \cdot 10^{-6}$ cm^{2}sec^{-1} of the soil solution [Bunzl (1991)]. If there are no further interactions between soil and radionuclide, migration rates of only a few centimetres per year are obtained. The diffusion within the particles themselves occurs even slower. Except for desert soils, this kind of transportation is commonly neglected compared to transportation by water.

Radionuclides, if soluble in the soil solution, may be transported deeper into the ground (transport by convection) by the rain or by artificial irrigation. The stronger the water current directed vertically is, the faster the radionuclide will be transported. The amount of the convection current is influenced by the porosity of the soil material. A radionuclide, insoluble in the soil solution, may be existent as precipitate, colloid, or adsorbed to soil particles. It can be transported if the particles are smaller than the pores that the water trickles through. Small particles, however, adsorbed to surfaces, can be filtered out to a considerable extent. Most radionuclides, especially those existing as cations, are adsorbed to the soil to a substantial degree. Such an adsorption occurs especially in loam, or clay minerals, sesquioxides, or in the humus. The distribution coefficient (K_d-value) is commonly reported as a measure of the sorption. This is the distribution of a radionuclide between soil and interstitial water:

$$K_d = \frac{Activity \quad of \quad soil \quad solids \quad [per \, unit \, of \, weigth]}{Activity \quad of \quad soil \quad solution \; [per \, unit \, of \, volume]} \qquad (5)$$

The K_d-coefficients for defined types of soil are outlined in [NCRP 76 (1984)]. The value of the coefficient characterizes the interaction of a radionuclide with components of the soil and depends on numerous parameters, the most important of which are: kind of radionuclide, its chemical formula, pH, redox-potential, concentration of complex forming ligands, type and kind of the minerals, dimension of the particles, and presence of colloids.

A typical depth profile for ^{137}Cs in forest soil is shown in Fig. 13. The ^{137}Cs-profile results from superimposition of the deposition from nuclear weapon tests in the 1960s with the fallout from the Chernobyl reactor accident in 1986. Since at this time the ratio of ^{137}Cs to ^{134}Cs from the fallout of the Chernobyl accident was 1.75, radiocaesium derived from nuclear weapon tests, however, contained no ^{134}Cs, it was possible to calculate the ^{137}Cs-profile originating from the Chernobyl fallout by measuring the ^{134}Cs-profile. Whereas the largest part of the nuclear weapon test fallout has disappeared from the upper soil layers, "Chernobyl caesium" is still found in the upper soil horizons. The data of [Straume et al. (1996)] on soil profiles of ^{129}I, ^{137}Cs, and plutonium from the Chernobyl fallout in various regions of

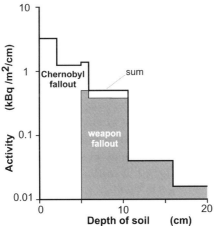

Figure 13: ^{137}Cs activity in the horizons of a forest soil 13 days after the accident at Chernobyl, based on [Bunzl (1991)].

Belorussia show remarkable variations. During 1993, seven years after the accident, 93 % of the plutonium was located in the first 5 cm of the ground, but ^{129}I and ^{137}Cs were found at most within the first 20 cm. Similar observations have been made by [Boone et al. (1985)] in soils near the Savannah River. Radionuclides deposited on the surface of arable soil are brought to deeper levels by ploughing. Repeated ploughing, however, may bring back the radionuclides to the surface.

[Cline et al. (1972)], over a period of 16 years, reported the behaviour of radioactive strontium and caesium in cultivated and fallow land areas. After 16 years ^{90}Sr was still completely available. The decline of activity was nearly exclusively due to physical decay, only 2 % was due to "harvesting". On the contrary, after eight years, only 70 % of the ^{137}Cs was found in the upper soil layers.

3.2.   Transfer of Radionuclides into Food-Chains

In the preceding sections the transfer through the mediums water, air, and soil was considered only to be caused by physicochemical processes without the participation of living organisms. By incorporation of radioactive matter into the organism of living beings the distribution, especially in water and soil, may be essentially altered.

3.2.1.   Transfer from Water to Marine Food-Chains

Radionuclides having entered rivers, lakes or oceans affect all of the organisms living in these ecological systems. Whereas freshwater, as a source of artificial irrigation or for use as drinking water for humans and animals, is to be taken into consideration separately, this point of view is inapplicable with regard to the oceans.

In the waters, basic elements like carbon, oxygen, calcium, nitrogen, etc. are normally dissolved or retained in sediments at the bottom as a reserve for living organisms [Reid et al. (1976)]. These nutrients are absorbed and metabolized by mobile or rooted plants using solar energy. The marine flora, predominantly of seaweeds (algae) which are abundant in various forms almost all nourish autotrophically. Numerous species, like the phytoplankton, float in the water, others are rooted to the bottom of the ocean (benthos). The immense production of organic matter of the algae forms the onset of the food-chain of animal organisms and is a resource for a lot of oceanic animals. Photosynthesis as an energy source restricts the space of life and development to the lightened water regions reaching down to 100 to 200 m of depth. While the phytoplankton species floating in the open water mostly are minute, algae and seaweed fixed on the bottom near the coasts can reach a length up to 300 m. The foodstuffs are adsorbed over the entire surface, the roots only serve as fastening organs. Phytoplankton is food for the zooplankton, as well as the basis of nourishment of life-forms that have a higher level of nutrition. It is also food for certain fish species and animals living near the bottom of the ocean. Sedimentation of excrements or decomposition of dead plants or animals started by certain organisms leads the elements, necessary for nutrition, back to their abiotic matrix, with the result that they are again available to the phytoplankton. Essential trace elements like Fe, Co, Mn, Zn, Cu, and numerous others are present in water and sediments in different amounts These trace elements are accumulated by the water organisms in various ways. Since the effluents of these trace elements into water may comprise radioactive isotopes of these elements, they have to be taken into consideration. [Onishi et al.(1981)], and [NCRP 109 (1991)], give a good survey of the transport of radionuclides in water.

The transfer of radionuclides from water into aquatic biota is characterized by the transfer coefficient, the bioaccumulation factor, *BF* [UNSCEAR (1977)] elsewhere also called concentration factor *CF* [Eisenbud (1987)], [Smith (1982)]:

$$BF = \frac{C_{biota}}{C_{water}} , \qquad (6)$$

where:

C_{biota} = radionuclide concentration of biota and tissue in *activity per kg fresh weight*,

C_{water} = radionuclide concentration of the water in *activity per L*.

The BF-values reported in the literature vary by some orders of magnitude. This fact is not surprising. The determination of transfer factors should

Table 9: Concentration factors in an aquatic environment, after [CEC (1979)].

	Sea Water					Fresh Water	
	Fish	Crustacea	Mollusc	Seaweed	Sediment	Fish	Sediment
^{3}H	1	1	1	1	0	1	0
^{14}C	5000	5000	5000	4000	100	5000	2000
^{90}Sr	1	10	10	10	500	30	2000
^{131}I	10	100	100	1000	100	30	200
^{137}Cs	50	30	30	30	500	1000	30000
^{239}Pu	10	100	1000	1000	50000	10	30000
^{241}Am	10	200	2000	2000	50000	30	30000

be carried out under equilibrium conditions, however in practice for the most part they are not given. The presence or absence of chemically related substances may considerably influence the numerical value of the transfer factors. Caused by competing bonds, as for instance, an influence on the transfer factor of further alkaline earths (e.g., strontium) is to be expected in the case of calcium concentration in water. This also becomes obvious for alkaline metals. The high concentration of potassium in sea water causes the bioaccumulation factor BF, for ^{137}Cs to be some orders of magnitude lower than in freshwater. In Table 9 the concentration factors for various radionuclides are given.

An important aspect for the determination of the transfer factors with respect to the human radiation dose is the question, whether this value relates to parts of organs and tissues which are important for human nutrition. The high transfer rates for instance for ^{90}Sr into the shells of crustaceans are in the first instance not of concern since they are not consumed by men. The same applies also for fish-bones. Accumulations, however, in not directly eaten parts of organs have to be considered, if they are introduced into the human food chain in an indirect way from animal food or fertilizers.

3.2.2.   Transfer from Soil to the Plant

Regarding plants, one has to differentiate between external contamination and internal content of radioactive substances. Both are only important for the food-chain if the whole plant is the basis of nourishment of other organisms.

In Fig. 14 the transfer of radioactive substances onto and within plants is demonstrated schematically. Radionuclides carried by the air or radioactive substances in the water of sprinkling systems are not only deposited on the soil itself, but also to a large extent, and depending on the season, onto the leaves of plants. This external contamination of plants is an important path of radiation exposure of man and animal for pastureland from short lived radionuclides such as ^{131}I. In this way radionuclides pass to grass, into cows, and into the milk. The short lived radionuclides usually only make a minor contribution to the total dose because enough tine usually passes before their direct intake by plants via the soil-root path. The external contamination of plants can be transferred to the ground surface and may enter the ground by washing or dropping off from the contaminated parts of the plants. Radionuclides, deposited earlier onto the soil, can be resuspended and deposited onto parts of the plants once more. By translocation, shifting of radioactivity within the plant may occur.

Plants mostly absorb foodstuffs by their roots. Trace elements and minerals are taken up preferably as ions [Wirth (1980)]. For the transfer to plant cells, these substances have to pass a boundary layer (plasmalemma) which only water, gases and organic molecules of a molecular weight of less than

500 daltons may pass relative easily. Ions are able to pass this barrier only by carriers present in the plasmalemma. These carriers are binding ions of the external solution, which transport them across the barrier and exchange them within the cell for plant ions derived from the respiration (H^+, HCO_3^-), being passed outwards as an equivalent. Because of the specification of these carriers the plant has the ability to absorb the essential elements preferentially. As a rule, these transport mechanisms are not so selective, that the absorption of superfluous or toxic substances is entirely excluded. In particular, these mechanisms are not suitable for preventing the absorption of the radioactive isotopes of an element necessary for the regular nutrition of the plant. The extent of absorption of a radioisotope from soil by a certain plant is called transfer factor T_F. T_F is defined as follows [Wirth (1980)], [NCRP 76 (1984)]:

$$T_F = \frac{C_P}{C_S} \qquad (7)$$

with

  C_P = activity of a radionuclide within a plant relative to the fresh weight,

  C_S = activity of a radionuclide in soil relative to the dry weight.

Some authors do not relate the activity in vegetation to the fresh weight of the plant, but to the dry weight.

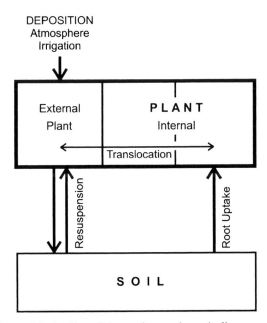

Figure 14: Transfer model of radionuclides to plants, schematically.

The extent of the transfer of a radionuclide depends on numerous parameters. Some of these parameters are related to the species of the plant. For example the exchange capacity of its roots determined by the production of H^+ and HCO_3^--ions is an important individual factor [Wirth (1980)]. By transporting ions outwards the plants are able to exchange adsorbed ions from the soil, to mobilize foodstuffs by acid digestion and to make them available for their roots. By excretion of organic chelating or reducing agents derived from their metabolism, plants are able to use even compounds that are difficult to dissolve.

Besides the species of a plant the soil composition is important for the extent of transfer. The transfer factor depends not only on the concentration of a radioisotope but also on the sum of all isotopes of this element prevalent in soil and, additionally, on the presence of chemically similar elements. In this way the natural occurrence of the stable elements strontium, calcium, caesium and cobalt essentially influence the absorption of their radioactive isotopes (carrier effect). Interactions with chemically similar elements may vary the uptake rate into plants. For instance at high concentrations of magnesium in soil [Cataldo et al. (1980)], the calcium and potassium accumulation in plants is inhibited, whereas the intake of zinc and manganese is augmented. As a consequence of the obviously not highly specified transport mechanisms, similar elements like potassium and caesium are influenced by each other. A further parameter is the availability of the elements for the plants in the soil, which is one of the reasons, for a distinction between mobile and reserve nutrients. Mobile substances are mostly water soluble or adsorbed but easily exchangeable, and are immediately available to the plants. The reserve nutrients in inorganic or organic compounds have to be disintegrated prior to becoming available. This disintegration process is initiated by weathering, prevalence of microorganisms in soil, or by the plants themselves. To what extent soils retain substances in a form available to the plants, depends on the ability for adsorption. The more finely grained the soil is, the larger is the surface for adsorption. Clay containing soils adsorb much more substances than sandy soils. The pH-value of the soil influences not only its ability for adsorption, but also the solubility of numerous compounds as well as the valence of ions. Water as a transport vehicle and as a solvent in soil provides for easier mobilization of reserve materials but also for dilution of water soluble compounds. Besides the rainfall, the water content of a soil depends on the sort, the structure, and the pre-treatment. Since plants are able to take up chelating agents too, the content of naturally and artificially added chelating agents, available to the plants, are of importance for the extent of the transfer of radionuclides.

The determination of transfer factors with respect to all these parameters is extraordinarily complicated. Since, for reasons of radiation protection, such experiments are mostly limited to controlled laboratory conditions. Problems will arise from transmitting the behaviour of the nuclides to the open ground, for instance a large field, or to undisturbed soils with different horizons. Furthermore, the chemical composition of the radionuclide containing compounds deposited onto the soil is scarcely known. For this reason, compromises have to be accepted for the calculation of radiation exposure using transfer factors. Generally, as a conservative assumption, the transfer factors for soluble compounds (available to plants), and only mature parts of plants destined for consumption should be considered for approximations of radiation exposures.

It is plausible that the depth distribution of the nuclide in the soil influences the uptake in plants by the roots. Immediately after the deposition of the radionuclide the main bulk of the nuclide is located directly nearby the surface of the soil. Depending on the type of soil and on the distribution of the microorganisms in soil, the concentration arrangement will adjust in the surroundings of the plant roots. Usually if the nuclides are not well fixed to soil, these concentrations decrease by an occurring dilution by more profound penetration or by taking away by water.

In publications of [Wirth (1980)] and [Matthies et al.(1982)], transfer factors for different radionuclides and foodstuffs are discussed in detail. Further papers concerning transfer factors are published by the following authors: [Römmelt et al. (1991)], [Cohen (1985a)], [Marckwordt et al. (1971)], [Nishita et al. (1961),(1958)], [D'Souza et al. (1971)], [Fredriksson et al. (1958)], [Garret et al. (1971)], [Keller (1990)], [NCRP 123 (1996)], [Sheppard et al. (1997)]. The knowledge of the order of magnitude of transfer factors is not only extraordinarily important with respect to precautions at the setting-up of nuclear installations but also is meaningful for the estimation of the radiation doses to be expected after nuclear accidents derived from ecological models (see 3.3).

3.2.3. Uptake in Animal and Human Bodies

As multifarious the fauna is, as numerous are the paths of transfer and the behaviour of radioactive substances in living organisms. Therefore in this chapter, only a rough survey can be given on the behaviour of radionuclides in animals playing a predominant role for the human food-chain.

Fig. 15 shows a survey on the transfer routes of radionuclides into the food-chain of animals and humans. By migration through the food-chain, radioactive substances may pass from the soil by way of consumable plants into the organism of animals or humans (ingestion). Whereas pasture cattle may take up radionuclides present in the upper layers of soil and upon the plants, for man this kind of transfer commonly should be excluded because of the cleaning of vegetable food. Transferred to the cattle, radioactive isotopes enter the human body by the intake of milk and meat. By this way considerable accumulations of radionuclides may occur. Since many animals form the basic nourishment for other animals, radionuclides are secondarily transmitted to these animals. This fact becomes obvious in the case of

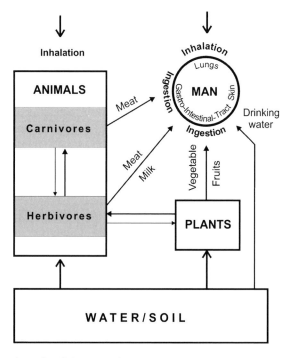

Figure 15: Migration of radioisotopes through the food-chain in the terrestrial and aquatic environment.

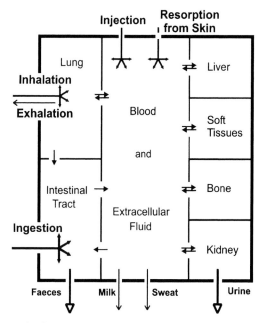

Figure 16: Behaviour of radionuclides in mammal animals or men.

predatory fish, their radioactivity does not start to increase considerably until radioactively contaminated water has entered the fish that they prey on. At this point in time the radioactivity in the water itself may have already decreased to zero again. In the case of ingestion, the transfer of the radionuclide may be limited to those parts of plants or of those organs of animals, which actually are consumed by humans or animals. Sometimes it is also important to consider parts of plants and animals that do not belong to nutrition in a natural way but which form useful by products of the agriculture and food industries, for example bone and fishmeal. However by migration through the food-chain, not only the ingestion is a noteworthy path, but also the inhalation (incorporation by lungs) and the percutaneous resorption (incorporation through skin).

The resorption of radioactive substances (see Fig. 16) by the lungs (inhalation) proceeds rapidly in general. Radionuclides having entered the lungs by inhalation, are firstly retained within the bronchi or alveoli of the respiratory tract. The extent and the place of the retention essentially depends on the particle sizes of the aerosols. Particles which are readily dissolvable or can be solubilized by biological complex-forming substances, are taken up into the blood stream within a few hours. Minor soluble particles are moved by means of mycociliary transport to the upper sections of the respiratory tract and from there they enter the digestive system. In the region of the alveoli they may be resorbed by macrophages and carried to regional lymph nodes or fixed in situ by biochemical reactions.

The transport mechanistics for radioactive substances in the body including their intake, biokinetics, distribution, uptake, retention, and excretion, are mostly characterized by their chemical and physical reactivity in a given physiological medium. The resorption of radionuclides by the digestive tract is not only marked by the physicochemical properties of the radioactive substance in interaction with the digestive mechanisms, but also by additionally occurring reactions with nourishing constituents, which may accelerate, delay, or even largely inhibit resorption. Table 10 gives a survey on the extent of resorption of various elements.

The further transport of the resorbed radionuclides mainly takes place in the vascular system (see Fig. 16). The amount of intake and the concentration in various organs or tissues depends on the organs themselves as well as on the specific properties of the radionuclide. ^{137}Cs, for instance, is distributed, because of its chemical relationship in a similar manner to potassium and is accumulated in muscle, the isotopes of iodine are stored in the thyroid gland, and the calcium isotopes and the chemically related alkaline earths are transferred into bone. The distribution of some elements after intravenous application are given in Table 11.

Table 10: Resorption of radionuclides from gastrointestinal tract of man, based on [Kriegel et al. (1985)].

Element	Resorbed Fraction (%)
H, C, N, O, S, Halogens, Alkalines, Noble Gases	up to 100
P, Hg, Ca	40 - 70
Tc, Tl, Re	40 - 50
Sr, Ra, Co, Cu, Ni	10 - 60
Mg, Al, Mn, Fe, Zn	- 10
Ba, Ru, Pb, Po, As	1 - 10
Ga, In, Cr(III), Nb	0.1 - 1
Rare Earths, Actinoides	< 0.1

Table 11: Distribution of radionuclides after intravenous application in relation to valence, based on [Kriegel et al. (1985)].

Chemical Valence	Distribution in Organism		
	Ubiquitous	**Hepatotrop**	**Osteotrop**
1	Li, Na, K, Rb, Cs	-	-
2	-	-	Be, Ca, Sr, Ba, Ra
3	Ru	La, Ce, Pr, Pm, Pu, Ga	Y
4	-	Hf, Th, Pu, U, Ce	Zr
5	Nb	-	P
6	Te, Po	-	Pu
7	Cl, Br, I	-	F

Numerous experiments with different animal species and studies on man with respect to the transfer and distribution of radionuclides in organisms have been published [Furchner et al. (1971ab), (1973)], [Stara et al. (1971)], [Liniecki (1971)], [ICRP 20 (1973)], [ICRP 30 (1979)], [ICRP 56 (1989)], [ICRP 67 (1993)], [ICRP 66 (1994)], [ICRP 69 (1995)].

From nuclear weapon tests and the development of the nuclear techniques the necessity arose to control the behaviour of the released radionuclides within the food-chain and their metabolism. Special emphasis is given to radioisotopes (see 3.4) with prolonged half-lives, and to those, which at their migration through the food-chain, have a chance to enter the human organism. Peculiar behaviour is found for radioactive iodine, which is, although short-lived because of its high mobility and degree of accumulation in the thyroid gland, of special importance for the internal radiation exposure.

3.3.   Ecological Models

In order to estimate, retrospectively or prospectively, the consequences of radioactive emission, numerous ecological models have been developed, which aim to quantify the transfer and spread of radionuclides through the food-chain to man. A survey was given by [UNSCEAR (1982)] and [NCRP 76 (1984)]. Some of the models are concerned with spreading from the place of emission, others preferentially deal with transfer through the food-chain.

The exactness of the prognoses depends on individual features. The more general such models are, the rougher are the prognoses for defined positions. Especially, these models are not suitable for a retrospective derivation of the original extent or the place of emission given the radioactive content in food or man.

Most of the models aim at defined locations, as for instance a milk farm for airborne radionuclides like ^{144}Ce, ^{137}Cs, ^{90}Sr, ^{54}Mn [Pelletier et al. (1971)]. The Columbia river near Hanford in the state of Washington, USA, is supposed to be the best studied river section in the world. Because of the effluents of the plutonium producing nuclear plants in its water streams, the behaviour of numerous radionuclides was observed and special spreading models have been developed and examined.[Price et al. (1985)], in 1985, detected in the river water, with regard to ^{137}Cs, ^{131}I, ^{89,90}Sr, and ^{60}Co, a relatively good agreement of the concentrations observed and those predicted from the model. Another example is a model for the resuspension of ^{137}Cs in the 30 miles zone around the Chernobyl reactor [Nair et al. (1997)].

The transfer routes in nature are complicated and depend on many parameters. This is the reason why for a special position, not only the meteorological conditions and sorts of soil, but also the season, the locally used production methods in agriculture, consumption, feeding habits, and the trade routes for food have to be taken into account. For realistic validation of radiation exposure of man, alterations in feeding habits must also be considered by knowledge of the situation as feedback mechanisms.

The RADFOOT-model by [Koch et al. (1986)], simulates the transport of various radionuclides from the agricultural production to the foodstuffs and calculates the radiation dose. Various types of soil, the intake of different vegetable and animal products are taken into account for the transfer. A similar computer model, "ECOSYS", was introduced by [Matthies et al. (1982)], which, after the Chernobyl disaster, forecasted realistic dose assumptions for the region of Munich. An advanced model by [Pröhl (1990)], [Müller et al. (1993)], [Poon et al. (1997)] evaluates doses occurring by a time dependent contamination of the food-chain after a nuclear accident. This model is suitable to be adapted to other geographical locations by entering data concerning the feeding habits and agricultural production methods of those locations.

3.4. Behaviour of Selected Radionuclides in the Environment

3.4.1. Behaviour of ^{106}Ru, ^{54}Mn, ^{60}Co, ^{65}Zn, ^{55}Fe, ^{22}Na, ^{241}Am, Np, Cf

The following chapter is concerned with the behaviour of selected radionuclides with importance for radiation protection. These radionuclides are naturally present, or released by nuclear industries. Literature for some of them with secondary importance is cited below.

[UNSCEAR (1977)], reported an accumulation of ^{106}Ru in the seaweed *Porphyra*, from which "laverbread" is produced, a nutrient rich in iron and iodine. In a later paper [UNSCEAR (1982)] it was reported that ruthenium may contribute to the dose in the lungs by inhalation. [Furchner et al. (1971b)] has examined the metabolism of mice, rats, monkeys, and dogs, [Stara et al. (1971)], that of rats and cats. [Masse (1982a)], gave a survey of the behaviour of ruthenium, including that of ^{54}Mn, ^{60}Co, and ^{65}Zn in the environment and in animals.

[Beasley et al. (1972)], reported on the ^{55}Fe-content in soil, fish, and humans at Rongelap after the nuclear weapon tests on the Marshall Islands and [Mattson (1972)], covered the ^{22}Na in the arctic food-chain "lichen-reindeer-man" interfered with by Soviet tests. A survey of the behaviour of radiocerium was given by [NCRP 60 (1978)], [Masse (1982b)] and [UNSCEAR (1977), (1982)]. A historical survey of americium was given by [Breitenstein et al. (1985)], and [Heid et al. (1985)]. The behaviour of ^{241}Am was detailed by [Stather (1982)]. [Palmer et al. (1985)], [McInroy et al. (1985)] and [Lagerquist et al. (1973)] have published data for the body burden from americium and techniques for its determination. [Harrison (1982)] supplied a contribution on intestinal uptake. [Durbin et al. (1985)] outlined a metabolism model for americium. [Voelz et al. (1985)] describes a follow-up study carried out with plutonium workers on the "Manhattan" project, which had incorporated plutonium and americium. [NCRP 90 (1988)] and [Nenot (1982ab)] reported on neptunium in detail, also considered californium, and [Benett (1976)] reported on the transuranic elements pathway to man.

3.4.2. Tritium

Hydrogen is bonded to oxygen (water) or to carbon in different compounds, one of the most important elements of living nature. This element forms only one radioactive isotope, ^{3}H, tritium (T). Tritium behaves chemically like the stable isotopes ^{1}H of hydrogen, despite the fact that, because of its larger mass, isotopic effects may occur. This is expressed, for instance, for tritium water (^{3}H$_2$O) by a boiling point increased by 1.5 °C, as well as a lowered vapour pressure [NCRP 62 (1962)].

Tritium (half-life = 12.3 y) is released into the environment by natural paths (see 1.3 and Chap. 4, 3.) produced by the interaction of cosmic radiation with oxygen or nitrogen [Smith (1982)], [Jammet et al. (1982)], as well as artificially from nuclear weapon tests (see 2.4 and Table 8) or from nuclear systems (see Chap. 3, 2.2) resulting from ternary fission of ^{235}U [Tadmore (1973)]. A survey of the behaviour of tritium in the environment was given in [NCRP 62 (1979)], [Jeanmaire (1982a)], and by [UNSCEAR (1962), (1977), (1982)].

In the atmosphere 5 km below sea level, tritium is rapidly transformed into HTO, in larger altitudes, however, it rather exists as HT. The residence time in the troposphere amounts to 20 to 40 days [Libby (1958)], [Barett et al. (1961)], [Bolin (1964)]. By raining, it reaches in the tritiated water the common water circuit and by that all ecosystems. The deposition of HTO is larger over sea than over land areas, since tritium is additionally exchanged between the air and the water surfaces.

In the ocean, tritium-water is very rapidly mixed with a thin warm-water layer, called the "mixed surface"-layer (see 3.1.2, Fig. 10), reaching down to a depth of 50 to 100 m. Below the thermocline, the time necessary for the mixing process decreases rapidly and the concentration of ^3H is weak.

The importance of hydrogen for living things has motivated a large number of studies on tritium metabolism [Pinson (1951)], [Pinson et al. (1957)], [Siri et al.(1962)], [Butler et al. (1965)], [Woodard (1970)], [Osborne (1966), (1972)], [Silini et al. (1973)]. The incorporation of tritium can occur by inhalation as well as by ingestion as tritium gas, water, or bound to organic matter. Introduced as tritiated water, the radionuclide enters within a very short time period the extra- and intracellular body water and biokinetically participates in the water system of the organism with a biological half-life of about 9.7 days in the case of humans. A small part is also bound organically. In compounds containing carbon, the half-life in the body may be extended considerably to about 450 days. In this case, the retention depends on stability of the organic compound and its metabolism, but also in this case, tritium may be substituted by stable hydrogen from body fluids to a considerable extent. Samples were taken by the US Food and Drug Administration while monitoring foodstuffs, especially from reactor adjacent areas. In 1975 and 1976, [Simpson et al. (1981)] observed tritium concentrations of 1.48 to 17.4 Bq/kg in food samples. During an observation period from 1975 to 1982, [Stroube et al. (1985)] established 3.7 to 7.4 Bq/kg in cereal products, 4.44 to 6.66 Bq/kg in fish and 2.2 to 5.18 Bq/kg in milk.

3.4.3. Radiocarbon

Carbon is, like hydrogen, a basic element for all living organisms. All radioactive isotopes, with the exception of ^{14}C (half-life = 5730 y), are far too short-lived to play an important role in the environment or in the accumulation within the food-chain.

^{14}C is naturally generated in the environment by the interaction of cosmogenic radiation with the atmosphere (see 1.2.1 and Chap. 4, 2.), and artificially released by nuclear weapon tests (see 2.4 and Chap.4, Fig. 4) and by emission from nuclear devices (see Chap. 3, 2.2). A survey of its behaviour was given in [NCRP 81 (1985)], [UNSCEAR (1962), (1977), (1982)].

Our planet contains $1.6 \cdot 10^{24}$ g of carbon, $4.1 \cdot 10^{19}$ g of which are exchangeable. About 95 % of the exchangeable part is located in the oceans alone, the rest is contained in the atmosphere and in the terrestrial biosphere. In the atmosphere carbon exists mainly as CO_2, whereas in the ocean the predominant inorganic compounds are 89 % HCO_3^-, 10 % CO_3^{2-}, and about 1 % dissolved as CO_2 and H_2CO_3. The next most frequent deposits are marine sediments (detritus) and dead organic material. Carbon in living nature is present in an overwhelming quantity as phytoplankton, synthesizing organic material in the upper levels of the oceans. It establishes the basis for life in the oceans. A further reservoir in the oceans are carbonates having been produced originally as $CaCO_3$ by biota in the surface layers. Terrestrial reservoirs worth mentioning are particularly woods, furthermore terrestrial animals, including humans.

^{14}C participates in the carbon circle and is, similarly to ^3H, highly mobile within the environment. During photosynthesis, the isotopic effect is noticeable by a slower reaction of the heavy ^{14}C compared to the lighter carbon isotopes. ^{14}C released into the environment (see Chap. 4, 2.2), enters the exchangeable carbon pools of the terrestrial biosphere, of the oceans, and of the sediments. In the stratosphere as well as in the troposphere, ^{14}C achieves an equilibrium state (half-life = 2 y, [Smith (1972)]) and circulates as CO_2, in balance with the carbon of the ecosystems. The medium duration of stay in the troposphere and in the "short-term" compartments of the biosphere (for instance annual plants) amounts to 7.5 years until the radioactive carbon enters the long-term compartments such as trees, animals, or oceans [UNSCEAR (1977)].

A man with a body weight of 70 kg contains about 14 kg of carbon and takes in 350 g of carbon per day [Jeanmaire (1982b)]. From this a biological half-life results of about 28 days. After incorporation of ^{14}C into the human organism as $^{14}CO_2$ (inhalation) a small part (only about 1 %) of carbon is retained in fat, proteins, carbohydrates, or as bicarbonate. If present in organic matter ^{14}C is transformed according to the respective metabolism of the compound. After conversion, being possibly very complex, ^{14}C is excreted to a large extent in the form of CO_2 or as urea. The biological half-lives depends on the nature of the compound. The longest half-life, for glycine for instance, amounts to 35 days [ICRP 10 (1968)], for KCN to 10 d, for methanol to 45 d, and for glucose up to 290 days [Jeanmaire (1982b)]. In each case the excretion proceeds multiexponentially.

3.4.4. Radioiodine

The biologically essential element iodine consists naturally, of only two isotopes occurring in appreciable amounts, the stable ^{127}I, and the radioactive isotope ^{129}I, a radioisotope with a half-life of $1.57 \cdot 10^7$ y (see Table 12).

^{129}I is found only as traces in air, water, and soil. It is generated by interaction of cosmic radiation with atmospheric xenon, or at spontaneous spallation of uranium or thorium [Edwards (1962)]. Previous to the nuclear weapon tests, in geological water samples from the year 1930, the relation ^{129}I/^{127}I was $5 \cdot 10^{-12}$ to 10^{-13} [Edwards et al. (1969)] and it increased during the tests in the 1950s to about 10^{-6} [Keisch et al. (1965)]. Among all known isotopes of this element, only those named in Table 12 are of importance for radiation protection. For intake into the food-chain or to the air used for respiration, only those are of interest where the half-lives are sufficiently long or the released activity amounts were large enough. From all radionuclides generated by nuclear fission, only the two products ^{131}I and ^{129}I commonly remain as long-term burdens. These two radionuclides enter the environment on a larger scale from nuclear weapon tests and accidents [UNSCEAR (1962), (1977), (1982), (1988)]. They play only a secondary role during normal reactor operation, however, the concentration of ^{129}I in the environment will be augmented by the release of reprocessing facilities in the future [Eisenbud (1987)], [NCRP 75 (1983)].

In nature, stable iodine is not distributed uniformly. The largest part is located in the water of the oceans. [Turekian (1969)] reported a concentration of 0.064 μg/g in sea water, i.e., a pool of about $6.4 \cdot 10^{14}$ g of stable iodine results from a total water volume of about 10^{22} g. A large part of the iodine is exchanged via evaporation between water and atmosphere.

Table 12: The most important radionuclides of iodine after thermal fission of ^{235}U or ^{239}Pu and activation.

Iodine-mass	Half-Life		Fission Yield (%)	
			^{235}U	^{239}Pu
125	59.4	d	activation	
129	$1.57 \cdot 10^7$	y	0.7	1.4
131	8.02	d	2.9	3.8
132	2.3	h	4.3	5.4
133	20.8	h	6.7	6.9
134	52.5	min	7.7	7.3
135	6.6	h	6.3	6.5
137	19	s	3.2	2.4

In the atmosphere the largest part of iodine is elemental iodine I_2, or iodine bound to aerosols. But also iodides, iodates, and organic iodine compounds like methyliodide are detectable. The average residence time of iodine in the atmosphere is between 15 days and two years [Allen et al.(1980)]. Iodine is dropped by wet or dry deposition onto the soil or onto the vegetation.

In soil iodine is concentrated preferentially within the first layers of the soil. The behaviour of ^{129}I in soil has been observed by [Boone et al. (1985)] near the Savannah region for 25 years. Within the first 30 cm of the soil the residence half-time was 30±6 years. The migration into the ground is of interest solely for ^{129}I, since all the other isotopes of iodine have only a short half-life. For those short-lived isotopes, direct deposition onto plants and resuspension is the starting point into the food-chain. Grazing animals, for instance, may take up these nuclides together with their food or by respiration of air. Exclusively in this way the transfer to the food-chain can take place fast enough to cause an internal radiation exposure of man. In 1953, for example, the presence of ^{131}I in the thyroid gland of cows was observed in the neighbourhood of the test site in Nevada [Van Middlesworth (1954)]. The availability of iodine from soil for the intake by roots depends on many parameters. Iodine, mostly occurring as iodide, is up to 10 % available for absorption by roots [Whitehead (1984)]. The average concentration in soil amounts to 5 $\mu g/g$ but in terrestrial plants and animals it is on average less than 0.5 $\mu g/g$ [Robl (1997)], [O'Dell et al. (1971)].

The intake of iodine by organisms can occur by ingestion as well as by inhalation. The absorption to the blood takes place very rapidly for three hours in the small intestine or in the lungs [Tubiana (1982)] and nearly at 100 %. In blood plasma iodine is quickly removed by the thyroid gland and the kidneys. Considerable accumulations occur also in the stomach, in the salivary glands, in mammary tissues and in their secretions, in which they can reach more than the 40-fold amount of that of the plasma concentration [O'Dell et al. (1971)]. The iodide concentrations in the other tissues, such as muscle and blood amounting to 1 to 4 $\mu g/L$ are extremely low [Tubiana (1982)]. The normal daily intake in adult humans ranges up to about 100-200 μg of iodine [Robl (1997)], [Tubiana (1982)]. At iodine deficiency, endemically present in larger regions of many countries, the daily supply amounts to only 30 to 50 μg of iodine.

The metabolic fate of iodine in the organism, however, is determined by the thyroid gland and its hormones. In regions with high iodine concentration in food, the uptake of the thyroid gland amounts to 20 %, at still sufficient supply to 50 %, and in iodine deficient regions it increases up to about 80 %. The gland is capable of concentrating iodine up to the 10000-fold amount compared to the concentration in blood. Iodine enters the thyroid gland from blood plasma as iodide (iodination). In the thyroid iodine is inserted into tyrosine by the peroxydase reaction (iodisation) leading to the formation of monoiodo- and diiodotyrosine. By coupling of the two amino acids, tetraiodothyronine (T4 ≡ thyroxine) and triiodothyronine (T3) are formed, both thyroid hormones, which bound to thyroglobulin are partially stored in the colloid of the thyroid follicles. About 80-90 % of the total iodine content of the organism is bound within the thyroid gland and only 10-20 % can be found externally, predominantly as hormone bound iodine. The concentration of hormone-bound iodine in blood and muscle amounts to 0.05 $\mu g/g$. The secretion of hormones into the blood circulation occurs after hydrolysis of the thyroglobulin. In blood, the hormones are to a large extent protein-bound (95-99 %). The level of hormones not associated with proteins is controlled by the pituitary and the hypothalamus. At decreasing hormone levels the thyroid gland is induced to increase the secretion of T3 and T4 by secretion of thyrotropin. From blood the hormones enter the cells, where they are metabolized while iodide is produced as one of the ultimate products. Free iodide is carried to the blood-iodide pool and may be again introduced into the thyroid gland. The excretion of iodide predominantly occurs by the kidneys. In rats, a non negligible secretion of iodide and thyroid hormones with the bile to the small intestine was observed [Berg (1976)]. The most important way with respect to the food-chain, is the fast and high excretion with milk. In milk, after a single oral administration of ^{131}I to cows, already after 30 minutes an increase is observed leading to a maximum of the concentration after 12 to 24 hours. The evidence for the high mobility of ^{131}I, despite its short half-life of only eight days, was the detection of ^{131}I in milk samples and in human thyroid glands in New York during a Soviet three months test series of 1961 [Eisenbud (1987)]. Although the test site was situated thousands of kilometres away, 3.7-7.4 Bq ^{131}I per litre milk and 18.5 to 25.9 Bq in the thyroid glands have been measured.

To protect the thyroid gland against excessive radiation doses by radioiodine, substances inhibiting the uptake of iodine can be administered. The absorption can be drastically diminished by thiocyanate, perchlorate, stable iodate, or iodide. Oral administration of 100 mg iodine (I⁻) [SSK (1989)] blocks the uptake of radioiodine within 30 minutes, 30 mg reduces the uptake considerably, and by a dose of 15 mg/d this suppression can be maintained [Tubiana (1982)]. A special risk exists during the foetal and perinatal period and during the period of childhood. The foetal thyroid gland starts concentrating the iodine from the 9th to the 11th week of gestation onwards. Furthermore there is an increased radiation exposure risk from ^{131}I by the growing of the gland (1 g for neonatal, 2 g after 6 months, 4 g after 4 years, 14 g after 14 years an 14 to 20 g for adults [Tubiana (1982)], [Brian et al. (1985)] and by augmented turnover, perinatally and during childhood. At iodine deficiency this risk is increased further by an augmented uptake.

[Bernhardt et al. (1971)] reported on actions diminishing the ^{131}I uptake in the human thyroid gland by interrupting or influencing the transfer from the path pasture-cow-milk. These precautions especially effect children with normally the highest milk consumption. At first, after the nuclear weapon test at the Nevada test site, such actions were undertaken in the surroundings and in Utah in 1962. After the "PINE-STRIPE"-event in 1966 (subterranean test) the farmers were instructed to feed the animals stored food and to detour fresh contaminated milk to dairies and to process it into milk products. Because of the physical decay of ^{131}I during storage and processing, the content of radioiodine was diminished drastically. Furthermore, with appropriate fertilization of the pastures, or by additional administration of stable iodine at a dosage of 4 g/d to the cows, the ^{131}I-concentration in milk can be reduced up to 50 %.

3.4.5. Krypton

Krypton (see Chap. 4, 5.3) is classed within the rare gases and usually not involved in biological processes, like for instance radon. The naturally occurring ^{85}Kr (half-life = 10.7 y) is generated from ^{84}Kr by neutron capture activation from cosmic radiation. In soil and in the oceans this isotope is simultaneously produced together with ^{129}I at the spontaneous or neutron induced fission of natural uranium.

At nuclear weapon tests or in nuclear reactors, ^{85}Kr is formed as a fragment with a probability of 0.285 % from ^{235}U, and of 0.144 % from ^{239}Pu [UNSCEAR (1977)]. 185 PBq of ^{85}Kr originate from nuclear weapon tests carried out from 1945 to 1962, and an additional 160 PBq from those taking place before 1980 [UNSCEAR (1982)]. Up to 1966, 555 PBq of ^{85}Kr have been released by the production of plutonium by the USA.

The utmost part of ^{85}Kr is produced in nuclear piles [Tadmore (1973)], [Smith (1982)] about 11.1 TBq [NCRP 44 (1975)] to 14.8 TBq [UNSCEAR (1977)] per MW electrical power. During regular reactor operation only a small amount (<1 %) escapes from the fuel elements (see Chap. 3, Table 11), the vast majority remains in the fuel rods until reprocessing (see Chap. 3, Table 12). It has been estimated that the reprocessing facility Allied Gull Nuclear Services near Barnwell, South Carolina, has released 592 PBq of ^{85}Kr per year [USAEC (1974)] and during the years 1966 to 1972 48.1 PBq [UNSCEAR (1977)]. In 1972 at Sellafield, formerly Windscale, Great Britain about 37 PBq [UNSCEAR (1977)], between 1980 and 1985, 230 PBq additionally whereas in France at Cap de la Hague 267 PBq [UNSCEAR (1988)] were emitted during the same period. The worldwide annual release for year 2000 is estimated to be 36.6 EBq with a total artificial inventory of 232 EBq [Nichols et al. (1971)], [NCRP 44 (1975)].

In the atmosphere $1.64 \cdot 10^{16}$ g of krypton are present [NCRP 44 (1975)], in the upper layer of the oceans down to 100 m depth (well-mixed region) about $6.7 \cdot 10^{12}$ g, with a total amount in the oceans, $4.7 \cdot 10^{14}$ g of krypton. The stable krypton pools and the exchange rates in relation to half-life of ^{85}Kr are a model for the final distribution of ^{85}Kr. ^{85}Kr released to the environment is mainly distributed within the atmosphere and only to a minor extent in the oceans or in the soil. The wash-out of ^{85}Kr by rainfall is negligible due to its poor solubility and the presence of a stable carrier. The same applies for the soil. Taking into consideration the long-term mixing process especially in the deep water, the oceans do not represent a sink for ^{85}Kr in the atmosphere. Lacking in a significant sink, ^{85}Kr will distribute in the atmosphere over the whole globe and decline with a half-life of 10.7 years.

Since krypton is deposited only to a negligible amount onto surfaces and not involved in biological processes, ^{85}Kr scarcely plays a part in the food-chain. Only the direct irradiation of man (external) and the burden from the inhaled krypton (internal) is to be taken into consideration. Krypton shows low solubility in blood but a larger one in fat, and it is rapidly diffusing in the tissues [NCRP 44 (1975)]. For the risk assessment for humans by ^{85}Kr, dose graduated environmental models [Machta (1974ab)] were developed which allow evaluations near the emission source (about 100 km).

3.4.6. Radon

Radon is classed within the rare gases and usually not involved in biological processes, like for instance krypton. Radon isotopes, being intermediates in the natural decay series, belonged to the first natural radionuclides to be discovered and investigated at the beginning of the 20th century. These naturally occurring radioisotopes, together with their radioactive progenies (see 1.2), are of fundamental importance for the radiation exposure. The resulting natural exposure of man amounts to about 55 % of the natural background radiation dose [Miller (2000)]. [Price (1997)] and [Lévesques et al. (1997)] reported data on radon in US-buildings in relation to the subsoil. In [UNSCEAR (1977), (1982), (1988), (1993)], [Eisenbud (1987)], and [SSK (1992ab), (2002)] detailed data were summarized.

Radon is produced in the soil and released partially into the open air. From loose soil, through the interstices between sand grains and from crevices, radon may pass to the atmosphere. ^{219}Rn (actinium emanation, actinon) has, because of its short half-life of 3.96 seconds, nearly no chance to pass through the deeper ground levels to the surface. ^{220}Rn (thorium emanation, thoron) with a half-life of 55.6 seconds, however, diffuses from a few centimetres of depth to the surface of the soil. With its physical half-life of 3.82 days ^{222}Rn (radium emanation, radon) reaches the ground even from larger depths, as for instance, from 1 m of depth at an amount of 50 % into the surface air [Israel (1962)]. [Wilkening (1956)], reported an emission rate from soil between $2.2 \cdot 10^{-4}$ to $5.2 \cdot 10^{-2}$ Bq·m^{-2}s^{-1}, and a total emission from ground of about $1.85 \cdot 10^{12}$ Bq·s^{-1}. The contribution from the oceans to the total radon content in air is about 100-fold smaller [UNSCEAR (1988)]. After having entered the air, radon can cover considerable distances within the lower troposphere. Impulses for the intermixing are the vertical temperature gradient, horizontal winds, and eddy flows [UNSCEAR (1977)]. With increasing sea level, the concentration of radon decreases, being about 50 % at 1000 m. Corresponding to the diurnal change of the temperature gradient, the concentration in the layers near the ground reaches its maximum in the morning and its minimum in the afternoon or evening. Especially high concentrations result from inversion weather conditions, whenever vertical intermixture is possible only to a limited extent. The progenies of radon are heavy metals, for instance polonium. In air these daughter products together with water vapour, or with other substances form aerosols, which are deposited on surfaces. In Germany, the mean radon concentration in outside air near the earth surface amounts to 14 Bq/m^3 [SSK (1992ab)]. A higher radon level may be found wherever rock formations with augmented uranium content occur or materials with increased radium content are deposited, disintegrated, dissolved, or processed (see Chap. 3, 1.2 and 1.3). Increased concentrations are also observed above waste dumps from uranium mines or phosphate deposits, for example in Florida, where 90 % of the American phosphate is mined [Eisenbud (1987)]. In case of water conducting strata in soil, radon can be dissolved at a considerable extent and thus reach the drinking water [SSK (1992ab)], [Hess et al. (1985)], from where it may be released into the air. During the processing of drinking water, the escaped radon may be followed by an increased radiation exposure by inhalation.

Radon, however is not only present in the outside air, but also indoors. It gets into the buildings by diffusion through walls, ceilings, and floors and may be exhaled by building materials which may contain significant amounts of uranium and radium. Besides that, radon gets indoors from the ground below the buildings if cellars and floors are not sufficiently sealed to prevent intrusion. Undergrounds with high uranium, thorium, or radium contents are located in Brazil in the state of Espirito Santo, and in Rio de Janeiro, in India at Kerala as Monazite sand, at Minas Gerais in Brazil as apatite or phosphates, in Germany at Schneeberg or at Jachymov (Joachimsthal) in the Czech Republic as pitchblende. In such regions, specially at the construction of buildings, special precautions are necessary to protect against the penetration of radon. [Nero (1985)], found a lognormal distribution for the radon concentrations with a geometric mean of 35.5 Bq radon per m^3 and an arithmetic mean of 61.4 Bq/m^3. From the data of [Nero (1985)], it can be concluded that in the case of detached houses the largest amount of radon, 85 % comes from the soil, 11 % from the outside air and only 2.8 % from the walls of the buildings. In the case of apartments the contribution from the floor is smaller but larger from the walls of the building. [Cohen (1985b)], found in houses near Pittsburgh an arithmetic mean of 233 Bq/m^3 of radon in the basement, 88.8 Bq/m^3 at the first floor and 74 Bq/m^3 at the second. The median values were considerably lower at 96.2, 44.4, and 35 Bq/m^3, respectively. Further data, for thoron too, were given by [Schery (1985)], and [Israeli (1985)].

Radon accompanied by its progenies, may enter man by inhalation partially as aerosols. The sections of the respiratory tract in which radioactive aerosols are deposited depends on the particle size. Large particles remain within the bronchi, smaller ones are transferred in the alveoli. Radon itself is, like krypton, soluble in body fluids and in fat. Radon is rapidly emanated from drinking water so that ingestion plays only a subordinate part [Cross et al. (1985)].

3.4.7. Radiocaesium

^{137}Cs is, with regard to the radiation burden of the environment, one of the most important radionuclides, which is generated during nuclear fission at a relatively large percentage of yield [NCRP 52 (1977)]. Because of its long half-life of 30 years and its high mobility in the environment, it is widely distributed and relatively quickly enters the food-chain. Main sources of ^{137}Cs were the nuclear weapon in the atmosphere tests (see 2.4 and Chap. 3, 2.1.3) [UNSCEAR (1977), (1982), (1988)], [NCRP 52 (1977)] and the Chernobyl reactor disaster in 1986 (see Chap. 3, 2.3.1.6) [UNSCEAR (1988)], [DOE (1987)], [Eisenbud (1987)], [AEN-NEA (1997), (2002)]. As further sources worth mentioning are the emissions at regular operation of reactors and reprocessing plants and accidental release from medical ^{137}Cs-systems (see Chap. 3, 2.2 and 2.6).

The knowledge of radiocaesium in the environment is due to investigations of fallout from the nuclear weapon tests. A total of 960 PBq ^{137}Cs [UNSCEAR (1982)] or 1.26 EBq [NCRP 52 (1977)], were originally injected into the stratosphere. The medium residence time within the stratosphere amounted to 6 to 12 months [NCRP 52 (1977)]. From the stratosphere caesium reenters the troposphere, especially by exchange in the late winter, and arrived at a maximum level in the air near the earth surface in springtime. At least in the temperate zones the depositions took place in the early growth period of the plants. In the northern hemisphere 3420 Bq/m^3 in all were deposited with main depositions at the northern temperate zones (40°-50°) of 5170 Bq/m^2, whereas in the southern hemisphere the concentration was only 860 Bq/m^2 [UNSCEAR (1982)]. The Chernobyl reactor accident led besides the emission of ^{137}Cs also to an emission of

[134]Cs. The radionuclides were transported by the air during a short time over large distances and deposited by rainfall onto country and sea areas, especially in the European regions [IAEA (1991)], [UNSCEAR (1988)].

Radiocaesium, deposited over the ocean surfaces, reached the sediments of the Atlantic Ocean only at a range of 3 % [Noshkin (1973)]. The concentration of activity in the surface water amounted to about 7.4 Bq/m^3 of ^{137}Cs, and was lower by a factor of 30 in the deep ocean [UNSCEAR (1977)]. The medium residence time in surface water was estimated to be 17 years for the Pacific [Hodge (1973)]. [Smith (1997)] developed as an example a model of fresh water for the contamination from Chernobyl-caesium in Great Britain.

Since the chemical properties of caesium are very similar to that of the essential element potassium, it is not surprising that caesium in biological systems, the human organism included, behaves like potassium. [137]Cs injected into the environment, may be transferred and accumulated on a large scale in terrestrial, as well as, in aquatic ecosystems. The vegetation, as basis of the food-chain, accumulates ^{137}Cs deposited on leaves and soil. The extent of transfer from the soil by the roots can vary considerably [UNSCEAR (1977)], [Wiechen (1972)]. [Konshin (1992ab)] tried to describe the behaviour in soil by a model and the following transfer from soil to grass and finally into the milk. The transfer to the herbivores takes place relatively fast, whereas the accumulation in carnivores reaches its maximum at a later time. For human nutrition, ^{137}Cs-suppliers especially are: milk, meat, fish [UNSCEAR (1977)], some sorts of berries from the woods [Berg (1991)], and mushrooms [Grueter (1971)], [Römmelt (1991)].

The human body contains about 1500 μg of stable caesium [ICRP 23 (1975)], [ICRP 30 (1979)], [Berg (1990)]. The daily intake with food amounts to 12 to 16 μg/d. From that, depending on the food composition, up to 80 to 100 % are absorbed. About 90 % of the excretion is by the kidneys with the urine and amounts to 9-10 μg/d. Despite the predominating excretion by kidneys, there is also an enteral circulation of Cs ions into the gastrointestinal tract with strong reabsorption. The decreasing of the reabsorption by the dye "Prussian Blue" is a chance to decorporate caesium [Müller et al. (1982)]. [Spiers (1968)], reported that about 80 % of the caesium is deposited in muscle and 8 % in bone. After a single application of caesium, 10 % of the retained ^{137}Cs was excreted with a biological half-life of 1 to 2 days [UNSCEAR (1977), (1982)], and the rest with a half-life of less than 50 d up to 200 days [UNSCEAR (1977)], [Henrichs et al. (1989)]. [NCRP 52 (1977)], reported a half-life of 15±5 d for children, and of 100±50 d for adults. [Clemente et al. (1971)], investigated 20 to 35 year old women and men, observed a shorter ^{137}Cs half-life for females (71.7 d) than for males (86.9 d). These half-lives are in good accordance with values of 83-85 d for

40 years old men, 60-66 d for 39 years old women, 52-55 d for 12 years old children, 37-43 d for ten-years-old and 37-43 d for six-years-old [Weng et al. (1973)]. In the case of pregnant women, showing similarly to children an augmented caesium turnover, [Lloyd (1973)] reported the half-life of less than half of that for non-pregnant females (49 d). Extensive surveys on the ^{137}Cs-content in human organism for many countries were given in [UNSCEAR (1982)], and on biokinetic models by [Melo et al. (1997)], [ICRP 56 (1989)], [Leggett (1986)].

Investigations of the behaviour of ^{137}Cs in the food-chains were carried out in numerous countries. For example [Kline et al. (1973)] investigated the ^{137}Cs-concentration in plants and soils of Puerto Rico. From 1959 to 1968 in Denmark [Aarkog (1971b)] studied the contamination of foodstuffs and [Simpson et al. (1977), (1981)], [Stroube et al. (1985)] from 1961 to 1982. [Hollemann et al. (1971)], [Salo et al. (1964)] investigated the behaviour of radiocaesium in an arctic food-chain (lichen → reindeer → meat → human). The ^{137}Cs-content in the human body (Lapland, Germany) caused by the nuclear weapon tests, has been determined by [Miettinen (1964)] and [Onstead (1960)]. Prior to the Chernobyl accident, [Marei et al. (1972)] reported on the caesium concentration in Belorusssia in the "soil→grass→milk→human"-path, [McAulay et al. (1985)] on ^{137}Cs in fish from the Irish Sea. [Takeshita et al. (1972)] investigated the ^{137}Cs-content in placentas as well as foodstuffs in Japan, and [Töttermann et al. (1972)] in foetal bone from 1962 to 1966 in Lapland. Results from studies in many countries of the world were quoted in the UNSCEAR-reports [UNSCEAR (1962), (1977), (1982), (1988)].

3.4.8.   Radiostrontium

Strontium, an element of the group of alkaline earths, was detected in a lead mine near the Scottish village of Strontian in the 18th century. In nature it occurs very rarely, and until 1940 it has not been often seen (see Chap. 4, 7.). By the discovery of nuclear fission, the radioactive isotopes of this element came more and more to the front, especially ^{90}Sr. This nuclide has because of its physical properties, i.e., long half-life and high energy released at decay, and because of its behaviour in living beings, a high radiotoxicity for most organisms.

In the biosphere strontium also behaves qualitatively like other alkaline earth elements like the essential element calcium. Nevertheless in nature there are considerable quantitative differences.

^{89}Sr (half-life = 50.5 d) and ^{90}Sr (half-life = 29.1 y) are formed as products resulting from nuclear fission (see 1.2.3, 2.3 and 2.4). The yield of ^{90}Sr varies with the kind of the split nucleus and with the neutron energy used [UNSCEAR (1962)]. At the fission of ^{235}U by thermal neutrons ^{90}Sr is generated at a yield of 5.77 %, at irradiation with fast neutrons at 4.38 %.

The fission by fast neutrons delivers, in the case of ^{239}Pu, a yield of 2.23 % and in the case of ^{232}Th even 6.8 % of ^{90}Sr.

By nuclear weapon tests in the atmosphere and by emission of reactors or reprocessing facilities radiostrontium was released into the environment. Whereas ^{90}Sr injected to the atmosphere by nuclear weapon tests (see. 2.4 and Chap. 3, 2.1.3) was distributed globally and deposited onto the surfaces of soil or of the oceans, the spreading by accidents or regular emission from nuclear plants was locally limited. From the nuclear weapon tests in the atmosphere about 460 PBq of ^{90}Sr were deposited within the northern hemisphere and 144 PBq in the southern hemisphere [UNSCEAR (1993)]. The activity deposited per unit area not only depends on the hemisphere, but also varies therein with the latitude. For 40° to 50° of northern latitude, for example, a maximum of 3.23 kBq of ^{90}Sr per m^2 was found. The seasonally varying rates of deposition have a maximum every springtime [HASL-329 (1977)], [Cambray et al. (1983)], [Larsen (1983),(1985)]. They have been declining towards zero since the end of the tests in the atmosphere in 1980, so that the deposition from the nuclear weapon tests may be considered to be terminated [UNSCEAR (1993)].

^{90}Sr deposited in the oceans [Noshkin et al (1973)] is located above the thermocline zone in the well-mixed layer. Only 1 % reached the ground sediments. In the well-mixed water layer the concentration amounted to about 7.4 Bq/m^3, below the thermocline it was lower at least by a factor of 30, and near the coast and in river deltas higher by a factor of 5. In marine biota (see Table 9) ^{90}Sr was, compared to the concentration in water, accumulated in calcified tissues of marine organisms, such as bones, scales, and shells. Typical concentration factors [UNSCEAR (1977)] for seaweed were, 100 for shrimps, and 2 to 10 for lobster. In muscle meat of fish, however, no significant accumulation was observed.

In a model, [Konshin (1992c)], has tried to describe the migration within the soil. The migration speed in soil depends on many factors (see 3.1.3). ^{90}Sr deposited on the bottom scarcely migrates into the ground. The most of the strontium, about 50 % [UNSCEAR (1977)] remained within the first 4 cm of the soil. Down to a depth of 30 cm nearly the total amount of deposited strontium was observed. By leaching out, it can partially be cleared to rivers and lakes.

Calcium, as a bioessential element of the animate nature, especially for mammalians, is submitted to homeostatic control thus deriving a considerable constant concentration in many organs and tissues [NCRP 110 (1991)], [Spencer et al. (1960)]. Therefore, often the activity concentration of strontium in organs, secretions, or tissues (e.g., bone, blood, milk) was related to the respective calcium concentration. At "steady state" conditions the Sr/Ca relation supplied with the diet (precursor) should be mirrored in the organs.

Therefore, [Comar et al. (1956)], defined a so-called *"OR"*-ratio, i.e., strontium (*[Sr]*) to calcium (*[Ca]*) observed concentration ratio, which has been used for many years, to predict and to describe the behaviour of strontium in organisms and in the environment:

$$OR = \frac{\dfrac{[\,Sr_{sample}\,]}{[\,Ca_{sample}\,]}}{\dfrac{[\,Sr_{precursor}\,]}{[\,Ca_{precursor}\,]}} \tag{8}$$

The uptake of strontium by plants can result from direct contamination, as well as, by the roots. Whereas the retention caused by external contamination reflects the actual deposition, in the case of the absorption by the roots, the cumulative deposition in soil is to be taken into consideration. The amount of exchangeable calcium in the soil solvent is an important factor and influences the transfer level of ^{90}Sr considerably [UNSCEAR (1962)]. Further influences onto the Sr-availability are the clay and the humic content, the pH-value, and the content of further electrolytes in the soil. The various plant tissues show a broad variation in the Sr/Ca relation. Commonly, in fruits the ^{90}Sr-concentration is lower than in leaves, stems, or roots.

With regard to the food-chain, the metabolism of ^{90}Sr in animals is of special interest, e.g. in order to estimate to which extent this nuclide is transferred to milk and meat. Simultaneously important findings are collected also for the strontium metabolism of man. Ingestion is the main route for the incorporation into the organism of mammalians. Strontium particles, inhaled from the air, may be absorbed by the lungs to a considerable extent, but play only rather a subordinate role for radiation exposure.

Table 13: Calcium and strontium absorption from intestinal tract, after [Thomasset (1982)].

Species	Age	Ca (%)	Sr (%)	References
Rat	35 d	92.7	71.5	
	74 d	99.8	88.0	Gran (1960)
	270 d	40.3	12.2	
Calf	2 d	99	94	
	60 d	85	77	Lengemann (1957)
	90 d	77	66	
	152 d	42	33	
Human	60 - 270 d	37	39	
	4 - 14 y	2	8	Bedford (1960)
	Adults	46	19	Spencer (1960)

The absorption of Sr from the intestinal tract (see Table 13), depending on species and age amounts to 20 to 100 %. Young growing animals have a larger turnover and absorb more strontium and calcium than adult ones. The individual absorption may vary considerably, since the intestinal absorption is influenced by the ingredients of the food. By addition of calcium-phosphate and sodium-alginate to the food of rats for instance, the absorption was diminished 7-fold [Slat et al. (1971)]. After a period of fasting, strontium was absorbed considerably better [Spencer et al. (1972)]. After the passage through the intestinal wall strontium circulates in blood and is bound there by 40 % to proteins, predominantly to albumin, whereas up to 60 % are fully diffusible [Berg et al. (1973)]. Within a few hours the concentration in the blood rapidly decreases [Spencer et al. (1972)], [Thomasset (1982)] by accretion and exchange with bone, by renal and faecal excretion, or occasionally by lactation. 99 % of the ^{90}Sr retained was found in the skeleton, first of all in the regions of growth, after increasing time also within the bone mineral. Since Europeans supply of 80 % their need for calcium by milk and dairy products, the secretion into milk is of particular interest. After a single administration of strontium to a milk cow, 0.5-2 % was secreted together with the milk about 48 hours later. At chronic administration under normal feeding conditions about 0.08 % of the daily intake was observed in one litre of milk [UNSCEAR (1962)].

In numerous countries, investigations of ^{90}Sr-contaminations have been carried out. During the growth period, the uptake of strontium into the skeleton is augmented with increased turnover. Because of the commonly very long biological half-lives of substances incorporated into the skeleton, ^{90}Sr content in bone is measurable long time after incorporation. From 1962 to 1965 and from 1966 to 1969 [Aarkog (1968), (1971a)] collected dental samples of children born between 1953 to 1963 in Denmark, Greenland, or on the Faeroes, in order to determine the ^{90}Sr-exposure resulting from the nuclear weapon tests. The largest concentrations of 888 mBq ^{90}Sr per g Ca were found after the numerous nuclear weapon tests in the atmosphere in the early 1960s for one year old children from the Faeroes, born in 1963. [Czosnowska et al. (1972)], investigated the excretion of ^{90}Sr and Ca with the urine in 4 to 7 year old children in Poland. At a medium daily intake of 459 mBq of ^{90}Sr an average excretion of 26 mBq/d by the urine was observed. From 1962 to 1971, [Kramer et al. (1973)] controlled the ^{90}Sr-content in metabolic non-milk diets, and found a time dependent ^{90}Sr-supply with a maximum value in 1964/65 of about 315 mBq/d. From 1962 to 1968 [Lalit et al. (1972)] analyzed milk samples in India, [Töttermann et al. (1972)] and [Salo et al. (1964)] examined the behaviour of radiostrontium in arctic food-chains. Results from studies in many countries were listed in the UNSCEAR-reports [UNSCEAR (1962), (1977), (1982), (1988)].

3.4.9. Plutonium

The widespread horror evoked by the dropping of the plutonium bomb onto Nagasaki on August 9th, 1945, has been engraved in the association with this element. In 1940 plutonium in the form of its nuclide ^{238}Pu was produced first by Mc Millan Kennedy, Wahl, and Seaborg by bombarding ^{238}U with deuterons. The large scale production of the isotope ^{239}Pu in the Manhattan-project led to the construction of a bomb. A survey of the behaviour of plutonium and further transuranic elements in the environment is given by [Thompson et al. (1972), (1975)], [Healy (1975)].

From the 15 isotopes known of this element in the reactor or at nuclear weapon tests five, ^{238}Pu, ^{239}Pu, ^{240}Pu, ^{241}Pu, and ^{242}Pu are frequently produced (see Fig. 4 and Table 7, 8). Plutonium decay, mostly emitting high-energy α-particles, ^{238,239}Pu at 5.5 MeV, ^{240}Pu at 5.2 MeV, and ^{242}Pu at 4.9 MeV [Metivier (1982)]. ^{241}Pu, however, is an exception, which decays by β^{-} emission to ^{241}Am, an α-emitting daughter of 5.49 MeV of energy [Stather (1982)]. In biological materials these α-energies of about 5 MeV cause extraordinarily high radiation doses, for instance about 5 Gy for a single disintegration in a 40 μm high cylinder with a radius of 0.01 μm [Metivier (1982)]. This is the reason for plutonium being highly radiotoxic. In aqueous solution plutonium can take on the formal oxidation states of +3 to +7, predominantly however state +4. Stable complexes of this state are formed for instance, with transferrin, citrate, or DTPA (diethylene-triamin-penta-acetic acid). Plutonium normally is stored as PuO_2 and is used in this compound in reactors. In special types of reactors metallic plutonium is utilized too.

In the sixties, especially before 1963, isotopes of plutonium (see Table 8) came globally into the environment by nuclear weapon tests. The release in the consecutive tests added a further 10 % [Harley (1971)]. The contribution by accidents (see Chap. 3, 2.3) was comparatively minor and of more local importance. Plutonium from the bomb tests, injected into the stratosphere, was distributed over the whole atmosphere by the mixing mechanisms and deposited in the form of oxide [Smith (1982)] onto the earth and the oceanic surfaces. The deposition rates of ^{239,240}Pu were compared with those of ^{90}Sr, showing that the Pu/^{90}Sr ratio observed of 0.017 was considerably constant within the stratosphere and in surface air [UNSCEAR (1977)]. Similar observations also applied for the deposition onto the soil [UNSCEAR (1993)]. Therefore, the distribution patterns onto soil in the southern and northern hemispheres resemble very much those of ^{90}Sr (see Table 14). The density of the deposition in the southern hemisphere was about four-fold lower than that in the northern part. Depending on the meteorological processes, the deposition was augmented within the respective temperate zones.

Table 14: Estimation of deposition (population-weighted) densities of major radionuclides produced in atmospheric weapon tests, after [UNSCEAR (1993)].

Radio-nuclide	Deposition Density (kBq/m²)				
	Northern Hemisphere		Southern Hemisphere		World
	40°- 50°	*Entire*	*40°- 50°*	*Entire*	
^{54}Mn	9.4	6.2	2.6	1.6	5.7
^{55}Fe	6.8	4.5	1.9	1.1	4.1
^{89}Sr	20	13	4.3	2.6	12
^{90}Sr	3.23	2.14	0.89	0.54	1.96
^{91}Y	25	17	5	3.3	15
^{95}Zr	38	25	8.3	5	23
^{95}Nb	64	43	14	8.5	39
^{103}Ru	28	19	6.2	3.8	17
^{106}Ru	24	16	6.7	4.1	15
^{125}Sb	2.9	1.9	0.79	0.48	1.7
^{131}I	19	13	4.2	2.5	11
^{137}Cs	5.2	3.4	1.4	0.86	3.1
^{140}Ba	23	16	5.1	3.1	14
^{141}Ce	21	14	4.6	2.8	13
^{144}Ce	48	32	13	8.1	29
^{238}Pu	$1.5 \cdot 10^{-3}$	$9.8 \cdot 10^{-4}$	$4.1 \cdot 10^{-4}$	$2.5 \cdot 10^{-4}$	$9.0 \cdot 10^{-4}$
^{239}Pu	$3.5 \cdot 10^{-2}$	$2.3 \cdot 10^{-2}$	$1.0 \cdot 10^{-2}$	$6.0 \cdot 10^{-3}$	$2.2 \cdot 10^{-2}$
^{240}Pu	$2.3 \cdot 10^{-2}$	$1.5 \cdot 10^{-2}$	$6.0 \cdot 10^{-3}$	$4.0 \cdot 10^{-3}$	$1.4 \cdot 10^{-2}$
^{241}Pu	$7.3 \cdot 10^{-1}$	$4.8 \cdot 10^{-1}$	$2.0 \cdot 10^{-1}$	$1.2 \cdot 10^{-1}$	$4.4 \cdot 10^{-1}$
^{241}Am	$2.5 \cdot 10^{-2}$	$1.7 \cdot 10^{-2}$	$7.0 \cdot 10^{-3}$	$4.0 \cdot 10^{-3}$	$1.5 \cdot 10^{-2}$

Plutonium deposited onto soil migrates, similarly to strontium, only very slowly into depth and remains in the upper 4-5 cm layer within the undisturbed ground at 40 to 70 % [Smith (1982)], [UNSCEAR (1977)]. At the cultivation of fields, a mixing occurs by ploughing and harrowing. From this down to a depth of about 30 cm, the distribution can be assumed to be homogeneous. Plutonium is supposed to be in the soil solution predominantly in the oxidation state +4 and is nearly insoluble because of polymerization and hydrolysis. However, reactions with the components of the soil are known, which increase the availability for the plants.

In the oceans plutonium, in contrast to other radionuclides, reaches not only the well-mixed layers above the thermocline (see 3.1.2), but also the deep water below that. [Noshkin et al. (1973)] reported that about 50 % of plutonium is located below the thermocline. The larger part of the deposited plutonium is very rapidly bound to insoluble, particulate matrices and finally ends at 2 to 36 % in the sediments [Smith (1982)], [UNSCEAR (1977)].

In marine food-chains the deposition of plutonium in sediments can be a starting point for reappearing in marine biota. Marine organisms and some species of fish, feeding on sediments or nearby sediments, incorporate plutonium in this way (bentic organisms or filter feeders). Phytoplankton and seaweed show in fresh water as well as in salt water concentrations of plutonium at the same order of magnitude as in the sediments. Typical concentration factors are (see Table 9) for edible parts of mussels and seaweed 1000, for crustaceans 100, or for fish 10. Since the plutonium concentration in bone is higher than in meat, the transfer to human is lower if only the meat is consumed.

Plutonium penetrates physiological membranes only with difficulty [Bair et al. (1974ab)]. By this fact, the transport by biological mechanisms and the accumulation within the food-chain is restricted. The absorption by roots of plants is very limited and subject to large variations. The concentration factors vary from $4 \cdot 10^{-8}$ to $3 \cdot 10^{-2}$ [UNSCEAR (1977)] at a mean of $1 \cdot 10^{-4}$.

Plutonium released into the environment apparently reaches humans more by physical transport mechanisms than by biological transport routes. Plutonium, resuspended by ground wind from soil, can reach the food-chain by deposition onto plants [Pinder et al. (1985)], [Cataldo et al. (1980)] or be inhaled directly with the respiration air. Especially in the case of graminivorous animals, for instance cows, an augmented incorporation not only by ingestion but by inhalation can take place. Into the human organism, the uptake of plutonium in measurable quantities mostly occurs more directly, by inhalation or by wounds at accidents, than by the detour of the food-chain, and less by intestinal absorption or resorption from the contaminated skin. Experiments on the absorption from the gastrointestinal tract in rodents showed that the resorbed part amounted to only between 10^{-4} % and 2 % and varied depending on the oxidation state, on the chemical compound, and the age of the animals [Metivier (1982)]. The citrate complexes were best absorbed by young animals. The values for resorption by the skin were in equal orders of magnitude. For men, [Bair et al. (1974ab)] reported intestinal absorption values to be between $2 \cdot 10^{-3}$ % and $1 \cdot 10^{-2}$ %. Further papers on plutonium metabolism, resorption, retention, and distribution in the organism have been published by the following authors: [Leggett (1985a)], [Bair (1979)], [Kocher et al. (1983)], [Larsen et al. (1981)], [Jee (1976)], [Harrison (1982)], [Sullivan (1979), (1980ab), (1985)], [Johansson (1983)], [Langham et al. (1950)], [Durbin (1975)], [Morin et al. (1972)], [Rosenthal et al. (1972)].

The inhalation is the most important path of incorporation for the population, as well as for workers in the nuclear industries. Plutonium is deposited as aerosol, mostly as oxide, according to the particulate size within the respiratory tract and thereafter into the pulmonary tissues. Part of it can be transported back to the upper respiratory tract by ciliary movements, enters the gastrointestinal tract and is mostly excreted. The part remaining in the lungs is accumulated by macrophages. By decomposition of cells and alveoli, plutonium may pass to the blood and be distributed onto the other body tissues, being available especially to liver and skeleton. A further part of the plutonium can be taken up by the lymph nodes. The clearance of the upper respiratory tract normally amounts to not more than 1 to 2 days [Metivier (1982)]. The biological half-lives of $^{239}PuO_2$ in the deeper regions of the lungs run up to 150 to 500 d in rats and to 300 to 300 d in dogs. For men, biological half-lives of 250 to 300 d after accidental release have been established. The fraction of plutonium entering the blood depends on the primary chemical compound of the plutonium inhaled. Which organ and to what extent it retains plutonium is influenced by the stability of the plutonium complex in blood. Predominantly, complexes of tetravalent plutonium with transferrin occur, and furthermore there are citrate complexes, which preferentially are retained within the bone. After inhalation of $^{239}PuO_2$, the relative concentration of plutonium in the thoracic lymph nodes in 7 to 9 years old dogs was 1400-fold higher than the value compared to lungs, followed by abdominal lymph nodes 100-fold [Metivier (1982)]. In other organs and tissues, the concentrations were less than in lungs (lung =1): in liver at 0.5, in spleen at 0.2, and in bone at 0.06. After inhalation of plutonium nitrate (see Table 15), by way of the blood, considerably more activity was retained in the liver and especially in the skeleton.

Table 15: Distribution of plutoniumnitrate after inhalation, after [Metivier (1982), [a][Ballou (1972)], [b][Morin et. al. (1972)].

Species	Time (d)	Retention of Initial Alveolar Burden			(%)
		Lung	Thoracic Lymph Nodes	Liver	Skeleton
dog [a] (^{239}Pu)	1	88	0.06	0.3	1.8
	30	32	0.4	9.0	43.0
	100	41	0.6	10.0	28.0
rat [b] (^{239}Pu)	1	78		1.8	3.2
	30	40		1.0	7.7
	90	15		0.3	4.4
rat [b] (^{238}Pu)	1	96		0.6	1.9
	30	53		2.2	18.1

The tissue distribution of plutonium after absorption by the skin and from the intestinal tract is very similar to that observed after inhalation, with the exception of the tissue of the lungs. After intravenous injection, plutonium nitrate is rapidly eliminated from the blood circulation, at 80 % by the liver, and at 6 % by the kidneys [Metivier (1982)]. With increasing time, however, more and more plutonium is transferred to the skeleton. At lower masses of plutonium injected, a larger part remains in its monomerous state and is under these circumstances preferentially stored in the skeleton up to 35 %. Similar observations have been made in the case of an application of mono-merous plutonium citrate complexes (44 % in bone, 6.2 % in liver). The biological half-lives in both organs are prolonged and are assumed to be 40 years for liver and 100 to 200 years for bone.

In 1945 and 1946, 18 fatally ill persons were injected with Pu-citrate or Pu-nitrate, respectively, in order to correlate the urinary and faecal excretion with the plutonium content of the organism. From these examinations the overwhelming part of knowledge about the behaviour of plutonium was derived directly for men [Langham (1959), (1950)], [Lagerquist et al. (1973)], [Popplewell et al. (1985)], [Durbin (1975)]. The excretion takes place by urine and faeces. Additionally the liver is able to secrete plutonium into the intestine via the gall bladder. [Leggett (1985b)], has developed a model describing retention, translocation, and excretion of plutonium in adult humans. Furthermore, much knowledge was derived from a 50 years long-term study of 26 workers [Hempelmann et al. (1973)], [Voelz et al. (1985), (1997)], having incorporated during their work 50 to 3180 Bq of plutonium in the years 1944/45 (Manhattan-Project).

Chapter 3

RADIONUCLIDES RELEASED INTO THE ENVIRONMENT

Dieter Berg

GSF - National Research Centre, Institute of Radiobiology, D – 85764 Neuherberg, Germany

1. CONTAMINATION OF THE ENVIRONMENT BY NATURALLY OCCURRING RADIONUCLIDES

1.1. Contamination without Human Influence

A part of naturally occurring radioactive materials (see Chap. 2, 1) existed before the origin of the earth. Others are produced in the atmosphere, the soil, and in the water by influences from the cosmos. From the fission fragments and activation products generated by the spontaneous nuclear reactions of uranium (see Chap. 2, 1.3.2, Oklo phenomenon) it becomes obvious that, in principle, nearly all radionuclides might occur. Short half-lives and low concentrations, however, hinder us in fact from identifying them in the biosphere.

Carbon, hydrogen, potassium, and also uranium are widely spread in the biosphere. For this reason it is not surprising that they play an important role in the development of life and participate in metabolism. In the Earth's history, geological and meteorological influences have always given rise to changes in the distribution in the biosphere. In 1980, for instance, after the eruption of Mount St. Helen in Fayetteville, augmented concentrations of uranium in rain and snow were observed [Essien et al.(1985)]. Also the organisms changed the distribution of natural radioactive substances by their metabolism. Radionuclides were partly concentrated in living matter by selective enrichment, bound and deposited after the death, ^{14}C in coal and oil for example.

R. Tykva and D. Berg (eds.), Man-Made and Natural Radioactivity
in Environmental Pollution and Radiochronology, 71-145.

1.2.   Contamination by Human Influences Excluding Nuclear Techniques

Men began to be involved in emission, spreading, and distribution of radioactive substances once they learnt to use fire, by introducing agricultural techniques, and by mining of minerals and ores. By the usage of fire, besides ^{14}C, also uranium/thorium are released. By agricultural techniques such as fertilizing and irrigation, radioactive substances are additionally carried to the arable soil. Mining of ores unknowingly brought radioactive substances to the biosphere, besides the desired metals.

1.2.1.   Combustion

The release of natural radioactive substances started with the combustion of wood. Since the discovery of coal, oil, gas, and peat as fuels at the beginnings of industrialization, the emissions have increased considerably by the augmented need for energy. In 1985 the global output of coal amounted to about $3.1 \cdot 10^{12}$ kg [UNSCEAR (1988), (1993)]. The main producing countries were China with a share of $0.81 \cdot 10^{12}$ kg, the United States with $0.74 \cdot 10^{12}$ kg and the former Soviet Union with $0.49 \cdot 10^{12}$ kg.

Coal contains the natural radionuclides [40]K, [238]U, and [232]Th, together with their progenies (see Chap. 2, 1.). The activity concentrations in coal, in fly- and bottom-ashes (defined below) are listed in Table 1. The enormous variations in the concentrations of activities are remarkable. Analyses of 800 coal samples from the United States showed [238]U-activity concentrations at a range of 1 to 1000 Bq/kg [UNSCEAR (1988)]. An intensive monitoring of coal from China, sharing 26 % of the world production, resulted in concentrations of still higher mean values [Pa Ziqiang (1993)]. Although the concentrations of activity from the global analyses of coal samples vary at a range of more than three orders of magnitude, their average values are rather consistent. For the annual production of 1 GW of electric energy about $3 \cdot 10^{9}$ kg of coal are necessary. A large part, about 40 % of coal production, is for the production of electric energy used in coal power stations [UNSCEAR (1993)]. Another 50 % is consumed in coke ovens or other industrial production facilities, and about 10 % is burnt as domestic fuel.

Looking at the release of radionuclides from coal, the whole fuel cycle has to be considered from the mining and the combustion of coal until the deposition or reutilization of the fuel ashes.

The fuel cycle starts with the hauling of coal. From the ventilation of mines, radon and thoron, the gaseous progenies from the natural radioactive decay series (see Chap. 2, 1.) are emitted with the exhaust air to the environment. In this process, a yearly release of 30 to 800 TBq of radon into the atmosphere from coal mines was estimated [UNSCEAR (1988)].

During the combustion in power plants at temperatures up to 1700°C the mineral substance of the coal is transformed to a glassy ash. Incompletely burnt organic matrix and heavier ash accumulates on the ground of the furnace as bottom ash or slag. The lighter components, the fly ash, together with volatile minerals and the flue gases reach the chimney, where they, depending on the quality of the filter system, escape more or less into the atmosphere. In the absence of organic matrix in the ash, the concentrations in ash and slag (Table 1) increase by about one order of magnitude, and are, compared with earth's crust, significantly higher. At a production of approximately 400 GW of electric energy per year by combustion of coal (Table 2)

Table 1: Concentration of natural radioactivity in coal, fly- and bottom ash from production of electrical power, based on [UNSCEAR [a](1982), [b](1988), [c](1993)], [d][Pa Ziqiang (1993)].

Radioactivity				(Bq/kg)	
	Coal			Fly Ash	Bottom Ash
	World Wide		from China		
^{238}U	[c]20	[b](1 -1300)	[d]36	[c]200	[a]48 - 100
^{232}Th	[c]20	[b](2 - 320)	[d]30	[c]70	[a]44 - 120
^{226}Ra		[a](7 - 100)		[c]240	[a]4 - 250
^{210}Pb		[a](10 - 50)		[c]930	[a]30 -3900
^{210}Po		[a](10 - 41)		[c]1700	[a]7 - 190
^{228}Th				[c]110	[a]90 - 560
^{228}Ra		[a](13 - 35)		[c]130	[a]20 - 67
^{40}K	[c]50	[b](1 - 800)	[d]104	[c]265	[a]240 -1200

Table 2: Releases of radionuclides from annual production of 400 GW electrical power in coal-, oil- or natural gas-fired power plants, based on [UNSCEAR [a](1982), [b](1988)].

	Release	(TBq)	
	Coal	Oil	Natural Gas
^{40}K	[a]1.6	[b]0.40 - 0.60	
^{238}U	[a]0.6	[b]0.08 - 0.12	
^{232}Th	[a]0.6	[b]0.04 - 0.06	
^{226}Ra	[a]0.6	[b]0.10 - 0.12	
^{210}Pb	[a]2.0	[b]up to 0.17	
^{210}Po	[a]2.0		
^{228}Th	[a]0.6		
^{228}Ra	[a]0.6		
^{220,222}Rn	[b]24.0		[b]800

about 1.6 TBq of ^{40}K, 0.6 TBq of ^{238}U, ^{232,228}Th, and ^{226,228}Ra, 2 TBq of ^{210}Pb, and ^{210}Po escape with the fly ash into the atmosphere. Besides this, an additional 24 TBq of ^{220,222}Rn will be released, since radon is not retained by filter systems.

Referring to emissions by the combustion of coal for domestic use, for cooking or heating, there are no published investigations. The [UNSCEAR (1993)] report assumed that about 0.7 TBq of ^{40}K, and 0.3 TBq of radionuclides from the ^{238}U and ^{232}Th decay series, excluding radon and thoron, are emitted by the smoke into the environment.

Globally, large amounts of ashes of $2.8 \cdot 10^8$ tons [UNSCEAR (1988), (1993)] result from the combustion of coal in power plants. The bulky part of the ash is used in cement and concrete factories, but also for road building within the asphalt or the filling materials, and in agriculture as a fertilizer.

The combustion of oil and gas results in far lower amounts of ash. The release of radionuclides into the atmosphere from the combustion of oil, however, hardly differs from that from the combustion of coal (factor 3 to 100 lower, see Table 2). In oil fired power plants, $2 \cdot 10^9$ kg of oil are necessary to produce 1 GW per year of electric energy. The levels of atmospheric releases from the annual production of 1 GW [UNSCEAR (1988)] with the use of oil shale from Colorado and according to investigations in France and in the USA amount to 200 MBq and 290 MBq of ^{238}U, 300 MBq and 250 MBq of ^{226}Ra, 150 MBq and 90 MBq of ^{232}Th, and 1000 MBq and 1500 MBq of ^{40}K respectively. With the use of natural gas, about $2 \cdot 10^9$ m^3 are necessary for the annual production of 1 GW electrical power. Because of that 2 TBq of radon are released. In Sweden and Finland also peat is used for the production of energy. Peat may contain ^{238}U-contents of 40 Bq/kg and in extreme cases even up to 10000 Bq/kg. In fly ash of peat-fired power plants additional ^{137}Cs was found from the fallout of the nuclear weapon tests [Mustonen et al (1985)].

The risks from the use of the various fuels for energy production were discussed by numerous authors, and were put together in the UNSCEAR reports. These evaluations commonly considered solely the radiation risks; health hazards from the output of non-radioactive pollutants such as arsenic or cadmium mostly are not taken into consideration. [Bertin (1982)], [De Santis et al. (1984)], [Cohen (1981), (1982), (1983), (1985c)], [Jacobi (1981)] and [Nakaoka et al. (1985)] compared the radiation risks resulting from the use of fossil fuels for energy production with those resulting from nuclear power.

1.2.2. Phosphate Industries

The phosphate and fertilizer industries use phosphate rocks as initial material. Phosphate rocks contain natural radionuclides ^{238}U, ^{226}Ra, ^{232}Th, and ^{40}K as well as progenies of the natural radioactive decay series. Table 3 gives a

Table 3: Activity of natural radionuclides in phosphate rocks from different countries, based on [UNSCEAR (1982)].

	Radioactivity	**Concentration**		**(Bq/kg)**
	^{232}**Th**	^{238}**U**	^{226}**Ra**	^{40}**K**
China	25	150	150	
Morocco	20 - 30	1500 - 1700	1500 - 1700	10 - 200
USSR	25 - 92	44 - 90	30 - 390	44 - 230
USA	10 - 78	150 - 4800	150 - 4800	48

survey of the activities from phosphate minerals originating from various countries. In the year 1982, mining amounted to $130 \cdot 10^9$ kg, with the main producers being China with a share of about 3.3 %, Morocco with 14 %, the former USSR with 19.3 %, and the United States with 37.6 % [FAO (1984)].

The phosphate deposits originate from sedimentary, volcanic, or biological processes. At present, sedimentary phosphate cover provides about 85 % of the demand for phosphate [UNSCEAR (1988)]. Phosphate rocks from sedimentary sources, e.g., from Florida or Morocco, tends to show higher ^{238}U- and ^{226}Ra-concentrations than those from volcanic origin. Apatites, phosphate rocks mined at the peninsula of Kola in Russia have concentrations of only 44 to 90 Bq/kg of ^{238}U and 30 to 70 Bq/kg of ^{226}Ra (see Table 3). Phosphate rocks of biological source in the form of Guano were developed from the excrements of sea birds in combination with limestone as tricalcium-phosphate.

The radionuclides listed in Table 3, together with their progenies, contribute to burdening the environment by sewage and exhaust air during prospecting, as well as during subsequent processing. The volatile radionuclides, e.g., radon and thoron, are released completely into the environment, at the latest during processing procedures. Thus, at a ^{226}Ra-concentration of 1500 Bq/kg, 1.5 MBq of ^{222}Rn escapes per ton of phosphate rock. Additionally, ^{210}Pb and ^{210}Po escape at the higher temperatures during the processing.

With the use of phosphate-fertilizers from phospate rocks, these radionuclides reach the surfaces of pastures and plough lands. Phosphate fertilizers contain per kg P_2O_5, 1700 to 9200 Bq of ^{238}U and 480 to 1700 Bq of ^{226}Ra [UNSCEAR (1988)].

1.2.3.   Mining Industries

The mining of mineral resources (see Table 4) – metals or non-metals – integrates in the subsequent processing also natural radioactive substances, because of their ubiquitous presence in the earth's crust. At the exploitation and processing they partly escape in an undesirable way into the biosphere, or are disengaged as waste product.

Table 4: Naturally occurring radioactivity in mineral resources, based on [NCRP 118 (1993)].

Mineral	Radioactivity Concentration of Mineral/Waste			(Bq/kg)
Natural Gas	2 -	54000	Rn	Gas (per m³)
Monazite	6000 -	20000	U	Sands
Zirconium		4000	U	Sands
		600	Th	Sands
	4000 -	7000	Ra	Sands
Tin	1000 -	2000	Ra	Ore or Slag
Titanium	30 -	750	U	Ore
	35 -	750	Th	Ore
Aluminium		250	U	Ore
	100 -	400	Ra	Bauxite, Limestone, Soil
	30 -	130	Th	Bauxite, Limestone, Soil
	700 -	1000	Ra	Tailings
Copper	30 -	100000	U	Ore
	20 -	110	Th	Ore

Table 5: Estimated annual atmospheric release of natural radionuclides from several mining (fire-clay, Al, Zn, Cu) and processing facilities, based on [UNSCEAR (1988)].

	Annual	Atmospheric		Release		(MBq/y)		
	^{238}U	^{234}U	^{230}Th	^{226}Ra	^{222}Rn	^{210}Pb	^{210}Po	^{232}Th
Fireclay Mine								
mining	< 0.1	< 0.1	0.1	< 2	1200000	< 10	< 7	< 0.1
kilns	5	5	< 6	< 8	-	< 200	130	< 1
Aluminium Mine								
Al-kilns	2.5	2.5	-	2	-	-	-	-
red mud	-	-	-	-	70000	300	340	-
reduction plant	-	-	-	-	-	1200	1000	-
Zinc Mine								
mining	0.004	0.004	0.004	0.003	8500000	0.01	0.006	0.002
mill	0.06	0.06	0.05	0.03	40000	0.05	0.07	0.02
smelter	10	10	5	7	-	20	2	3
Copper Mine								
mining	-	-	-	-	240000	-	-	-
mill	10	14	-	7	70000	30	0.07	-
smelter	400	400	700	60	-	7000	7000	500

For instance zirconium sands or monazite sands mostly contain ^{238}U and ^{232}Th at activity concentrations of more than 500 Bq/kg [UNSCEAR (1988)]. However, in such mines radon is normally not a problem. Table 5 gives a survey of the yearly emission into the atmosphere by reference factories, of ceramics-, aluminium-, zinc-, and copper-industries.

Products made from clay or loam, e.g. bricks or pottery, are widely used. In the USA a fireclay pit was chosen as a reference of radioactive emissions. On mining of clay a yearly release into the atmosphere may by expected of as much as 1 TBq of ^{222}Rn (see Table 5). During the processing in the kilns ^{210}Pb and ^{210}Po also escape at temperatures above of 1100 °C.

In the aluminium industries in a reference facility the yearly emissions were estimated as following: the most important contributions yield from ^{222}Rn, ^{210}Pb, and ^{210}Po.

In the zinc industries, from mines in the USA, estimated from a reference factory, 8-9 TBq of ^{222}Rn were emitted into the atmosphere, and from the mills about 40 GBq yearly. The emission of other radionuclides however was of minor importance.

In the copper industries it was estimated that from the mines and mills mostly only ^{222}Rn escapes whereas the melting facilities emit ^{238,234}U, ^{230,232}Th, ^{210}Pb, and ^{210}Po. In East-Germany, in the regions of Saxony and Saxony-Anhalt, copper- and silver-mining, as well as smelting were carried out for about 800 years. [Schubert (1998)] investigated old waste dump grounds and realized, that partially the dump areas were polluted and should not be used unrestrictedly. [Keller (1993)] found exhalation rates of 1 Bq/(m²·s) of ^{222}Rn from waste dumps in nearby Schneeberg at the Erzgebirge and concentrations in the air in several houses of up to 50 kBq/m³.

1.3.   Pollution after the Introduction of Nuclear Techniques

1.3.1.   Radium

After the discovery of radioactivity by Henri Becquerel in 1886, the systematic search for radioactive "elements" began. Thus, in 1828 Berzelius discovered thorium, and in 1898 in Erlangen G.C. Schmidt identified the radioactive properties of thorium [Cotton et al. (1962)]. In the same year in Paris Marie and Pierre Curie isolated polonium and radium (^{226}Ra), and the radioisotope ^{224}Ra was discovered by Soddy and Rutherford in 1902.

For the production of radium on an industrial scale the Curies used the waste dumps of a paint factory of uranium based colouring materials at Joachimsthal which proved to be particularly effective. The first definition of the unit of the radioactivity was derived from radium: 1 g radium corresponds to one "curie"-unit, which represents about $3.7 \cdot 10^{10}$ nuclear transmutations per second.

Radium, as one of the first radioactive elements, was immediately and widely introduced into medicine and laboratory procedures. From 1912 to 1940, radium and its decay product radon found use in medicine, sometimes however, with dubious quack applications.

Beginning about 1913 ^{224}Ra was administered to arthritic patients [Wick et al.(1993)]. Intravenously injected, ^{224}Ra caused later on, instead of curing the original disease, grave diseases, such as bone or breast cancer, leukaemia, and others [Nekolla et al. (1995)], [Wick et al. (1995)].

A survey of the history of ^{226}Ra was given by [Pratt (1993)], [Landa (1993)] and [Adams (1993)]. [Rundo (1993)] has presented a survey of the detection methods of radium in the human organism, which started on a large scale with investigations of the incorporated radium in dial painters in 1930. Since 1920, radium was used in luminous paint, leading to considerable incorporations for numerous dial painters and subsequently to grave radiation damages such as tumours of bone and the nasal cavities. Approximately 60 deaths, caused by this incorporation, are known. In the meantime, the radium in luminous paints was replaced by promethium or tritium. Clock faces and indicating scales, e.g., in airplanes, contained 3.7 to 11.1 kBq ^{226}Ra and 74 kBq respectively. ^{226}Ra is used in similar orders of magnitude in antistatic devices, as well as in smoke and fire detectors [UNSCEAR (1977)].

Additionally ^{226}Ra was used in gynaecology, especially for irradiation of uterine carcinoma. During the use of encapsulated irradiation sources, leaks often occur, leading to contamination of persons and buildings. All to often, radium sources were lost or showed up in normal waste depositions, dumps, or incinerators. [Villforth et al. (1969)] have investigated 415 such incidents in which radium plays a part.

1.3.2. Tritium, Krypton and Thorium

Numerous articles for daily use contain radioactive substances, e.g., luminous paints, electronic and antistatic devices, or gas and smoke detectors. Today commonly, on luminous dials, instead of ^{226}Ra, tritium or promethium are used. On dials for instance, between 37 to 925 MBq of ^{3}H are present, in "sealed tubes", sometimes even up to 1.11 TBq [UNSCEAR (1977)]. Smoke detectors contain ^{85}Kr amounts of up to 259 MBq and electron tubes between 37 and 185 kBq.

Thorium has been used up to very recently as "Welsbach mantle" in gas lamps, producing luminescence, in amounts of 250 to 400 mg of thorium per gas mantle. During operation, the secondary products ^{224}Ra, ^{212}Bi, and ^{212}Pb are released. Furthermore, thorium is used as an additive in lamps and in special sorts of glass, in the glaze of pottery, and in welding rots.

Many of these commodities are discarded because of defects, reaching normal household trash and thence dumps, which generally are not equipped for the deposition of radioactive substances.

1.3.3. Uranium

Uranium, discovered in 1789 by Klaproth, is distributed ubiquitously; however, it occurs in exploitable amounts, especially as uranium pitchblende or as carnotite. Up to now only chemical toxic, but not radioactive, effects have been found in the kidneys [Leggett (1989)] and the reproductive system [Domingo (1994)]. [Seeber et al. (1998)] investigated the uranium intake in adults from regions of East Germany, polluted by uranium mining, and observed daily intake of 0.8 to 4.7 μg. Uranium was first used in homeopathic medicines for the treatment of diabetes [Stannard (1988)]. In the ceramic-, porcelain-, and glass-industries, uranium compounds were used to create yellow to orange hues.

This element, however, achieved really great importance by the discovery of nuclear fission in 1938/39. For construction of the first atomic bomb, starting in 1942 with the "Manhattan project", large amounts of uranium were needed. For the production of electric energy from nuclear processes, the need for uranium as fuel was considerably augmented. Naturally occurring uranium contains 99.8 % ^{238}U, 0.72 ^{235}U, and 0.0058 % ^{234}U (see Chap. 2, Table 1), all being radioactive nuclides. Therefore, mining and processing of uranium, cause these radionuclides together with their progenies to be released into the environment. In Fig. 1 the fundamental steps of the nuclear fuel circuit are outlined.

During the opencast or underground mining of uranium ores the radioactive noble gas ^{222}Rn carries the highest risk of release of radioactivity to the environment. Its emanation from the ore may result in high radon concentrations in the mines, leading to considerable exposure for the miners. Particularly in underground mining, attention has to be paid to good ventilation.

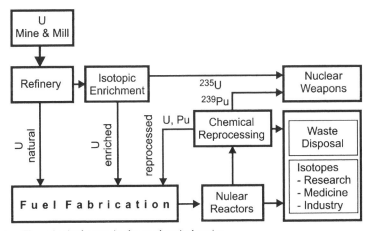

Figure 1:   The principal steps in the nuclear industries.

The worldwide yearly production of uranium, at the end of the 1980s, was estimated to be about 50000 t [UNSCEAR (1993)] for which, it was necessary to extract about 20 million tons of ore. In Table 6 the production rates of some important producer countries are listed.

The amount of radon emission, as the most important radionuclide source from mines, was estimated from data measured in Australian, Canadian, and German mines. The standardized emission was situated between 1 and 2000 GBq per ton of the produced uranium oxide (U_3O_8) with a mean of about 300 GBq per ton [UNSCEAR (1993)]. From a yearly production of 1 GW of electric output, there results an emission of nearly 75 TBq of radon from the uranium mines (see Table 7).

After mining, the uranium ore is transported to the so-called uranium mills, where the uranium is extracted and concentrated to U_3O_8 precipitates, called "yellowcake". The processing of the ore starts with mincing. From the resulting product, which has a sand-like consistency, uranium is extracted by means of acids or bases and is concentrated by ion exchange or solvent extraction. Although 95% of the uranium may be extracted [Eisenbud (1987)], a small part still remains, depending on the ore and the extraction method. Above all, all progenies from the uranium decay series are left in the resulting tailing. The long-lived progenies of the uranium-radium series, ^{230}Th and ^{226}Ra (see Chap. 2, Fig. 1a), are the starting point for the continuous exhalation of ^{222}Rn from such slag dumps. Additionally, precautions have to be taken to prevent radium or thorium from reaching the biosphere by wind forwarding or erosion, for instance into the drinking water. For this reason, wastes from mills are in fact deposited in open slag heaps, but behind dikes or dams, and are finally covered with solid matter or with a layer of water. Measurements at open dumps [UNSCEAR (1993)] showed exhalation rates of about 1 Bq·m^{-2}·s^{-1} of ^{222}Rn from a content of 1 Bq of ^{226}Ra per g of dumped material. At dumps of uranium mills in Australia, Canada, and East Germany, emission rates of ^{222}Rn of 0.1 to 43 Bq·m^{-2}·s^{-1} were measured at a mean of 10 Bq·m^{-2}·s^{-1}. This corresponds, in standardization of the yearly energy production, to a release of about 24 TBq ^{222}Rn per 1 GW·y (see Table 7). By covering the dumps, the emission can be considerably diminished.

The uranium concentrates from the mills are further purified in refineries and converted to metals, passing through different intermediate compounds like UO_3 (orange oxide) and UF_4 (green salt). The refining processes use procedures that lead to dry, powdery uranium compounds. By suitable filter systems, the release of uranium containing dusts into the environment may be avoided.

Table 6: Production rates of uranium in some countries during the years 1986 until 1989, based on [UNSCEAR (1993)], [Barthel (1993)], and [Lange et al. (1991)].

Country	Annual Production of Uranium				(t)
	1985	1986	1987	1988	1989
Canada	10880	11720	12440	12400	11000
South Africa	4880	4602	3963	3850	2900
German Dem. Rep.	4470	4086	4059	3924	3800
United States	4300	5200	5000	5050	4600
Namibia	3400	3300	3500	3600	3600
Australia	3206	4154	3780	3532	3800
France	3189	3248	3376	3394	3190
Niger	3181	3110	2970	2970	3000
Gabon	940	900	800	930	950

Table 7: Releases of radionuclides in effluents from a model uranium mine, mill, fuel conversion, enrichment, and fabrication facility normalized to an annual energy production of 1 MW, based on [UNSCEAR (1993)].

	Releases of Radionuclides								($kBq \cdot MW^{-1} \cdot y^{-1}$)	
	Mine	Mill	Mill Tailings		Conversion		Enrichment		Fabrication	
			a1	b2	a	b	a	b	a	b
	$\cdot 10^6$	$\cdot 10^6$	$\cdot 10^6$	$\cdot 10^{-3}$					$\cdot 10^{-3}$	
^{222}Rn	75	3	20	1	-	-	-	-	-	-
^{238}U	-	0.4	-	-	130	94	1.3	10	340	170
^{235}U	-	-	-	-	6.1	4.3	0.06	0.5	1.4	1.4
^{234}U	-	0.4	-	-	130	94	1.3	10	340	
^{228}Th	-	-	-	-	0.02	-	-	-	-	-
^{230}Th	-	0.02	-	-	0.4	-	-	-	-	-
^{232}Th	-	-	-	-	0.02	-	-	-	-	-
^{234m}Th	-	-	-	-	130	-	1.3	-	340	170
^{226}Ra	-	0.02	-	-	-	0.11	-	-	-	-
^{210}Po	-	0.02	-	-	-	-	-	-	-	-

[a] airborne, [b] aquatic, [1] in operation, [2] abandoned

After the purification processes in the refineries, the uranium finally arrives at the enrichment facilities in the form of UF_6. There, ^{235}U is enriched out of the other uranium isotopes by means of gaseous diffusion, centrifugation, or other aerodynamic, or electromagnetic methods. Highly enriched uranium is necessary for the production of nuclear weapons and as a fuel for certain nuclear reactors. Strengths of 2 to 5 % are necessary for light-water moderated and light-water cooled reactors (PWR's and BWR's) and for graphite moderated and gas cooled reactors of the type AGR. This strength is not necessary for reactors of the type GCR nor for the type of heavy-water moderated and cooled reactors (HWR's). The resulting uranium, depleted during the production, is applied in shielding and for tank penetrating projectiles.

The emissions reaching the environment from the enrichment facilities and the fuel element production of uranium fuel rods are minute compared to the preceding processes (see Table 7). They merely contain long-lived uranium isotopes together with the short-lived decay products of ^{238}U. The long half-life of ^{230}Th prevents a renewed formation of substantial amounts of ^{226}Ra and its decay products, e.g., ^{222}Rn (see Chap. 2, Fig. 1a).

2.   CONTAMINATION BY ARTIFICIALLY PRODUCED RADIONUCLIDES

2.1.   Emissions during Production and Tests of Nuclear Weapons

Major contaminations of the environment by artificially produced radionuclides started immediately after the discovery of nuclear fission in 1938/39. At the end of the 1940s, by producing and testing of nuclear weapons (see Table 8), a considerable amount of radioactive material was released. For a long time, consideration was seldom given to the environment. During the nuclear arms race, in the 1960s, test explosions reached their maximum. The atmospheric bomb tests having polluted the whole earth by

Table 8:   Beginning and latest nuclear test, number of atmospheric (ATM), subterranean shots (SUB), and estimated total yield of atmospheric tests, based on [a][UNSCEAR (1982)], [b][Greenpeace (1996a)] and [c][Van der Vink et al. (1998)].

Country	Year		Number			Tot.Yield (Mt)
	Begin *a*	**Last** *b*	**ATM** *a*	*b*	**SUB** *b*	*a*
United States	1945	1993	193	215	[b]815	138.6
Soviet Union	1949	1990	142	207	508	357.5
United Kingdom	1952	1991	21	21	24	16.7
France	1960	1996	45	50	160	11.9
China	1964	1996	22	23	20	20.7
India	[bc]1974	1999	-	-	[c]6	-
Pakistan		1998				

their fallout. The worldwide rapidly increasing concentrations of the radioactive pollution caused the "Super Powers", to cease at least from the atmospheric testing. But instead, more underground tests were carried out, continuing until very recently. The last known tests were performed by India and Pakistan. Although underground tests as a rule emit far fewer radioactive substances into the biosphere, at some explosions radionuclides in considerable amounts have escaped. After the beginnings of energy production by means of nuclear techniques, major emissions mainly occurred from accidents.

2.1.1. Emission of Radionuclides during the Production of Fissible Materials

For manufacture of nuclear weapons the fissile isotopes ^{235}U, ^{239}Pu, ^{238}U are used together with the radionuclide ^{3}H as "intensifier", which is needed for the amplification of the neutron flux by means of the thermonuclear reactions ^{3}H + ^{2}H → ^{4}He +n + 17.6 MeV or ^{3}H + ^{3}H → ^{4}He + 2n + 11.3 MeV. These reactions are the main component of thermonuclear devices as well. Production of nuclear weapons needs on the one hand mining of natural uranium and enrichment of ^{235}U from the natural abundance of 0.7 % (see Chap. 2, Table 1) up to 90 % by gaseous diffusion or centrifugation, and on the other hand the nuclear production of ^{239}Pu. For nuclear breeding of ^{239}Pu, ^{235}U as reactor fuel is enriched only to a small extent of several percent. ^{239}Pu is generated in the reactor by β-decay from ^{239}Np following the reaction ^{238}U(n,γ)^{239}Np (see Chap. 2, Fig. 4). By reprocessing of the fuel, ^{239}Pu and tritium are recovered, however, ^{85}Kr escapes into the atmoshere. From the atmospheric ^{85}Kr-stock [UNSCEAR (1993)] estimated a total production of 200 tons of plutonium for the manufacture of nuclear weapons. From the knowledge of the (published!) number of devices a mean of 5 kg per device can be deduced. To maintain the tritium concentrations in the weapons, the half-life of ^{3}H being 12.3 y, a production totalling 290 kg of tritium (about 100 EBq) since the 1960s was assumed [UNSCEAR (1993)].

Data on the release of radioactive substances from weapon producing plants into the environment have been partially subject to military secrecy until now. Weapon producing plants in the USA are situated in Albuquerque, Oak Ridge, Richland, and Savannah River. In the former Soviet Union those in Chelyabinsk 40, Tomsk 7, and Krasnoyarsk 26 have to be included.

2.1.1.1. *Hanford*

In the USA near Hanford a plant for the production of nuclear weapons was built. In 1944 two reactors and in 1945 a third one started with the production of plutonium [UNSCEAR (1993)]. The reprocessing of the fuel rods was put in to operation immediately after the completion of the necessary facilities in 1945. Consequently the release of radioactive material into the atmosphere as well as into the Columbia River began. At premature reprocessing and extraction, ^{131}I escaped into the atmosphere, instead of waiting for a nearly complete decay of ^{131}I after removing the fuel rods from the core. Between 1944 and 1956 20 PBq of ^{131}I were released [Cate et al. (1990)], [UNSCEAR (1993)]. The maximum effluent into the atmosphere of 18 PBq of ^{131}I occurred in the years 1944 to 1946; from 1947 to 1956 only 2 PBq of ^{131}I escaped. [Shipler et al. (1996)], [Heep et al. (1996)], [Walters et al. (1996)], [Ramsdell et al. (1996)], and [Farris et al. (1996)] investigated airborne releases and effluents into the Columbia River and the resulting radiation doses. Between 1944 and 1992, one of the airborne nuclides ^{131}I, ^{90}Sr, ^{103,106}Ru, ^{144}Ce and ^{239}Pu, namely the ^{131}I transferred to cow's milk, a significant exposure pathway for the thyroid gland of the inhabitants adjacent to the Hanford area. The thyroids of children received doses ranging from 0.7 mGy to 2.3 Gy. From the nuclides ^{24}Na, ^{32}P, ^{65}Zn, ^{76}As, and ^{239}Np released into the Columbia River, ^{32}P and ^{65}Zn prevailed in fish. By consumption of fish, adults might have been exposed to as much as 5 mSv to 15 mSv.

2.1.1.2. *Rocky Flats*

Rocky Flats is a nuclear weapon plant 16 miles northwest of Denver. In the evening of September 11th, 1957, in the production plant a glove box caught fire, probably by spontaneous ignition of metallic plutonium. The fire spread to the waste air filters, where an explosion occurred, leading to the breakage of the filters. Plutonium from contaminated filters was released into the atmosphere. It was not possible to measure the amount of plutonium then released, since the explosion also destroyed the measuring device. [Mongan et al. (1996a)] later on reconstructed a release of 1.9 GBq of plutonium.

In 1958 the storage of oil and solvents, contaminated by plutonium and depleted uranium, began using containers deposited in the open air at the so-called "Pad 903". Until 1967, 5237 drums accumulated at the storage area, 3572 of those containing plutonium. In January 1964, leakage by corrosion of the drums was discovered. By this, oil contaminated by plutonium entered the soil [Webb et al. (1997), [Iggy Litaor (1999)], [Hulse et al. (1999)], followed by wind transportation of plutonium out of the dump, especially during cleaving and excavation works for rehabilitation. [Mongan et al. (1996b)] reconstructed a release of about 0.26 TBq of plutonium for the years 1960 to 1970. [Ripple et al. (1996)] specified the emissions for plutonium and enriched and depleted uranium from the plant for the time period 1953 to 1989. On May 11th, 1969 [Poet et al. (1972)], plutonium was released into the atmosphere by a fire. Probably plutonium escaped in the form of highly insoluble PuO_2-particles. Measurements of the soil in the neighbourhood of the plant showed indeed the presence of plutonium, however the distribution pattern was inconsistent with the meteorological conditions during the fire. Up to now, the exact amount released at this accident remains unknown.

2.1.1.3. *Oak Ridge*

In 1943 at the Oak Ridge National Laboratory a radiochemical factory was built as a pilot project for the processing of irradiated fuel. In November 1959, a chemical explosion occurred in one of the hot cells. No person was injured, however considerable damage resulted by contamination of the building, of the adjacent road, and of the surfaces of the building. The explosion took place after decontamination of an evaporator. Probably residues of the phenol containing decontaminating agent had remained in the evaporator. When later on nitric acid was heated in the evaporator, these residues induced an explosion [King et al. (1961)], [Eisenbud (1987)], which broke a door of the cell leading directly outside the building. About 15 g of plutonium, about 37 GBq, escaped out of the cell together with small quantities of ^{95}Zr and ^{95}Nb. In addition to the chemical factory plutonium also entered the air ventilation system of the adjacent building housing an air cooled graphite moderated reactor. The primary contamination in the hot cell was 0.370 to 3700 kBq/100 cm^2; in some spots at the road and the surfaces of the building more than 1.85 kBq/100 cm^2 were deposited, mostly however less [Eisenbud (1987)].

2.1.1.4. *Chelyabinsk-40/65 (Mayak/Ozyorsk)*

In May 1946, between Chelyabinsk and Ekaterinburg (Sverdlovsk) in the South Ural (see Fig. 2) in a town with the code name Chelyabinsk 40, later Chelyabinsk 65 and today Ozyorsk, the USSR started, the construction of facilities for uranium production for nuclear weapons. In 1948 at this centre, called "Mayak" (lighthouse) by the physicist Kurchatov, the first graphite moderated uranium reactor was activated and the radiochemical laboratories were put in operation. In order to make up for the loss of time gained by the USA in the nuclear weapon techniques race, radiation protection in the USSR was neglected in the first years and remained incomplete for a long period. The radioactive waste resulting from the process of plutonium extraction, especially in the radiochemical plant, has repeatedly been the cause of serious problems caused by inadequate storage of radioactive waste.

Techa River contamination

From 1949 to 1956, 7.6·10^7 m^3 of moderate and highly radioactive liquid waste at a total of 100 PBq of β-activity was injected into the river-basin Techa-Iset (see Fig. 2), about 98 % thereof in the years 1950 and 1951 [Akleyev et al. (1994)], [N 1140-501 (1991)], [Kossenko (1996)], [Yachmenov et al. (1996)]. The single deliveries varied considerably with regard to activity and radionuclide composition. However, on an average they consisted of 20.4 % of ^{98,90}Sr, 12.2 % of ^{137}Cs, 26.8 % of radioisotopes of rare earths, 25.9 % of ^{103,106}Ru and 13.6 % of ^{95}Zr+^{95}Nb. The radionuclides were spread all over the Techa River, were washed downstream to the Iset River, and precipitated into sediments in the river-bed, of the embankments, and the flooded areas. At the time of the maximum injections in the years 1950 to 1951, the dose rate from external radiation exposure near the injection site or in villages nearby was increased considerably by the depositions. The usage of river water as consumable and utility water led to a considerable incorporation of radionuclides by numerous villagers. 124000 people were exposed to enhanced radiation, 28000 of them persons from riverside villages, whose main drinking water source was the river. They received effective doses of 35 to 1700 mSv. 7500 people from 20 villages had even to be evacuated. For lowering the dose of the population, water pipes and wells were established, highly contaminated embankment areas were ploughed, and the agricultural usage of certain areas was restricted or even forbidden. By these steps incorporation and external exposure was lowered considerably. [Kozheurov et al. (1994ab)], [Degteva et al. (1994), (1996)], and [Mokrov (2002)] [Kossenko et al. (1994), (1996)] reported on the radiation doses and the following damages.

After 1951 the effluents into the Techa River were reduced to a great extent. Since then the concentration of ^{90}Sr and ^{137}Cs in the water has been decreasing, being mainly determined by washing out of the radionuclides absorbed to the sediments and flooding zones. During a measuring campaign at the Techa River (Muslyumovo) in 1992, ^{90}Sr and ^{137}Cs-determinations in soil, water, air, sediments, and foodstuffs [Winkelmann et al. (1994)] as well as in humans [König (1994)] were carried out. In the flood plane area of Muslyumovo 3 to 3.6 MBq of ^{137}Cs per m^2 and concentrations of 4.7 to 5.6 kBq/kg of ^{90}Sr were measured, in other places 15 to 70 kBq/m^2 of ^{137}Cs and 46 to 200 Bq/kg of ^{90}Sr. The ^{90}Sr-concentration in the water amounted to 7.4 to 18 Bq/L and 0.7 to 1.5 Bq/L in milk of cows from that area. [Kryshev et al. 1998)] analyzed a series of foodstuffs from the contaminated areas and observed the following concentrations of ^{90}Sr and ^{137}Cs (Bq/kg):

potatoes	0.2- 6.7 of ^{90}Sr	and	0.5- 3.8 of ^{137}Cs,
grain	0.5-12.6	and	0.3- 2.9,
milk	0.2- 6.3	and	0.2- 4.5,
beef	0.2- 1.7	and	0.3- 2.6,
fish	7 - 480	and	2 -32,
mushrooms	400- 1100	and	110-1600, and
berries	700-16000	and	150.

Kyshtym accident

In the early fifties in Chelyabinsk-40, radioactive waste from a production plant for plutonium situated near Kyshtym, was stored in water-cooled tanks [UNSCEAR (1993)]. In a 300 m³ tank containing 70 to 80 tons of radioactive waste in form of nitrate or acetate-compounds, refrigeration was interrupted by a defect in the cooling system. Within the tank, at first water was evaporated and the dried residues remained. The radioactive material got heated by the radioactive decay and reached a temperature of 330° to 350°C. On September 29th, 1957 the tank exploded with an estimated power of 70 to 100 tons of TNT [Romanov et al. (1990)]. 90 % of the material in the tank containing a stock of 740 PBq radioactive waste was deposited close nearby, whereas 10 %, 74 PBq, descended further away from the explosion centre (see Fig. 2) [Buldakov et al. (1990)], [Burnazyan (1990)], [Romanov et al. (1990)], [Nikipelov et al. (1990)], [Trabalka et al. (1990)], [Ternovskij et al. (1990)], [Yachmenov et al. (1996)]. Except for caesium, the waste had a composition of fission products resulting after one year of storage of about: 66 % $^{144}Ce+^{144}Pr$, 24.9 % $^{95}Zr+^{95}Nb$, 5.4 % $^{90}Sr+^{90}Y$, and 3.7 % $^{106}Ru+^{106}Rh$. Beyond those nuclides the waste additionally contained

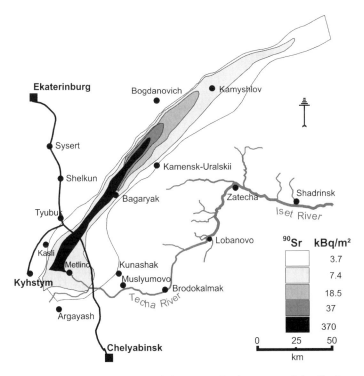

Figure 2: The Techa-Iset River area and the contamination trace of the Kyshtym accident based on [Akleyev et al. (1994)], [Yachmenov et al. (1996)].

about 0.036 % of ^{137}Cs and traces of ^{89}Sr, ^{147}Pm, ^{155}Eu, and ^{239,240}Pu [Buldakov et al. (1990)]. The radioactive cloud reached a height of about one km. The wind caused a rapid spreading out to a track 300 km long. Within 11 hours nearly all the raised material was deposited on the earth or plant surfaces again. Redistribution occurred only within the first days, the radioactive material being resuspended by the leaves of trees and ground surfaces and then transported further. Still 30 years thereafter, no further redistribution has been observed [Ternovskij et al. (1990)]. The contaminated area extended to 15000 to 23000 km^2, inhabited by about 270000 persons [Nikipelov et al. (1990)], [Buldakov et al. (1990)]. This area was graded into three zones relating to the ^{90}Sr deposition: more than 40 MBq/m^2 of ^{90}Sr, between 4 and 40, and between 0,07 and 4 MBq/m^2. Near the centre of the explosion, the concentration by ^{90}Sr even amounted up to 150 MBq/m^2 [Buldakov et al. (1990)]. Within the first 10 days 1154 persons in the zone with the highest contamination were evacuated; because of monitoring ^{90}Sr in agricultural products in areas of more than 150 Bq/m^2 ^{90}Sr [Romanov et al. (1990)], later on an additional 10730 persons had to be removed [Buldakov et al. (1990)]. Resettlement started after eight months. The health effects of the afflicted inhabitants were reported in a study of [Kostyuchenko et al.(1994)].

Wind Dispersion from the Lake Karachay

In springtime 1967, in the same region [UNSCEAR (1993)] a further accident happened as a result of a disposal of liquid radioactive waste into Lake-Karachay [Aarkrog et al. (1992)], [Nikipelov et al. (1992)], [Trabalka et al. (1990)]. This lake contained about 4.4 EBq of ^{90}Sr and ^{137}Cs. Radioactive dust, about 22 TBq of ^{90}Sr and ^{137}Cs [Akleyev et al. (1994)], from the desiccated shore zones of the lake were resuspended by wind and forwarded over a distance of 75 km onto an area of 1800 km^2, partially on the same track of the 1957 accident. The maximum contamination deposited amounted to 0.4 MBq/m^2 at a ^{137}Cs/^{90}Sr ratio of 3:1.

2.1.1.5. Tomsk 7

Seversk, formerly named Tomsk 7, was, similarly to Ozyorsk, a closed city situated about 15 km northwest of Tomsk ashore of the Tom river, a tributary of the Ob river. This place was chosen to construct a plant for the production of nuclear weapons. For that purpose, nuclear reactors, a factory for chemical reprocessing of uranium and plutonium, a uranium enrichment facility, and installations for the storage of radioactive waste were built. The first of the five reactors was started in 1955 and it was in operation until 1990. The other reactors became critical the first time in 1958, 1961, 1965, and 1967 respectively. The first reactor was equipped with an open chilling system. The cooling water was drained off directly into the Chemilshikov

River, a tributary of the Tom. Because of the systematic release of contaminated cooling water, radioactive substances entered the river system by this way. In sediment samples from the Chemilshikov Channel not only the short-lived radionuclides ^{51}Cr, ^{58}Co, ^{65}Zn were found, but also ^{137}Cs. Near the village of Chemilshikov on the right site bank of the Tom river even ^{239}Pu was discovered.

Some additional accidents occurred with release of radioactive substances. The most severe accident taking place on April 6th, 1993 [IAEA (1998e)], when by mistake uranium, niobium, zirconium, and ruthenium were fed into a 34 m^3 large tank, already filled with a mixture of paraffin and tributylphosphate. The tank, containing a volume of 25 m^3, about 8773 kg of uranium and 310 kg of plutonium (710 TBq), exploded. As a result the building was damaged and a share of 4.3 TBq from a total of 20.7 TBq of long living fission and activated nuclides [Yablokov (1994)] was distributed to the northwest across an area of 120 km^2.

During the production over a period of 30 years, 40 EBq of waste accumulated [Yablokov (1994)]. Liquid waste was deposited into sandy layers at a depth of 320-460 m and also pumped into an open basin [Bøhmer et al. (1995)]. The storage in open reservoirs led to similar problems as in the case of Lake Karachay. Therefore in 1991 covering up of the open reservoirs was started.

2.1.1.6. *Krasnoyarsk-26*

The town of Zheleshnogorsk, until 1994 Krasnoyarsk-26, is situated 50 km north of Krasnoyarsk on the bank of the Yenisey River and was also called Dodonovo, Devyatka, Town 9, City No 26, or Atomgrad. All important production plants were installed underground, at a depth of 250 to 300 m [Bøhmer et al. (1995)]. They consisted of plutonium producing reactors with radiochemical installations for reprocessing. The first reactor was put into operation in 1959, the second in 1961, and the third in 1964. It has been estimated that these reactors were able to produce about 1.5 t of plutonium a year [Bøhmer et al. (1995)]. The first reactors were both equipped with an open cooling system. The cooling water was taken from the Yenisey River and until 1991 drained off directly. A contamination of the river water resulted as far as 500 km downstream. During the dry months of the summer, injected radionuclides were relatively fast adsorbed to sediments. Numerous studies showed that the contaminations contained also plutonium.

The reprocessing was started in 1964. The resulting liquid wastes, containing $4 \cdot 10^7$ m^3 about 26 EBq [Yablokov (1994)], were stored either in deep-lying cavities in the rock or 20 km to the north at a depth of 190 to 470 m. During the transport of radioactive substances from the radiochemical factory by means of a pipeline, sprung leaks had evolved. This led not only to contamination of the surrounding earth, but also secondarily to a contamination of plants and surface water, which was drained to the Yenisey River.

2.1.2. Nuclear Weapon Test Sites

Locations of nuclear explosions in the atmosphere, underground, and under water performed by different countries are shown in Fig. 3.

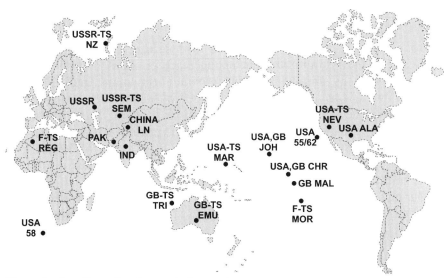

Figure 3: Locations of nuclear explosions:

ALA: Alamogordo	**CHR**: Christmas Islands	**EMU**: Emu & Maralinga
JOH: Johnston Islands	**LN**: Lop Nur	**MAL**: Malden Islands
MAR: Marshall Islands	**MOR**: Mururoa & Fangataufa	**NEV**: Nevada
NZ: Novaya Zemlya	**REG**: Reganne	**SEM**: Semipalatinsk,
TRI: Trimouille Islands (Monte Bello)	**TS**: Test site	

F: France, *GB*: United Kingdom, *IND*: India, *PAK*: Pakistan, *USA*: United States of America, *USSR*: Union of Soviet Socialist Republics

2.1.2.1. *Los Alamos*

As a result of the feverish efforts during the "Manhattan-Project", on July 16th, 1945 at 5.30 local time the first nuclear explosion test took place in the desert of New Mexico at Alamogordo near Los Alamos (see Fig. 3). This was the only test, before three weeks later, on August 6th the devastation of the Japanese town of Hiroshima by detonation of a uranium bomb was made public. Three days later Nagasaki suffered a similar fate by the first plutonium bomb ignited.

At the first test, the TRINITY-Test in 1945, the bursting charge was detonated on top of a 30 m high steel tower, the explosive force was 19 kilotons compared to TNT. Radioactive fragments were transported by the wind as far as 40 km and caused skin burns in cows [Lamont (1965]. Minute particles even reached the state of Indiana [Webb (1949)], [Eisenbud (1987)], [Stannard (1988)]. The subsequent tests were transferred into the Pacific region, to the Marshall Islands.

2.1.2.2. *Test Sites Marshall-, Christmas-, and Johnston-Islands*

The Marshall Islands, located in the central Pacific zone (Micronesia) (see Fig. 3 and 4), were until the end of World War II under Japanese administration. In 1947, they were entrusted to the USA by the "United Nations Security Council". [Simon (1997)], [MSC (1978)], and [Deines et al. (1991)]. [Schultz et al. (1991)] summarized the fate of these islands, the performed nuclear weapon tests, and the following environmental programs initiated by several organizations. [Stannard (1988)], [Eisenbud (1997)] outlined the events during the test "BRAVO", and [Eisenbud (1990)] described the environmentally relevant results during the "CASTLE"-series.

In 1946, the military governor of the Marshall Islands decreed that the population of the Bikini-Atoll had to leave their island temporarily, in order to enable the USA to carry out atomic bomb tests for the "benefit" of mankind. The navy evacuated 166 inhabitants to the atoll of Rongerik, from where they had later to be resettled again because of unsatisfactory food and water supply. The operation "CROSSROADS" started and during these experiments the inhabitants of islands of Enewetak, Rongelap, and Wotho were evacuated too. The spectacle of the "CROSSROADS"-explosions were the third ("ABEL") and fourth ("BAKKER") atomic blasts and were observed by 42000 soldiers, pressmen, and politicians. About a hundred ships took part, among others, the aircraft carrier "Saratoga", the battleships "Arkansas" and "Nagato", and the cruisers "Prince Eugene" and "Sakawa".

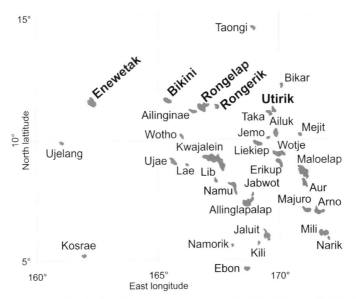

Figure 4: The Marshall Islands with the test sites Enewetak, Bikini, Rongerik, Uterik.

The "CROSSROADS"-series were followed by the operations "SAND-STONE" comprising five tests and "GREENHOUSE" comprising four tests near the atoll of Enewetak. The test "MIKE" in 1952 on the island of Elugelab was the first thermonuclear test and, left behind a crater a half mile deep. After this explosion in the radioactive fallout two new elements, namely einsteinium and fermium as well as two new radionuclides of plutonium were discovered, ^{244}Pu and ^{246}Pu. Fortunately, the most of the fallout of this experiment descended in the open Pacific [Simon et al. (1997a)], [Eisenbud (1990)]. In 1954 the first blast "BRAVO" out of the test series "CASTLE" was fired on the Bikini-Atoll. The explosive force of "BRAVO" was three times higher than expected and the wind carried radioactive fallout to the neighbouring atolls of Rongelap, Rongerik, and even to the far more distant atolls of Uterik. Furthermore the crew of the Japanese tuna fishing boat "Lucky Dragon", being 135 km east of Bikini, was considerably exposed to radioactive fallout. In 1957 during the operation "REDWING", 17 tests, some of them hydrogen bomb tests were performed near the atolls of Bikini and Enewetak. In 1958 on the Marshall Islands the last test series took place during the operation "HARDTACK I" comprising 35 detonations. During these tests and in the following years, foodstuffs, the marine food-chain, and whole body activities in humans were measured and the radiation doses were calculated. The results have been published in numerous papers [Simon et al. (1997a)], [Black et al. (1986)], [Cohen et al (1985d)], [Beasley et al. (1972)], [Musolino et al. (1997)].

In 1962 the USA continued their tests in the atmosphere outside the Marshall Islands near the Christmas Islands (24 tests), the Johnston Islands (12 tests) and in some other locations (3 tests) in the middle Pacific [Simon et al. (1997b)] (see Fig. 3). Thereafter, for economical and for additional security reasons, the nuclear tests and experiments were transferred to the Nevada test site. A stocktaking of the radiological conditions at the Marshall Islands and the eventuality of a repopulation of the atolls of Bikini under the present circumstances has been examined by the [IAEA (1998a)].

2.1.2.3. Test Site Nevada (USA)

In 1950 an area within the United States was destined for nuclear test explosions. From 1951 until the test stop agreement in 1963 on the Nevada test site (see Fig. 5), 90 % between 1953 and 1957, about 100 explosions at a bursting force up to one megaton were carried out in the atmosphere [UNSCEAR (1993)], [Stannard (1988)]. More than half of the external γ-dose from the test series resulted only from three events, "HARRY" in 1953, "BEE" in 1955, and "SMOKY" in 1957, in the three adjacent places, at St. George in Utah, and at Ely and Las Vegas in Nevada [Anspaugh et al.

(1986)]. During several explosions of the 30 subterranean tests at this site [UNSCEAR (1993)], between 1961 and 1980, ^{131}I emissions into the atmosphere took place: during "DE MOINES" in 1962 1.2 PBq of ^{131}I, "BANDICOOT" in 1962 0.33 PBq, "PIKE" in 1964 130 TBq, "RED HOT" in 1966 7 TBq, "PIN STRIP" in 1966 7 TBq, and during "BANBERRY" in 1979 3 PBq of [131]I. After the "BANBERRY" shots [Brown et al. (1973)] recovered ^{131}I in the human food-chain. At the very same area, for instance also experiments for the peaceful utilization ("SEDAN", "CABRIOLET", "SCHOONER") were carried out, followed also by escape of radioactivity [Black et al. (1986)]. The project "SCHOONER" in 1968, a nuclear "excavation" experiment, detonating 108 m under the ground, caused a 3990 m high cloud with a diameter of 2420 m and leaving behind a crater with a radius of 130 m [Gudiksen (1972)]. From this experiment in the tuff rock, besides [141]Ce, [132]Te and among others, [187]W was generated as an activation product too [Chertok et al. (1971)].

[Thompson et al. (1996)] tried to estimate the radiation doses for the population in the neighbourhood of the test site and in 353 places in Utah, Arizona, California, and Nevada. [Whicker et al. (1972), (1996)] and [Kirchner et al. (1996)] found that, just like at Hanford, ^{131}I contributed the major share to radiation exposition via ingestion by milk.

Figure 5:   The Nevada test site.

2.1.2.4. *Test Site near Semipalatinsk*

The Russian nuclear weapons development program began in 1939, for based on the fundamental works of Y. Zeldowich and Y. Khariton, and was intensified under the leadership of V. Kurchatov from 1943 onwards. The first European nuclear reactor, moderated by graphite, similarly to the "Hanford 305 Design", was put in to operation at the Kurchatov Institute in Moscow in 1946.

The first large site for nuclear test explosions of the former Soviet Union was selected in Kazakhstan (see Fig. 3 and 6) about 120 km northwest of Semipalatinsk in the Semipalatinsk district ("oblast") comprising an area of 18500 km^2. The town of Kurchatov is located near the northeast corner, at the Irtysh River. There the scientific and technical staffs had been accommodated. The area, surrounded by a steppe-like region with continental climate, is influenced by strong seasonal winds from predominantly northwest during wintertime and from northern directions during summertime. Here, at 7 a.m. on August 29th, 1949, at an altitude of 30 m, the first Soviet atomic bomb was ignited. The plutonium bomb had an explosive force of 22 kt (compared to TNT). The radioactive cloud reached an altitude of nine km. At the time of the test, unstable weather prevailed with showers and strong winds, running up to speeds of 60 km/h at a height of six km. The cloud drifted away to the Altayskiy kray [Shoikhet et al. (1998)] and then reached populated areas about 150 km away after 2.5 hours, so that inhabitants there were exposed to considerable doses [Kiselev et al. (1994)], [Loborev et al. (1994ab)]. In the Uglowskoye county ("rayon") at some villages, e.g. in Topolnoye,

Figure 6:   The former Soviet Union test site of Semipalatinsk (TS) and the Altai trace from the first nuclear explosion on August 29th, 1949.

Belenkoye, and Laptev Log, effective doses were estimated ranging from 1.5 to 0.6 Sv resulting from both external and internal exposure, and in the more distant county of Rubzowsk, e.g. in Veseloyarsk still 0.5 Sv [Shoikhet et al. (1998)].

Subsequently up to 1989, on the test site 456 further nuclear tests took place, the number varying dependent on the source of citation, and whether military or civil experiments were included. [Gorin et al. (1993)], [Dubasov et al. (1994abc)], and [Mikhailov et al. (1996)] published information on the different tests and experiments. In the period from 1949 to 1962, 124 nuclear explosions were performed in the atmosphere, 25 near the ground plane at a height of 30 to 40 m, eight below, and 91 above a height of 10 km. On August 12th, 1953 a sort of "fusion bomb" was tried out, followed on December 22nd, 1955 by the first thermonuclear bomb at an explosive force of 1.6 Mt. All together, during the tests, about 6.7 PBq of ^{137}Cs and 3.7 PBq of ^{90}Sr were released into the atmosphere [Algazin et al. (1995)]. Outside this area 33 further explosions took place in Kazakhstan and two in Uzbekistan.

The fallout clouds had drifted to northeast into the district of Semipalatinsk, of East Kazakhstan, of Pavlodarsk, and to the Altayskiy kray. [Logachov et al. (1993)], [Izrael et al. (1994)], [Gabbasov et al. (1995)], and [Zinchenko et al. (1997)] tried retrospectively to calculate the extension of the tracks of the fallout from the tests. These tracks have been communicated in a report by [Voigt et al. (1998)] quoted from [Dubasov et al. (1993)].

Along the tracks contamination by deposition at the vegetation and at the soil occurred. Estimating the radiation doses, especially that during the "hot testing phase", one has to rely on retrospective reconstruction calculations due to lack of sufficient measurements. During the 1990s the soil and food stuff, besides measurements of the dose rate, were examined for the assessment of the radiation risk [Voigt et al. (1998)], [Semiochkina et al. (1998)], [Yamamoto et al. (1996)], [Shebell et al. (1995), (1996)], and [Hill et al. (1995)]. [Voigt et al. (1998)] concluded that the contamination with ^{90}Sr and ^{137}Cs, apart from in some hot spots, is relatively low and amounted to only 1-2 orders of magnitude above the global fallout level. Similar results were found by [Hill et al. (1995)], who additionally carried out whole body measurements for the determination of the ^{137}Cs concentrations in the body. [Yamamoto et al. (1996)] and [Shebell et al. (1995)] investigated soil samples from the test site itself and found partially increased amounts of ^{239,240}Pu, ^{137}Cs, ^{60}Co and ^{152,154}Eu. [IAEA (1998b)] reported that the remaining radioactivity from nuclear weapon tests is low, except for some points at the testing site, e.g. "Ground Zero" and the lake Balapan. The contamination levels have been as follows:

– villages outside of the testing site 5-100 Bq/kg for ^{137}Cs, 5-63 Bq/kg for ^{90}Sr, 0.2-0.4 Bq/kg, for ^{238}Pu and 0.2-0.7 Bq/kg for $^{239+240}$Pu, with the exception of the village of Dolon with 30-250 Bq/kg for plutonium

– at the testing site itself, a wide variation of range occurred, depending on the place, up to 50000 Bq/kg for ^{137}Cs, up to 11000 Bq/kg for ^{90}Sr, up to 6000 Bq/kg for ^{238}Pu, and up to 14000 Bq/kg for $^{239+240}$Pu.

2.1.2.5. Test Site on Novaya Zemlya

The former Soviet Union concluded to move the nuclear test site from Semipalatinsk to Novaya Zemlya. Novaya Zemlya, comprising two large islands, which are separated by a narrow strait, the Proliv Matochkin Shar, is 900 km long with an area of about 82180 km^2 (see Fig. 3). Numerous smaller islands belong to it, another about 1000 km^2. The main part of the northern island but also a part of the southern is covered by glaciers. After the evacuation of the people, the nenets-families, formerly living there, two larger settlements have been built up as military and fleet bases, one with 4000 inhabitants on the Belochaya bay, and another lying nearby the strait of Matochkin Shar. In 1954, at the establishment as a testing site for nuclear explosions, the islands were put under military administration. There have been two test areas at Novaya Zemlya, one located at the Chernaya bay on the southern island, and the other near the Matochkin Shar strait on the northern island.

All explosions carried out in the atmosphere were detonated at the northern testing ground. The southern territory was used merely between 1973 and 1975. On Novaya Zemlya 132 bombs detonated, 86 of them in the atmosphere during the period from 1957 to 1962, further 43 underground explosions from 1963 to 1990. In the period of 1955 to 1961, three explosions were ignited at sea, two of them at the Barents Sea in the vicinity of the Matochkin Shar strait.

The first test in the atmosphere on Novaya Zemlya took place on September 24th, 1957. Up to the end of 1957, further 23 bombs exploded, until the opposition of the Russian engineers increased. Andrey Sakharov criticized the tests in the range of megatons to be without scientific sense. In the following years no further Soviet tests were carried out until September 1961. During the Cuba Crisis however, the test resumed and from September until November 1961 within short intervals, one by one, 24 bombs with an explosive force in the range of megatons each 20 to 30 Mt were exploded. On October 30th, 1961, the strongest bomb ever detonated, a hydrogen bomb with 58 megatons was ignited. It exploded at an altitude of 365 m above the ground, and the mushroom cloud rose up to a height of 60 km. Since August 1962 to the end of the year, 32 further tests followed, many of them also in the megaton range, until in 1963 the nuclear test stop agreement for experiments in the atmosphere became effective. Thereafter, until 1990, on the northern area of Novaya Zemlya only underground tests took place at a depth of 300 to 400 m. At the southern testing area, between 1973 and 1975, only seven tests were carried out. After an agreement between the USA and the USSR in the year 1976, to restrict the explosive force to 150 kilotons, on the southern territory no further tests have been performed.

Although the emission of fallout should be avoided by the transfer of the tests to subterranean area, at about 100 Soviet subterranean explosions, radioactive gases were released into the atmosphere anyway [Nilsen et al. (1994)]. Indeed, after a subterranean test on June 2nd, 1967 on the northern area, radioactive fallout was proved in northern Europe. In the Chernaya bay, outside the southern testing site, probably by explosions at sea, increased ^{137}Cs and ^{239}Pu concentrations in sediments have been measured.

For the civil sector 115 nuclear explosions were triggered, in order to create cavities, to dig canals, for seismic trials, or even to blow out large fires. At the northern test site on Novaya Zemlya 41 such civil explosions were ignited.

2.1.2.6. *Test Sites at Trimouille Is./Monte Bello Is., Emu/Maralinga in Australia and Malden Islands in the Pacific Region*

The decision of Great Britain to produce nuclear weapons was made in 1947. The first plutonium producing reactor at Windscale, now called Sellafield, became critical in 1950 and started the production of plutonium in 1952. Because of the dense settlement no nuclear explosions in Great Britain could be carried out, the first testing explosion was realized at the Trimouille Island near the Monte Bello Islands (see Fig. 3) northwest of Australia. Under the code name "HURRICANE", on October 3rd, 1952, the first British nuclear device, a plutonium bomb, was ignited inside a ship ("HMS Plym") 2.7 m below the water line. The ship was riding at anchor about 365 m away from the shore. Two further explosions in the atmosphere followed in 1956, the strongest with an explosive force of 98 kt.

Between 1953 and 1963, nine more nuclear weapon tests and other experimental nuclear explosions were performed by Great Britain in South Australia (see Fig. 3 and 7) at Emu and Maralinga [Symonds (1985)], [Haywood et al. (1992)]. The first two tests were near Emu, the other at Maralinga. The radioactive material released comprised, among others, in some cases also plutonium and uranium. As a consequence of these tests, as recently as 1992 significant activities of caesium, plutonium, americium, cobalt, strontium, and uranium were still found at the test site, so that in some areas restrictions for access have been necessary. Studies on the bioavailability of the radionuclides and an estimation of the radiation doses were carried out by [Stradling et al. (1989), (1992)], [Harrison et al. (1989ab)], [Popwell (1989)], [Haywood et al. (1990), (1992)], [Johnston et al. (1992)].

Furthermore, in the year 1957 , under the code name "GRAPPLE", thermonuclear devices exploded on the Malden Islands (see Fig. 3) and further six explosions took place between 1957 and 1958 on the Christmas Islands (see Fig. 3), mostly in the range of megatons. Additional tests were performed together with the USA at the test site of Nevada.

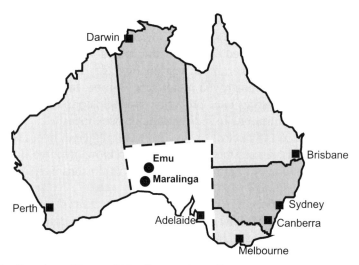

Figure 7:   Test sites of Emu and Maralinga on Australia.

2.1.2.7. *Test Sites at Reganne and near the Mururoa Atoll*

On February 13th, 1960 at Reganne (see Fig. 3), in the western Sahara in Algeria, France detonated the first nuclear bomb and later on carried out a further 16 nuclear tests at this place. Since 1966 the blasts were transferred to the pacific region to the atolls of Mururoa and Fangataufa. As a whole 45 tests in the atmosphere were performed, the last in the year 1974. From 1975 to December 1995, a further eight subterranean explosions were triggered, the last one on December 12th, 1995. The radiological conditions resulting from the French bomb tests at the atolls of Mururoa and Fangataufa were examined by a group of international experts and the results were published by [IAEA (1998cd)].

2.1.2.8. *Test Site at Lop Nur*

The first Chinese nuclear bomb, a ^{235}U-bomb at an explosive force of 22 kt, was detonated on October 16th, 1964 at the testing site of Lop Nur in the Xinjiang Province (see Fig. 3). In June 1967, the 6th test, the first Chinese hydrogen bomb was exploded there at an explosive force of 3.3 Mt. Altogether, 22 tests in the atmosphere and 20 underground were carried out (see Table 8), the last test on August 16th, 1996.

2.1.2.9. *Nuclear Test Explosions from Other Countries*

On March 24th, 1993 the then president of South Africa, De Klerk, announced in parliament that South Africa had produced nuclear weapons. From this development, the only South African test happened in the Indian Ocean on September 9th, 1979. On this date the satellite Vela 6911 detected a double lightening, typical for a nuclear explosion, over the Indian Ocean at the Prince Edwards Islands near South Africa. This observation was controversial for a long time, until April 1997, when the South African Foreign Minister Azis Pahad confirmed the test.

On May 11th, 1998 in the neighbourhood of the India-Pakistani frontier [SZ (1999)], in the province of Rajasthan, in the Thar desert a series of three tests was ignited followed two days later by two further subterranean shots (see Fig. 3). This represented India's second nuclear test series since May 5th, 1974.

As a counter action Pakistan also announced a series of six tests in total and the experiments followed on May 28th and 30th, 1998 in the province of Baluchistan in the vicinity of the Afghan border.

2.1.2.10. *Nuclear Explosions in the Open Sea outside the Testing Sites*

With the operation "WIGWAM", the USA performed a nuclear test, which was carried out on May 5th, 1955 about 800 km west of San Diego (see Fig. 3) at a water depth of 610 m. The bursting force was 30 kt. At this point, the ocean has a depth of 4870 m, deep enough to avoid damage in the deep sea or at the seabed. The place was chosen because it is located far enough away from the normal shipping routes, because there was no fishing area, the weather was settled, and the sea currents and temperature gradient were known. The bursting charge was hanging on a cable, fastened to an unmanned boat. The aim of the experiment was to find out, how strong the shock waves were, and what is the impact of the total energy of the explosion remaining in the water. Should it be possible, for instance, to destroy a submerged submarine from a "surface boat" by usage of nuclear weapons without self-exposure to danger?

In the same region of the Pacific Ocean, about 640 km west of San Diego, "SWORDFISH", within the scope of the "DOMINIC" I series, was detonated on May 11th, 1962, testing a nuclear submarine weapon. "DOMINIC" was the last of the American test series in the atmosphere before the test stop agreement.

In the Southern Atlantic, 45° of southern latitude, out of the way of the shipping lines, on August 27th, 28th, and on September 6th, 1958, three tests performed by the USA in the atmosphere took place under the code name "ARGUS". These operations were the first to use a ballistic missile with a nuclear bursting charge launched from a ship to an altitude of approximately 483 km before being detonated.

2.1.3.   Global Effluents to the Biosphere by Fallout from Nuclear Tests

The radionuclides produced and released during nuclear weapon testing are compiled in Chap. 2, Table 8. They were distributed all over the globe by the transport mechanisms described in Chap. 2, 3. Since a large part of the experiments and nuclear tests was performed in the northern hemisphere, this part of the world was the most affected by the deposition of the radioactive fallout. In Chap. 2, Table 14 the activities of the most important radionuclides deposited with the soil are summarized. From that it becomes obvious that in the northern hemisphere the deposition had been about 4 to 5 times higher than in the southern.

The spectrum of nuclides essentially contributing to the radiation exposure of humans [UNSCEAR (1977)] comprises the following:

– ^{131}I burdens the thyroid gland after inhalation, especially in foetuses from the 9th to 11th week of gestation and in children.
– ^{137}Cs leads to a radiation exposure of the whole body from outside by deposition on surfaces and from inside by uptake with foodstuffs (milk, meat products, mushrooms) (see Fig. 8, Table 9).
– ^{90}Sr, after incorporation, especially with milk products, leads to an exposure of the skeleton and the haematopoietic tissue (see Table 9).
– ^{14}C burdens, after uptake into the organism, gonads, lungs, and the bone marrow.
– ^{106}Ru and ^{144}Ce burden the lungs.
– Plutonium isotopes are mostly a factor of importance in the vicinity of a testing site, elsewhere the deposition was low.

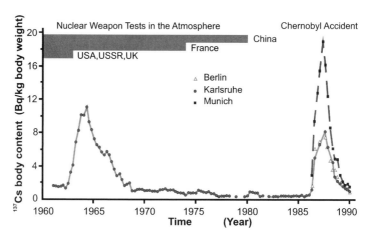

Figure 8:   Whole body content of people from reference groups from three towns in Germany: Karlsruhe [Doerfel (1987)], Berlin [Schmier et al. (1988)], and Munich [Berg et al (1986),(1987a-c)].

Table 9: Concentration of ^{90}Sr and ^{137}Cs in milk, daily intake from the diet, ^{90}Sr in bone, and ^{137}Cs in the whole body (WB) in the years 1974, [a]1975, or [b]1977, based on [UNSCEAR (1982)]: *C*: children 5-19 a old, *A*: adults.

Location	Milk (Bq/L)		Intake (Bq/d)		Incorporation (Bq/kg)		
	^{90}Sr	^{137}Cs	^{90}Sr	^{137}Cs	^{90}Sr - Bone[1] C	^{90}Sr - Bone[1] A	^{137}Cs - WB[2]
Northern Hemisphere							
Canada	0.2	0.3	0.3	-	70	60	-
Denmark	0.2	0.3	0.4	0.6	52	52	0.72
Faeroe Islands	0.9	9.4	0.6	8.8	-	-	-
Finland	0.2	1.0	-	-	-	-	2.00
Inari	-	-	-	-	-	-	160
France	0.3	0.4	0.4	0.6	-	-	1.34
Germany West	0.3	0.7	0.4	0.6	[b]52	[b]30	0.7-1.22
(Berlin)	0.2	0.3	0.2	0.6	-	-	-
Greenland	-	-	0.3	2.0	-	-	-
India	0.1	-	0.2	-	-	[a]85	-
Italy	0.3	-	-	-	-	-	-
Japan	0.2	0.3	0.3	0.6	57	41	0.80
Netherlands	0.2	0.3	-	0.3	-	-	-
Nepal	-	-	-	-	[a]150	110	-
Norway	0.3	2.2	-	-	110	89	-
Poland	0.3	0.8	-	-	-	-	-
Sweden	0.2	0.4	-	-	-	-	1.64
Switzerland	0.2	0.4	-	-	-	-	0.60
USSR	0.3	0.7	0.5	0.6	110	53	-
Murmansk	-	-	-	-	-	-	520
United Kingdom	0.1	0.3	-	-	-	-	0.66
USA	-	0.1	-	0.5	-	-	-
New York	0.2	-	0.4	63	44	-	-
San Francisco	0.05	-	0.1	-	26	26	-
Southern Hemisphere							
Argentina	0.11	0.4	0.08	0.4	[a]36	[a]36	-
Australia	0.20	0.3	0.10	-	37	37	0.72

[1]standardized to calcium, [2]standardized to body weight

Short-lived radionuclides are of importance only for a few weeks following the test explosion. The ratio of the expansion velocity (a measure of radionuclide mobility in the environment) to the physical half-life determines whether radionuclides can penetrate into the food chain or they have already decayed to a sufficiently low level. They mostly have no major importance with respect to the global distribution, but are only of local interest.

In Table 9 the activity concentrations in milk for ^{90}Sr and ^{137}Cs and their daily intake though food are compiled for different countries for the year 1974. Furthermore Table 9 shows values for ^{90}Sr-activity in bone for children (C) and adults (A), and data on the ^{137}Cs-content in the whole body (WB). The ^{90}Sr-concentrations are standardized with respect to the calcium content of the samples, and the ^{137}Cs-values, originally related to the body potassium, with respect to the body weight (2 g potassium per kg body weight). Contaminations measured in the northern hemisphere are higher than those of the southern. Especially in regions which are located far north of or near test sites, higher concentrations could be proved. Particularly in the case of consumption of reindeer meat, in parts of Finland, Norway and the former Soviet Union, the ^{90}Sr content in bones as well as the ^{137}Cs in the whole body increased considerably. In 1964 in the Federal Republic of Germany (see Fig. 8), after the end of the tests in the atmosphere, ^{137}Cs amounted to 12 Bq/kg in humans and dropped until 1986, prior to the reactor accident in Chernobyl, to about 0.5 Bq/kg body weight. Since the uptake of calcium by the bone, the mineralization rate, in the case of growing organisms, is extraordinarily high, the calcium like ^{90}Sr is built in to an increased extent in children during the period of growth. Until 1980 about 460 PBq of ^{90}Sr from the tests in the atmosphere entered the northern hemisphere and further about 144 PBq the southern. Cumulatively deposited ^{90}Sr reached its maximum of 355 PBq in the northern hemisphere in the year 1966 and in the southern hemisphere with 108 PBq [UNSCEAR (1982)] in the year 1974. The maximal concentration of radionuclides in the organism should, however, have been reached only some years after the end of the test series in the atmosphere.

The emissions of plutonium by the nuclear weapon tests globally amounted to about 0.33 PBq of ^{238}Pu, 11 PBq of $^{239+240}$Pu, and 140 PBq of ^{241}Pu (see Chap. 2, Table 8). There are additional an 0.51 PBq [Bunzl et al. (1994)] by the burning up of the satellite SNAP-9A. Until 1978 from decay of ^{241}Pu also 3.1 PBq ^{241}Am were generated [Bunzl et al. (1994)], which also could be proved in environmental samples. In Germany the $^{239+240}$Pu-concentration for foodstuffs and organs per kg fresh weight according to [Bunzl et al. (1994), (1983)] has been as follows: in potatoes 0.04-0.12 mBq/kg, in wheat 0.04-40 mBq/kg, in beef liver 0.07-2.46 mBq/kg, and in human liver, depending on age, between about 6 and 21 mBq/kg (about 6 mBq/kg at age of 20-39 y, 17 mBq/kg at age of 40-59 y, 21 mBq/kg at age of 60-79 y).

2.2. Emissions by Nuclear Installations at Standard Operation

2.2.1. Emission from Nuclear Reactors

The worldwide importance of nuclear energy is demonstrated by the fact that about 17 % [Poong et al. (1998)] of the total electrical energy produced originates from nuclear techniques. In 1994 the world wide leading producers were the USA with 73 GW·y, followed by France with 39 GW·y, by Japan with 29.5 GW·y, by Germany with 16.3 GW·y, and by Canada with 11.6 GW·y. In some countries, in Belgium, Bulgaria, France, Hungary, Lithuania, Slovakia, Sweden, and Switzerland the contribution amounts even to more than 40 %. In 1995 globally 437 units in total were operating and 39 were under construction.

The hazard to the environment from emissions of nuclear plants or research reactors during normal operation usually at most affects the population living directly in the neighbourhood. When problems arise, however, further sections of the population may be affected by emitted radionuclides, which, due to their half-life and mobility in the environment, may spread widely and get into the food-chain (see Chap. 2, 3.2). Tables 10 and 11 give a survey of the release of radioactive matter into the environment at normal action of the reactor. Table 10 shows the means of the emissions into the environment, standardized to the produced energy over the years 1985-1989. The kind and amount of the emission are dependent on the design of the reactor. The light-water reactors, LWR's, water cooled and moderated, are subdivided into pressurized water reactors (PWR) and boiling water reactors (BWR). Some LWR's, however, use graphite as a moderator (LWGR). PWR's and BWR's are those reactor types, which at present produce most of the electric power. Other reactor designs, like heavy-water reactors (HWR), graphite moderated, gas cooled reactors (GCR), or fast breeders (FBR) hold an essentially lower share in power production. Concerning the emission it is appropriate to group radionuclides with similar radiobiological behaviour in the following way: the inert gases contained in off-air, tritium in airborne or liquid effluents, ^{14}C in waste air, ^{131}I in waste air, radioactivity adsorbed to particles in waste air, and "others" contained in the liquid releases (see Table 10).

Radioactive noble gases occur by nuclear fission and from the radioactive decay of fission products in the fuel. Furthermore, noble gases result by neutron activation. In this way also ^{41}Ar may be produced in the air around the reactor vessel, at FBR's within the reactor container from the argon buffer gas for the sodium coolant, or at gas cooled reactors out of argon impurities of the cooling gas. At GCR's and LWGR's (see Table 10) the emissions of noble gases, standardized to the energy production, totalling to about 2000 TBq/(GW·y) are higher by about one order of magnitude than at other types of reactors. The percentage distribution of the activity of the different radioactive isotopes is summarized in Table 11 for BWR's and PWR's based on data from American nuclear plants [UNSCEAR (1993)]. Some radionuclides with very short half-lives or minor shares are not included in the list. For that reason, the sum of the percentiles is smaller than 100 %. In BWR's the ^{133}Xe at a range of 90 % is the main contaminant of the noble gas, in PWR's amounting to only 32 %, but there is additional ^{135}Xe at a range of 20 % and $^{88+85}$Kr of 15 %. Until the year 1989, altogether approximately 3.16 EBq of radioactive noble gases were released into the environment.

Table 10: Averaged normalized releases of radionuclides (1985 - 1989) from different reactors types (PWR, BWR, GCR, HWR, LWGR, FBR), their electrical energy production in the year 1989, and the estimated sum of worldwide emission up to 1989, based on [UNSCEAR (1993)], [b][Poong (1998)].

	Releases of Radionuclides				(TBq·GW^{-1}·y^{-1})		Sum-1989 (TBq)
	PWR	BWR	GCR	HWR	LWGR	FBR	
[a]*Noble gases*							
	81	290	2100	190	2000	150	3160000
^3H							
airborne	2.8	2.5	9	480	26	96	65800
liquid	25	0.79	120	370	11	2.9	71700
^{14}C							
airborne	0.12	0.45	0.54	4.8	1.3	0.12	1140
^{131}I							
	0.0009	0.0018	0.0014	0.0002	0.014	0.0009	44.7
[a]*Particles*							
airborne	0.002	0.0091	0.0007	0.0002	0.012	0.0002	17.7
[a]*other*							
liquid	0.045	0.036	0.96	0.03	0.045	0.028	759
Electrical Energy Production (GW·y)							(GW)
1989	135.9	44.23	8.13	10.33	11.99	0.81	211
[b]**1990**	-	-	-	-	-	-	219
[b]**1994**	-	-	-	-	-	-	243
Estimated total Energy Production (GW) up to 1994:							3000

[a]composition see Table 11

Table 11: Composition of radionuclides in noble gases, in airborne particles, and in liquid effluents, based on [a][UNSCEAR (1993)], [b][Bonka (1982)].

	Distribution (%)			Distribution (%)	
	PWR	BWR		PWR	BWR
			[a]*Noble Gas*		
^{41}Ar	0.2	0.9	$^{85+85m}$Kr	6.7	4.1
^{88}Kr	8.1	0.2	$^{133+133m}$Xe	32.3	89.9
$^{135+135m}$Xe	20.2	3.9	^{138}Xe	16.5	0.2
			Sum	84.0	99.2
			[b]*Airborne particles*		
^{51}Cr	5.0	10.0	^{54}Mn	2.0	2.0
^{58}Co	10.0	10.0	^{60}Co	30.0	25.0
^{59}Fe	2.0	2.0	^{65}Zn	-	15.0
^{89}Sr	0.1	0.1	^{90}Sr	0.1	0.1
^{95}Nb	1.0	1.0	^{95}Zr	1.0	1.0
^{124}Sb	20.0	2.0	^{123m}Te	2.0	-
^{134}Cs	8.0	10.0	^{137}Cs	17.0	20.0
^{140}Ba	2.0	2.0	^{238}Pu	0.004	0.002
^{239}Pu	$4\cdot10^{-5}$	$2\cdot10^{-4}$	^{241}Pu	0.2	0.08
^{241}Am	$2\cdot10^{-4}$	$8\cdot10^{-5}$	^{242}Cm	0.05	0.02
^{244}Cm	0.004	0.002	*Sum*	100	100
			[a]*Liquid effluents*		
^{24}Na	0.18	86.52	^{51}Cr	4.24	5.78
^{54}Mn	2.90	1.35	^{55}Fe	8.56	0.50
^{57}Co	0.08	0.00	^{58}Co	40.96	0.77
^{59}Fe	0.35	0.45	^{60}Co	11.62	2.87
^{65}Zn	0.03	0.36	^{89}Sr	0.15	0.06
^{90}Sr	0.02	0.01	^{95}Nb	0.87	0.00
^{95}Zr	0.51	0.00	^{99}Mo	0.11	0.00
$^{103+106}$Ru	0.20	0.00	^{110m}Ag	3.76	0.02
^{124}Sb	1.31	0.00	^{125}Sb	6.01	0.00
^{131}I	3.59	0.09	^{133}I	1.16	0.01
^{134}Cs	3.46	0.35	^{137}Cs	7.16	0.79
^{140}La	0.13	0.00	^{144}Ce	0.09	0.00
^{239}Np	0.03	0.01	*Sum*	97.50	99.96

Tritium is generated as well at the nuclear fission itself, but also by capture of neutrons and decay processes of the fission and breeding material, within the moderator, or from other construction materials of the reactor. The standardized release of tritium (see Table 10) is highest for HWR's with 480 TBq/(GW·y) in airborne effluents and 370 TBq/(GW·y) in liquid effluents. In all, until the year 1989, the emissions of tritium in airborne and liquid effluents were estimated to be 137.5 PBq compared to 150 PBq tritium generated naturally per annum (see Chap. 2, Table 3).

^{14}C is produced by fission, by neutron capture in the fuel rods, in the cooling agent, and also on the outer surface of the reactor pressure vessel (see Chap. 2, 2.3). The amount of ^{14}C released as CO_2 or with the waste water, respectively, is dependent on the reactor design, and is for HWR's at 4.8 TBq/(GW·y) (see Table 10) about one order of magnitude higher than from other reactor types. The liquid releases are lower by about three orders of magnitude [Bonka (1982)]. Compared to the 1.4 PBq of ^{14}C (see Chap. 2, Table 3) generated every year, the amount released from nuclear power stations until the year 1989, was estimated to amount to 1.14 PBq in total.

Radioactive iodine isotopes are generated as fission and as decay products from the nuclear fuel and from uranium impurities on the surface of the fuel rods. They may diffuse into the cooling agent by leakage of the cladding of the fuel rods. [Bonka (1982)] calculated the emission rates for reactors during normal operation and found the emission of iodine in FBR's to be particularly low. Examinations by [Tichler et al (1988-1992)] of American nuclear plants established, that in 1988 about 36 GBq of radioiodine were emitted by PWR's and about 89 GBq by BWR's. In PWR's the share of ^{131}I was 16 %, 44 % for ^{133}I, and 26 % for ^{135}I. In BWR's the shares were 55 %, 31 %, and 6 % respectively. The major part of the iodine emission of about 50 % was exhausted as elementary iodine (I_2), partially absorbed to aerosols. Table 10 shows, that, with the exception of the LWGR's (14 GBq), all types of reactors emit about 1-2 GBq/(GW·y) of ^{131}I into the atmosphere.

Airborne radioactive particles are, in this context, understood as aerosols, which are contaminated by accumulation of radionuclides. Aerosols may derive from materials within the nuclear plant or from contaminated liquids. Especially solid radionuclides, deposited on aerosols, may result from the decay of noble gases, activation of corrosion products or structure material, but also from fission and decay within the fuel, and from uranium contamination on the surface of the fuel rods. The emissions are minute at intact fuel rods (see Table 10) and were supposed to be about 18 TBq totally up to 1989. [Bonka (1982)] has estimated the composition of the "radioactive particles" (see Table 11). The main part of the emissions in PWR's and BWR's is ^{60}Co at 25 to 30 %. After that, in PWR's there follows ^{124}Sb, ^{137}Cs, ^{58}Co, ^{134}Cs, and in BWR's ^{137}Cs, ^{65}Zn, ^{58}Co, ^{51}Cr, and ^{134}Cs. The contribution of plutonium, and americium is extraordinarily low. At FBR's, the main contaminants are ^{24}Na (33 %), ^{137}Cs (33 %), and ^{134}Cs (16 %).

The liquid effluents, tritium unconsidered, from HWR's, LWGR's, and FBR's are lower by a factor of 10 to 20 than from PWR's, BWR's, or GCR's (Table 10). The main part of radioactivity in the waste water of PWR's consists of the isotopes of cobalt, ^{58}Co, and ^{60}Co, by more than 50 % and in BWR's by 87 % (see Table 11).

2.2.2. Emission from Reprocessing Facilities

By reprocessing of burnt up fuel elements, uranium and plutonium are extracted which can be reused, in addition, fission and activation products are separated. For each ton of 3.5 % enriched ^{235}U PWR fuel, 25 % of the ^{235}U and 98 % of the ^{238}U remain after burning and about 9 kg of plutonium, with a calorific value of about 9 kt, is generated. Uranium and plutonium may be repeatedly reprocessed into fuel elements depending on the design of the reactor.

Reprocessing was first started in the USA. During the World War II the two national reprocessing facilities "Hanford" and "Oak Ridge" (see 2.1.1) and in the 1950s the factories near "Savannah River", South Carolina, and "National Reactor Testing Station" near Idaho Falls, Idaho were put into operation. All these plants were built for the production of plutonium for nuclear weapons, for military purposes. Since 1966, running of private reprocessing facilities was attempted. This failed, because of the complicated technology and the authorization procedures. President Carter put an end to these efforts, because of security concerns connected with the uncontrolled transfer of weapon-grade plutonium.

In the former USSR, also for military interests, a series of reprocessing plants were constructed immediately after the end of the World War II (see 2.1.1). Factories for the reprocessing of burn up fuel rods from civil usage are in operation in France at Marcoule and at Cap de la Hague, in Germany at Karlsruhe, in Japan at Tokai-Mura, in India at Tarpura and Trombay, in Russia at Kysthym-Ozyorsk, and in Great Britain at Sellafield and at Dournreay [UNSCEAR (1993)]. In the plants of Sellafield, Marcoule, and Trombay there was a reprocessing capacity of altogether 2150 t of metallic fuel in 1989 (1500, 60, 50 t/y). The reprocessing capacity of fuel as oxide at Cap de la Hague, Kysthym, Tokai-Mura, Tarapur, and Karlsruhe amounted to 1145 t/y in total (400, 400, 210, 100, 35 t/y) and to 6 t fuel from fast breeders in Marcoul and Dounreay every year. The plant at Karlsruhe was operating only for test purposes from 1971 to the end of 1990.

The low emissions at standard operation from nuclear plants can be attributed to the dense cladding of the fuel, which securely encloses the major part of the radioactive material. At reprocessing, however, the covers have to be removed to recover the fuel. In order to avoid the release of radioactive substances into the environment during unwrapping and in the further course of the extraction, rather great efforts are required. Furthermore, during each phase of the reprocessing procedure, it must be ensured, that no criticality of the fuel, leading to disastrous explosions, is achieved.

Before the transport of the fuel elements to reprocessing can be realized, the hot rods, removed from the reactor core, have to be in the interim stored in a cooling pond for at least five months or more, so that the short-lived radionuclides are allowed to disintegrate. After the transport to the plant, they are stored for about two additional years, to reach a ^{131}I concentration as low as possible by the further decay.

Mostly, reprocessing is performed by the so-called "PUREX"-process [Eisenbud (1987)]. The procedure starts by cutting the rods into small pieces and removing the wrapping material, mostly zirconium or stainless steel 304, by dissolved ammonium fluoride, or sulphuric acid, respectively. The fuel itself is dissolved in nitric acid. The uncladding is the first critical action with respect to the release of radioactive substances, the second is the dissolving of the fuel by nitric acid. On this occasion gaseous and volatile fission products may escape, including iodine, tritium, krypton, xenon, ruthenium, tellurium, and caesium. The dissolver off-gas is purified by caustic scrubbing, drying, and by highly effective filtration, to remove radionuclides, especially the iodine, but also to recover the nitric acid. The plutonium, the other transuranium elements, uranium, and their fission products are separated from the dissolved fuel by extraction with the organic complexing agent tributylphosphate dissolved in kerosene. From the organic matter plutonium and uranium are extracted by nitric acid [UNSCEAR (1993)] and the so-formed nitrates are further purified and reconverted to oxides for further use as fuel. The remaining waste in the solution is highly radioactive. It is concentrated, then stored in subterranean tanks as long as the heat from the radioactive decay has decreased, and the remaining waste is then occasionally solidified.

In Table 12 the emissions of the most important radionuclides from the three reprocessing plants Cap de la Hague, Sellafield, and Tokaimura in the years 1985 to 1989 are compiled [UNSCEAR (1993)]. The worldwide emissions until the year 1989 were estimated from the produced amount of energy and the fraction of reprocessed fuel rods. The noble gas isotope ^{85}Kr yielded with 1.3 EBq the largest contribution to the emission to the environment, succeeded by ^{3}H with 55 PBq, ^{137}Cs, with 40 PBq, ^{106}Ru with 16 PBq, and ^{90}Sr with 6 PBq. Whereas ^{85}Kr exclusively escapes into the atmosphere, most other nuclides are emitted into the environment with the liquid waste, especially ^{3}H, ^{137}Cs, ^{106}Ru, and ^{90}Sr. [Woodhead (1973)], [McAulay et al. (1985)], and [Hunt (1980)] carried out radioactivity measurements in the waters

Table 12: Airborne and liquid effluents (1985-1989) from the reprocessing plants, estimated worldwide (WORLD) effluents up to 1989, and normalized (NORM) to fuel reprocessed, based on [UNSCEAR (1993)], [Lange et al. (1991)]:
CDLH: Cap de la Hague,   SELLA: Sellafied,   TOKA: Tokaimura.

	Effluents		(TBq)		**NORM**
	CDLH	**SELLA**	**TOKA**	**WORLD**	$(TBq \cdot GW^{-1} \cdot y^{-1})$
			^{85}Kr		
airborne	203300	202600	47500	1318000	12300
			^{131}I		
airborne	-	0.013	0.0	3.70	0.0007
			^{129}I		
airborne	-	0.104	0.004	0.41	0.0057
liquid	0.26	0.620	0.0	3.90	0.032
			^{3}H		
airborne	99	1380	15.4	5890	41
liquid	14100	8455	1074	49100	643
			^{14}C		
airborne	-	29.4	-	243	2.0
liquid	1.4	11.0	-	54	0.54
			^{137}Cs		
airborne	-	0.022	-	4.7	0.0015
liquid	87.1	396.6	0.0	39600	13
			^{90}Sr		
liquid	308	104.6	0.0	5840	11
			^{106}Ru		
liquid	910	179.7	0.0	15900	39

around Sellafield and proved ^{137}Cs in fish. [Tadmore (1973)] investigated the release of ^{85}Kr and ^{3}H from reprocessing plants and [Rabon et al. (1973)] performed ^{137}Cs-monitoring in game.

On September 30, 1999, an accident happened at the conversion testing facility at Tokaimura (Japan) leading to serious consequences for the staff [Völkle (2000)], [IAEA (1999)], [IPSN (1999)]. Two workers poured 16 kg instead of the advised 2.3 kg of uranium enriched to 18.8 % ^{235}U into a precipitation tank. This started a chain reaction, which could only be stopped after 20 hours on October 1st by removing some cooling water. Radioactive noble gases and iodine isotopes, produced during $5 \cdot 10^{17}$ to $5 \cdot 10^{18}$ nuclear fissions, were released into the environment through the ventilation system. In the vicinity of the perimeter fence contamination of soil samples of 0.45 Bq/kg ^{131}I, 0.16 Bq/kg ^{133}I and of green vegetables with maximally 37 Bq/kg were measured. However, no increased contamination of other agricultural products or seafood was determined. [Fischer and Kirchner (2000)] compared this accident, which caused the death of one worker, with earlier severe accidents in the USA, Great Britain and Russia.

2.3.   Pollution of the Environment from Accidents

2.3.1.   Accidents in Reactors
In the past a series of troubles in nuclear reactor facilities have arisen. Especially accidents, concerning the reactor core may lead to uncontrolled release of radionuclides into the environment at a considerable amount. Many of the current reactors are equipped with a so-called containment, a shield, which is believed to prevent the release of radioactivity and to protect the reactor core from influences from outside. Table 13 lists accidents after reactor core damage, which became public. Some of these accidents will be discussed in the following sections.

Table 13: Reactor accidents that involved core damage, based on [Eisenbud (1987)].

Year	Location	Type		Contamination
1952	Canada	NRX	Experimental	None
1955	Idaho, USA	EBR-1	Experimental	Trace
1957	Windscale, UK	Windscale	Military	^{131}I,^{137}Cs,^{89}Sr,^{90}Sr
1957	Idaho, USA	HTRE-3	Experimental	Slight
1958	Canada	NRU	Research	None
1959	California, USA	SRE	Experimental	Slight
1960	Pennsylvania, USA	WTR	Research	None measured
1961	Idaho, USA	SL-1	Experimental	^{131}I
1963	Tennessee, USA	Orr	Research	Trace
1966	Detroit, USA	Fermi	Experimental	None outside plant
1969	St. Laurent, France	St. Laurent	Power	Little
1969	Lucens, Switzerland	Lucens	Experimental	None
1979	Harrisburg, USA	TMI-II	Power	Slight ^{133}Xe,^{131}I
1986	Chernobyl, USSR	RBMK-4	Power	^{131}I,^{134,137}Cs,^{90}Sr,Pu
1992	Sosnovy Bor, RUS	RBMK	Power	Slight

2.3.1.1. *Accident at Windscale in October 1957*

Windscale, now named Sellafield, is located in the northwest of England close by the shore of the Irish Sea. In this area there are two air-cooled, graphite moderated reactors, operated by natural uranium, designed to produce plutonium. In October 1957, after a misconducted standard routine for carrying off the "Winger" energy in the graphite [Eisenbud (1987)], the core of one of the reactors was destroyed. This led to release of fission products to the environment.

The design of the reactor with the graphite moderation and the usage of metallic uranium paired with inadequate instrumentation, were of decisive importance for the accident. At the bombardment of the crystalline graphite by neutrons, the carbon atoms are broken out of their normal position in the crystal lattice. The resulting vacancies in the normal crystal structure have effects on the physical properties of the graphite, resulting in an expansion up to 3 %. The permanent irradiation after 10^{21} neutrons per cm^2 finally leads to breakage of the crystal structure [Harper (1961)], [Wittels (1966)]. To avoid damages the original molecular formula is restored by an annealing process. During this process, interstitially vagabonding carbon atoms, emitting thermal energy ("Winger" energy), diffuse into the vacant positions of the crystal lattice. As a result up to 500 cal/g graphite [Eisenbud (1987)] are released, sufficient to cause the temperature to increase to a dangerous level, especially if when this process is performed rapidly.

The accident at Windscale was caused during the effort to carry off this energy released from the graphite. During the usual tempering process, which ran too fast, a local overheating in the core occurred. The local temperature rise remained undiscovered, because of the insufficient instrumentation of the reactor core [UKAEO (1957), (1958)]. Subsequently, the fuel unit was apparently damaged by the overheating, and the metallic uranium and the graphite started to react with the air. In the morning of October 12th, a fire broke out, raging for three days UNSCEAR (1993)]. The first hint as to an accident was produced by the observation of increased β-activity in a dust collector installed outdoors. Visual inspection from a plug hole at the front side of the reactor showed that the uranium cartridges of about 150 channels were red-hot [Eisenbud (1987)]. Because of the deformation already having occurred the cartridges could not be discharged, but the fuel rods from neighbouring channels could be removed. Thus the fire was interrupted and the extent of the accident limited. Since all efforts to extinguish the smouldering fire failed, on the following day the graphite was cooled by flooding the core with water. The reactor became cooled down in the evening of October 13th. At this accident the most essential part of the reactor core was destroyed and radionuclides were released into the environment [Dunster et al. (1958)], [Clark (1974)], [Crick et al. (1984)].

The emission of radionuclides took place mainly during two periods:
- when the air current across the core was restarted, to cool the reactor and
- when water was pumped into the reactor to extinguish the fire [Crick et al. (1982)]

The releases to the environment were estimated to be 13 EBq for noble gases [NTIS (1987)] and 740 TBq for ^{131}I, additionally there were 22 TBq of ^{137}Cs, 3 TBq of ^{89}Sr, 0.3 TBq of ^{90}Sr [Stannard (1988)], 3 TBq of ^{106}Ru, and 1.2 PBq of ^{133}Xe [Clark (1989)]. First reports stated, that, at the time of the accident, the reactor had been used only for the production of plutonium. In a later report, however, it was published, that the reactor also was used as a neutron source for the activation of ^{210}Po from bismuth. ^{210}Po together with beryllium in nuclear weapons usually serve as neutron starting source. Hence 8.9 TBq of ^{210}Po were emitted into the environment.

The radioactive fallout for the most part was forwarded south-south-east in the direction of London town and eventually even to Belgium, before it drifted north to Norway. At the time of the accident and immediately thereafter ^{131}I was the main component of contamination of the pasture land. The order to restrict milk consumption diminished the intake of ^{131}I from the food-chain (pasture → cow → milk → human) and reduced the ^{131}I-uptake into the human thyroid gland [Clark (1989)]. In settlements in the neighbourhood of the reactor, in Leeds and London, measurements of the thyroid gland in adults and children were carried out. The doses of the thyroid were about 10 mGy in adults and 100 mGy in children, living nearby Clark (1974)]; and 1 mGy in adults from Leeds and 0.4 mGy from London and about twice that much in the case of the children [Baverstock et al. (1976)]. [Chamberlain (1986)] and [Jakeman (1986)] investigated the contamination of long lived radionuclides (^{90}Sr, ^{137}Cs) in the vicinity of Sellafield, [Black (1984)] and [Bobrow (1986)] examined the possibly resulting health effects.

2.3.1.2. *Accident in the Reactor SL-1 at Idaho Falls in January 1961*

The low-power reactor SL-1 of the Army at Idaho Falls USA, was a directly cooled boiling water reactor with enriched uranium as fuel formed as platelets, cladded with aluminium [Eisenbud (1987)]. It was planned for a nominal thermal output of 3 MW. After an operation time of more than two years, the reactor was shut down on December 23rd, 1960 and was intended to restart after a service period of 12 days on January 4th [Buchanan (1963)], [Horan et al. (1963)]. The accident happened on January 3rd, during the night-time. Then the service staff consisted of three military employees, two of them licensed reactor operators and one undergoing training. They were staying at the top of the reactor near the driving mechanism of the control rods. Caused by manually removing of one or even more of the central rods of the reactor core beyond that extent put down in the instructions for maintenance, the reactor went into prompt critical condition, thus finally producing an explosion. It never became known, for what reason the rod was removed by the service crew. Two of the operators died immediately by the violence of the explosion.

First indications that within the reactor core something might have been out of order resulted from vague radiation and exceeding heat, raising alarm. The health physicist and some men from the fire department rushed to the reactor and registered at arrival dose rates up to 2 mGy/h in the vicinity of the reactor building. Invading the building, they neither could discover the three operators on duty, nor identify the source of the fire. In the corridors however, dose rates up to 5 Gy/h were measured. After searching for one and a half hour, the operators finally were found, two of them dead and the third one yet alive, but dying soon after from a head injury.

From the reactor building, equipped with no containment for the core, from the total 37 PBq of medium to long-lived radionuclides about 0.37 TBq were released into the environment and at a rough estimate an additional 0.37 TBq of ^{131}I, 19 GBq of ^{137}Cs and 3.7 GBq of ^{90}Sr [NTIS (1987)], [Eisenbud (1987)].

2.3.1.3. *Accident within the Fermi Reactor at Monroe/ Michigan in October 1966*

At the west bank of lake Erie, near Monroe, Michigan USA, there was a sodium chilled fast breeder called "Enrico Fermi-reactor" at an electrical power of 66 MW and operated with uranium oxide, enriched to 25.6 % [Eisenbud (1987)]. The core was loaded the first time in 1963. In the primary circuit, liquid sodium was used as coolant, which, because of it's high boiling point, permits to running the reactor at high temperatures but low pressure. The heat coupling to the steam production for the turbines took place by an intermediate loop, since on the one hand the primary coolant sodium, passing the reactor, becomes highly radioactive by activation, while on the other hand a reaction of water and sodium had to be excluded at all costs.

On October 5th, 1966, at 27 MW electrical output the monitors, installed into the ventilation system of the building, gave rise to an alarm. Although the reason for this was not obvious immediately, the building was isolated automatically, and the reactor was shut down manually. At the time of the alarm, nobody was present in the reactor building.

Measurements established that 740 TBq of fission products had entered the primary sodium circuit. Indeed radioactive noble gases, but no iodine, were detected. Out of the containment, fortunately no contamination was observed neither by solid nor by gaseous particles.

After draining off the sodium from the primary circuit to inspect the reactor vessel, later on it came to light that the coolant current had been blocked by pieces of zirconium, detached from the flow guide. By it, two fuel assemblies were overheated and damaged. The reactor could be repaired and started up again in October 1970.

2.3.1.4. *Accident at the Experimental Nuclear Power Plant near Lucens in January 1969*

On January 21st, 1969, in the Swiss experimental nuclear power plant, 2 km away from Lucens, not far from the lake of Geneva, a grave accident happened, leading to the total loss of the reactor [Fritzsche (1981)], [Miller (1975)], [LCI (1979)]. The reactor, designed and built in the early 1950s, cooled by CO_2, and moderated by heavy water, was operated by uranium enriched to 0.96 %, and had an output of 30 MW of thermal and 5-6 MW of electrical power. The reactor complex was installed into three neighbouring submountain caverns to ensure, that, by filtration and absorption of radioactive material by the surrounding rock, in case of an accident no radioactive material may be released outwards.

After an operation of about 70 days at full output, the reactor was shut down in October 1968, to carry out maintenance and repair works. As had already been the case, difficulties resulted at the CO_2-blowers, constructed especially for this type of reactor. At various occasions, a considerable increase of humidity in the CO_2-gas was observed. Thus possibly water might have entered the reactor cooling circuit. Therefore, extensive tests were performed before the reactor was restarted on January 21st, 1969. During the maintenance however, corrosion damage having taken place at certain fuel and zirconium rods, caused by the invading water, had not been found as was detected much later by minute examinations. After the start up, by these corrosions a blocking of the coolant circuit resulted, which was not registered by the instrumentation. By local overheating, a melting of the fuel cladding and the fuel in one element resulted and breakage of the appropriate pressure tube occurred. The cooling agent CO_2 expanded into the moderator vessel, filled with heavy water, leading to a break at its rupture disk, so that CO_2, and 7.4 t of heavy water together with fission products first entered the reactor cavern. When unexpectedly radioactivity, especially that of iodine, spread into the other caverns too, the ventilation system was turned off, to avoid emissions from the facility into the open air. Although by the waste air stack, 13 noble gas isotopes and 31 further radionuclides in the form of aerosols had escaped into the environment. From their emission, maximum whole body dose for the common population in the most unfavourable point was estimated to be 0.6 μSv in the case of the rare gases and 0.01 μSv in the case of the aerosols. Later on, by expulsion of tritium contaminated D_2O over a range of many weeks, 3.7 TBq of tritium might be released by the ventilation system into the open air and about 0.37 TBq as liquid effluent.

2.3.1.5. *Accident at Three Mile Island, Harrisburg in March 1979*

The reactor block 2 at Three Mile Island is a pressurized water reactor with an electrical power of 850 MW [Eisenbud (1987)]. It is located outside of Middletown in Pennsylvania USA, not far away from the capital Harrisburg and was put into service for the first time in December 1978. The accidental happenings in this reactor block began on March 28th, 1979 at four a.m. by the failure of the feedwater pumps, which supply the reactor's steam generator with water. This breakdown led to the automatic shut down of the reactor and the steam producing system. The residual heat, produced by the radioactivity in the reactor core, increased the temperature and pressure within the reactor coolant circuit. To diminish the increased pressure, a relief valve was opened in the pressurizer. This valve, however, which should be opened only for some seconds, jammed and remained open. Thus reactor cooling water escaped and the seriousness of the trouble started to progress by the steady loss of coolant. Fourteen seconds after the accidental switch off of the reactor, the emergency feed water pumps automatically started. The operator in the control room in fact noted the start of these pumps, but missed the two signalling lamps on the panel, indicating that emergency feed water valves were closed, so that no water could reach the reactor coolant circuit. The supervision system of the reactor correctly reported an emergency. Two minutes later, the high-pressure injection pumps started working at a flow rate of 3785 litres per minute. One of the operators, however, completely turned off one of these two pumps and reduced the flow of the other to less than 378 litres per minute. As a result, steam bubbles started forming. Instead of activating the emergency high pressure pumps again, respectively increasing the flow, the operators opened the reactor coolant letdown system, whereby additional water flowed off from the coolant circuit. These series of wrong decisions led to a continuously further decreasing water level and by it, to a further elevation of the temperature in the reactor core. The fuel claddings became damaged and the fuel partially melted, so that radioactivity entered the coolant and, furthermore through the cooling system via the relief valve, also reached the reactor building. The temperature increased up to a point where the zirconium alloy cladding started reacting with the water steam. By this process hydrogen developed, partially invading the reactor building. After some days, enough hydrogen had accumulated in the building in order to induce an explosion, which, however, caused almost no damage. By means of a backup valve the jammed valve could be closed, so that further loss of coolant was avoided. After 6.5 hours the reactor core was completely covered with water again.

Three hours after the beginnings of the accidental happenings, in the reactor auxiliary building an augmented level of radiation was registered by the monitors, an endangered area was established and half an hour later a common emergency was declared. A further hour later a measuring team monitored 10 to 30 μGy/h at the terrain outside. The dramatic incidents, following the mastering of the accident in the first days, also put a light on the difficulties in the cooperation of different responsible departments concerned with the emergency: The Pennsylvania Emergency Management Agency in the office of the governor, the Department of Energy, the Nuclear Regulatory Commission (NCR), the Bureau of Radiological Health of the Food and Drug Administration, the Pennsylvania Bureau of Radiation Protection, and an emergency team from the Brookhaven National Laboratory.

The Department of Energy sent a helicopter for radiation monitoring. Within the first hours 34 thermoluminescence dosimeters provided information on the γ-radiation level within a 15 mile's radius. Additional dosimeters were installed on the third day after the onset of the accident. Besides, during the first days a considerable number of food samples were collected. This was completed by whole body measurements of 700 persons. All these measurements proved to be negative increased radiation exposures.

In such a situation, the trouble management depends essentially on the interpretation of the registered data and a functioning communication of the official departments in coordination with the media. For instance, on Friday, March 30th, discrepancies arose, when radioactive gases, accumulated in the makeup water storage tank, should be transferred to a delay tank outside of the containment. The operator was not sure if during the transfer gases might be released into the atmosphere too, and demanded a helicopter to carry out radiation measurements near a waste air exhaust. Simultaneously, an NRC-specialist had estimated a dose rate of 12 mGy/h by ground contamination in case of a fault arising in the relieving valve of the delay tank. At the same moment the helicopter registered a dose rate of 12 mGy/h. The officials wrongly supposed this value to be measured really on the ground and concluded the valve of the delay tank to be defective. In fact the gases came not from the valve of the tank, but from the waste air exhaust due to small leakages in the transport system only. This misunderstanding led to a wrong estimation of the danger, so that NRC-officials demanded the evacuation of the population within a distance of 10 miles downwind of the reactor. The management of the Pennsylvania Bureau knew the correct facts of the helicopter measurements and from this they concluded that evacuation might not be necessary, but could not reach the bureau of the governor at this time, because of the overburdening of the transmission circuits of communication. Later, the governor was assured that no evacuation was necessary. It was ruled however, that anybody living inside a five miles zone in the downwind area should stay inside their houses for the next half hour. Shortly after 12 o'clock noon after consulting the chairman of the NCR's, the governor recommended, that pregnant women and children of preschool age from five miles around Three Mile Island should leave the region, and that all schools within this zone had to be closed.

A further series of misunderstandings occurred on Friday afternoon, when a 28.3 m³ large gas bubble inside the reactor container was discovered. The bubble consisted of hydrogen, which had been built by reaction of the hot steam and the zirconium cladding of the fuel rods. An explosion was considered, a misjudgement, which publicly was not or only half-heartedly admitted, because not enough oxygen could accumulate in order to build an explosive gas mixture. This was known to the NRC-experts, but no appropriate statement was made before the beginning weekend on March 31st. All these events are described in detail by [Eisenbud (1987)], [Kemeny (1979)], [Rogowin et al. (1980)], [Nyas (1981)], [Gerusky (1981)], and [Fabrikant (1983b)]. As a consequence of the accident about 370 PBq of noble gases were released from the containment, 2.1 PBq of ^{85}Kr over a longer period. Furthermore about 0.63 TBq of ^{131}I escaped to the environment, but ^{137}Cs and ^{90}Sr could not be detected [NTIS (1987)]. The radiation doses for the population two miles around the reactor were estimated by [Gerusky (1981)] to be 0.2-0.7 mSv.

2.3.1.6. *The Chernobyl Disaster in April 1986*

Preliminary remarks to the accident

The nuclear power plant "Chernobyl" is located 100 km north of Kiev, in the borderland to the Ukraine and Byelorussia, northwest of Kiev's water reservoir, on the banking area of the River Pripyat, flowing into the water reservoir near the town of Chernobyl (see Fig. 9). The construction of the plant started in 1970 intending a final capacity of 6 GW of electric power. The reactor blocks 1 and 2 were put in operation in 1977 and in 1978, the blocks 3 and 4 in 1981 and 1983. Two further were intended to be started in 1986 and 1987. These RBMK-reactors (**R**eactor **B**olshoi **M**oshtshnosty **K**ipyashtshiy = high power reactor of boiling water type) are graphite moderated, light water cooled, boiling water reactors of the pressure tube type, at an electric output of 1000 to 1500 MW, and are operated on uranium as fuel, enriched to 2 %.

The accident in block 4 happened on April 25th, 1986 at 1.23 a.m. and resulted in the destruction of the reactor core and part of the building. Without doubt, this has been the most disastrous of all accidents ever taken place in nuclear industries until now and at the same time the most expensive in industrial history. For the first time, people had to be evacuated from their homes on a large scale, partially far away from the place of the disaster. The consequences of the accident were strengthened by the obscuring tactics of the then Soviet leadership. Many days passed before it was admitted that a grave accident had happened. The fall of the Soviet Union in the years thereafter and the economical and administrative difficulties arising thereby in the successor states complicated the steps in the struggle against the consequences of the accident.

In August 1986, Russian experts very frankly reported at the Vienna Conference [USSR (1986)] on the type of reactor involved on the causes of the accident, and on the actions to reduce the consequences of the accident. Further numerously detailed reports on the accidental events are given by [NTIS (1987)], [SSK (1987), (1996ab)], [NEA-OECD (1996)], and [HSK (1986)]. In April 1996, on the 10th anniversary of the accident, scientists from different countries reported at a meeting [IAEA (1996a)] on the health effects for the population and on damages to the environment.

The accident took place in the course of a procedure with one of the two turbine generators. In this experiment, it should have been determined if, at a power breakdown followed by a shut down of the reactor, the rotational energy of the turbines is sufficient to guarantee the power supply of the four main feed pumps for a short time, that is for 40 to 50 seconds, until the emergency power unit might take over the supply. One year before, in block 3 a similar experiment had failed and therefore it was to be repeated under improved conditions. This time it was intended to keep the reactor in operation during the experiment. Such a test arrangement alone would not have caused the accident. But additionally, apparently vastly unknown special physical features of the RBMK-reactors, together with operational shortcomings, led to the disastrous accident.

The course of the accidental events

On April 25th, 1986, block 4 was to be shut down for a routine revision. It was decided to repeat the failed experiment of the previous year with the bridging of the power supply of the main feed pumps. At 1.00 h, the reactor output started to be decreased. Until 13.05 h, the reactor output dropped to 50 %. One of the two turbines was turned off. At 14.00 h the emergency cooling system was put out of action in order to avoid that during the experiment, water for cooling might be fed by emergency signals. Meanwhile, because of augmented energy demand, the load dispatcher in Kiev asked for continued operation of the reactor at one turbine. It was forgotten, however, to put the emergency cooling systems back into operation again. Only in the evening at 23.10 h, was the experiment to be continued by starting to run down the reactor to 25 % of its output. At 00.28 h on April 26th, however, due to troubles in the regulatory system the output dropped unexpectedly to below 1 %. The reactor, however, was, according to the instructions, not allowed to be run below 20 % of output. In order not to break off the experiment, it was attempted to raise the output by extending the control rods. Indeed, by this action the output increased to about 7 %, an output at which range the reactor, according to our present knowledge, was not designed to be operated; because of the withdrawn control rods the necessary allowance for an immediate shut down of the reactor was exceeded.

According to the schedule of the experiment, at 1.03 h the four circulation pumps were switched on and for stabilization of the reactor output, the control rods were extended still more. The plant was in an extremely labile state. At 01.19 h the water supply was increased and warning signals on water level and pressure, which would have led to emergency shut down, were shunted. At 01.22 h, the operator achieved, that the water supply was increased to two thirds of the necessary flow and stabilized, although the controller action was not adjusted for such a small throughput. At 1.23 h, the real experiment started by closing the quick-acting gate valves of the turbine generator. The reduction of the throughput and heating of the water caused a positive reactivity coefficient, for which compensation was attempted by sending in control rods. About 30 seconds later, the output was increasing further, since the automatic control system could not stop the increase in efficiency. 36 seconds after the experiment began, the shift foreman ordered the shut down of the reactor by the emergency switch. But within only a further four seconds, the output rose to nearly 100-fold of the nominal power of the reactor. The quick-action release system of the control rods, however, needed 18 to 20 seconds to become effective. At this moment, outside the reactor building, two explosions, accompanied by ejection of material, were observed within an interval of 2 to 3 seconds. Glowing lumps of material shot through the roof into the air, fell onto the roof of the engine room, and started a fire. The fast increase of the output within the tenth of a second heated the fuel such that enclosed gases together with evaporating fuel caused a pressure increase, which fragmented the fuel. Hot debris touched water, thus producing steam. The fuel element channels did not stand up to these stresses from pressure and temperature. The reactor cover plate, 1000 t of weight, was lifted and the pipes connected with the plate were torn off. The fuel loading machine dropped onto the reactor core and caused further devastation. After the fragmentation of the fuel and interaction with water by means of a zirconium-water reaction, hydrogen developed, probably leading to the second explosion.

Today it is considered to be true that the accident paradoxically was ultimately caused by the emergency shut down of the reactor. By sending in the practically completely withdrawn scram and control rods, the reactivity of the reactor by the faulty conception of the rods was for a short time not lowered, but augmented. This circumstance had been noticed already in 1983 at the plant of Ignalia in Lithuania, but obviously had not become general knowledge.

Clearing and Safe-Guarding of the Reactor Ruin

Since no catastrophe- or emergency-schedules existed, at the time from April 26th, to May 2nd, all actions to minimize the endangering of the population, the environment, and the staff, had to be improvised. First of all, the authorities concentrated on safe-guarding of the devastated reactor with the following focal points: prevention of a renewed fission reaction, extinguishing the fires, preventing a further heating of the fuel elements, shielding the direct radiation, and minimizing the further release of radioactive substances.

For cooling, the fire brigade pumped, immediately after the accident, 200 to 300 tons of water per hour from the stock tank of the intact block 3 into the destroyed reactor core. Since the cooling of the graphite blocks failed and contaminated water flowed out of the plant, this action was stopped 10 hours later. The graphite heated up further and began to burn. The conventional fires, ignited up to an area of 150 m by hot fragments from the reactor core, posed no difficulties to extinguish, although the fire-fighters were exposed partially to high irradiation. The graphite fires, however, caused considerable problems, since there were only a few experiences from which to draw knowledge, even on an international level.

It was decided to cover the reactor core with various materials through the gravely damaged roof. In order to prevent further fission reactions, military helicopters started as quickly as possible to throw 40 t of boron carbide as neutron catcher through the roof. Furthermore 500 t of dolomite were discharged, to absorb the developing heat by its decomposition and to quench the graphite fire by the developed CO_2. In order to lower the temperature in the core the heat should be absorbed by 2400 t of lead at melting, thus simultaneously contributing to the protection against the γ-radiation. The dropping of about 1800 t of sand and clay as a filtration and covering material served to diminish the emission of radioactive substances. By the covering, however, the temperature increased and thus also the emission of fission products into the atmosphere. This could only be stopped by May 6th, after installation of a nitrogen refrigeration system underneath the reactor cavity. Additionally a tunnel was dug beyond the reactor core, thus creating the possibility of installing a further nitrogen cooling system, together with a concrete slab, to avoid contamination of the ground water by liquefied radioactive material. For safe-guarding of the reactor ruin, a "sarcophagus" was built, the quality of which, because of the difficult circumstances during the stage of construction, left much to be desired. Prior to closing the roof, highly contaminated material scattered around, e.g., also ejected nuclear fuel from the roof of block 3, was thrown back into the "sarcophagus" by 600000 "liquidators".

36 hours after the accident, the evacuation of the 45000 inhabitants of the town of Pripyat was started and completed within one and a half hours. At a radius of 30 km around the reactor, a restricted zone was established, which is still maintained today. To avoid further evacuations of the general population, the administration tried to decrease the external radiation exposure by ploughing of the land patrially down to depth of 50 cm. Nevertheless, a further 90000 inhabitants had to be evacuated, some from far away from the reactor site.

Release of Radioactive Material into the Environment

The core charge of the reactor consisted of about 200 t of uranium, 4.5 kg/t thereof ^{235}U, furthermore of 2.6 kg/t of ^{239}Pu, 1.8 kg/t of ^{240}Pu, and 0.5 kg/t of ^{241}Pu. Related to the running time of the reactor at the moment of the accident, the core inventory of fission and activation products is summarized in Table 14 for the most important nuclides. How much of the inventory really was ejected from the core or escaped during the further course of events remains still controversial or has been corrected repeatedly. Radioactive material, escaped during the explosion, reached altitudes of partially more than 1500 m. Already at this stage, all noble gases might have escaped completely from the core. The release of iodine was estimated to be half of the inventory and the release of caesium one third. Less volatile elements such as zirconium, ruthenium, barium, cerium, and strontium, but also the transuranic elements, fortunately were released to an essentially smaller percentage, to about 2 to 6 % of the inventory.

Dispersion of the Chernobyl Fallout

The dispersion and deposition of radioactive substances are governed by the prevailing meteorological conditions, particularly by the wind and rain-fall conditions. On April 25/26th at the place of the accident, faint winds from changing directions prevailed together with ground fog. At a height of 700 to 1500 m, however, south to southwest winds predominated at velocities between 8 to 10 m/sec. Consequently due to the explosion and the high temperatures, the first cloud of fallout passed to a large altitude, and drifted to the north. On April 28/29th, the fallout reached Scandinavia and was registered first in Sweden [Snihs (1996)]. This cloud was mainly responsible for the deposition in Scandinavia, the Netherlands, and in Great Britain. By change of the weather condition, the wind went round and the radioactive clouds of the plume of the burning reactor now was passed to the west and finally to the south and burdened Central Europe, the Balkans, and the northern Mediterranean. Fallout reached the southern part of Germany on April 30th, 1986 [GSF (1986)]. Data on the spreading of the fallout, deposition onto the ground, transfer to the food-chain and into the human body are compiled for many countries in [UNSCEAR (1988)].

Deposition of the Chernobyl fallout

During the first days and weeks following the accident, the concentration of radioisotopes of iodine in air and their deposition, followed by their penetration into the food-chain, have been of special importance for the burdening of the thyroid glands. At this point of time the decision, a prophylaxis for blocking the thyroid gland by the administration of iodine pills, was strengthened by the fact that only few reliable ^{131}I-measurements in the relevant areas

Table 14: Estimated release of radioactive material from the Chernobyl reactor accident (a) first estimation [USSR (1986)] (b) corrected estimation, based on [Krüger et al. (1996)].

Radionuclide	Radioactivity		(Bq)
	Release		**Inventory by Accident**
	(a)	**(b)**	**(b)**
Noble Gases			
^{85}Kr	$3.3 \cdot 10^{16}$	$3.3 \cdot 10^{16}$	$3.3 \cdot 10^{16}$
^{133}Xe	$1.7 \cdot 10^{18}$	$6.5 \cdot 10^{18}$	$6.5 \cdot 10^{18}$
Volatile Material			
^{131}I	$2.7 \cdot 10^{17}$	$1.8 \cdot 10^{18}$	$3.6 \cdot 10^{18}$
^{125}Sb		$2.9 \cdot 10^{15}$	$1.9 \cdot 10^{16}$
^{134}Cs	$1.9 \cdot 10^{16}$	$5.0 \cdot 10^{16}$	$1.5 \cdot 10^{17}$
^{137}Cs	$3.7 \cdot 10^{16}$	$8.6 \cdot 10^{16}$	$2.6 \cdot 10^{17}$
More Persistent Materials			
^{95}Zr	$1.4 \cdot 10^{17}$	$1.4 \cdot 10^{17}$	$4.4 \cdot 10^{18}$
^{103}Ru	$1.2 \cdot 10^{17}$	$1.2 \cdot 10^{17}$	$4.1 \cdot 10^{18}$
^{106}Ru	$5.9 \cdot 10^{16}$	$2.5 \cdot 10^{16}$	$8.5 \cdot 10^{17}$
^{140}Ba	$1.6 \cdot 10^{17}$	$1.6 \cdot 10^{17}$	$2.9 \cdot 10^{18}$
^{141}Ce	$1.0 \cdot 10^{17}$	$1.2 \cdot 10^{17}$	$4.3 \cdot 10^{18}$
^{144}Ce	$8.9 \cdot 10^{16}$	$9.0 \cdot 10^{16}$	$3.2 \cdot 10^{18}$
^{89}Sr	$8.1 \cdot 10^{16}$	$8.0 \cdot 10^{16}$	$2.0 \cdot 10^{18}$
^{90}Sr	$8.1 \cdot 10^{15}$	$8.0 \cdot 10^{15}$	$2.0 \cdot 10^{17}$
Transuranic Elements			
^{238}Pu	$3.0 \cdot 10^{13}$	$3.3 \cdot 10^{13}$	$9.3 \cdot 10^{14}$
^{239}Pu	$2.6 \cdot 10^{13}$	$3.4 \cdot 10^{13}$	$9.6 \cdot 10^{14}$
^{240}Pu	$4.1 \cdot 10^{13}$	$5.3 \cdot 10^{13}$	$1.5 \cdot 10^{15}$
^{241}Pu	$5.2 \cdot 10^{15}$	$6.3 \cdot 10^{15}$	$1.8 \cdot 10^{17}$
^{242}Pu	$7.4 \cdot 10^{10}$		
^{241}Am		$4.9 \cdot 10^{12}$	$1.4 \cdot 10^{14}$
^{243}Am		$2.0 \cdot 10^{11}$	$5.7 \cdot 10^{12}$
^{242}Cm	$7.8 \cdot 10^{14}$	$1.1 \cdot 10^{15}$	$3.1 \cdot 10^{16}$
^{244}Cm		$6.3 \cdot 10^{12}$	$1.8 \cdot 10^{14}$

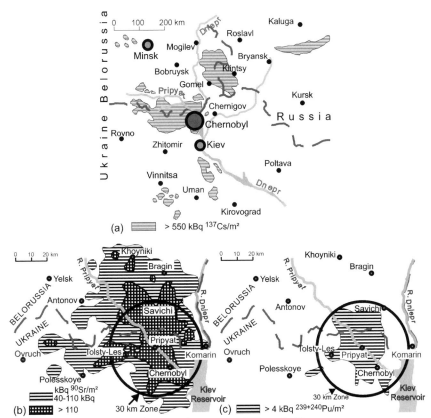

Figure 9: Surface contamination near the reactor of Chernobyl by
(a) ^{137}Cs in Belorussia, Russia, Ukraine, (b) ^{90}Sr and (c) $^{239+240}$Pu.

around the reactor were available. The insufficient information politics of the
USSR made this decision not easier and had the consequence that the pills
were distributed either not sufficiently or much too late. This is why until
today a reliable dosimetry is missing for the thyroid dose in relationship to
the diseases of the thyroid gland, which are increased in children [GSF
(1997)], [WHO (1996)], [Kinzelmann (2003)]. [Buglova et al. (1996)],
[Gavrilin et al. (1999)] in fact have published a map for the ^{131}I-deposition
based, however, more on reconstruction than on exact measurements. From
the map follows, that high ^{131}I concentrations must have been present in
areas north of the 30 miles zone, for instance in the region of Gomel at a
concentration of 1.85 to 5.55 MBq/m^2 and in the neighbourhood of the town
of Chechersk even up to 18.5 MBq/m^2.

The ^{137}Cs, ^{90}Sr and $^{239+240}$Pu depositions from the accident in the former
Soviet Union are outlined in Fig. 9a-c [IAEA (1991)], [NEA-OECD (1996)].
While caesium, because of its larger volatility (see Fig. 9a), has been

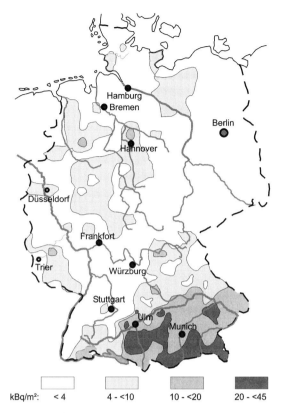

kBq/m²: < 4   4 - <10   10 - <20   20 - <45

Figure 10: Soil deposition of ^{137}Cs for Germany, based on [SSK (1987),(1996a)], [Bayer et al. (1996)].

transported to much further distances, the radionuclides of strontium (see Fig. 9b) and plutonium (see Fig. 9c) remained predominantly in the immediate surroundings of the reactor. The map of caesium depositions demonstrates that apart from the 30 km zone, larger likewise burdened areas of more than 550 kBq/m² of ^{137}Cs are located in the regions Kaluga-Bryansk, Gomel-Klintsy-Mogilev, and west of Chernobyl, close to the frontier area with Belorussia and the Ukraine. A part of these areas loaded additionally with ^{90}Sr, had to be evacuated, and even today are prohibited zones. Further maps relating to contaminations in the region of Rovno in the Ukraine can be found in [Likhtarev et al. (1996)].

Because of the weather conditions, the fallout of April 27/28th, 1986 was passed to Western Europe and then washed out of the atmosphere on April 30th, until May 1st, especially in the area of the Alps, so in Bavaria, Austria, and East and South Switzerland [De Cort et al.(1990)], [Raes et al. (1989)]. Besides ^{131}I, ^{134}Cs and ^{137}Cs in surface air also ^{133}Te, ^{133}I, ^{140}Ba, were traced, and were deposited to the soil [GSF (1986)]. The ratio of ^{137}Cs to ^{134}Cs being 2:1. In fact ^{90}Sr and ^{239}Pu were detected too, but without any importance for the radiation dose.

In Fig. 10 the deposition for the Federal Republic of Germany is outlined. The pollution from the accident was highest south of the River Danube, especially towards the Alps, and reached values up to 45 kBq/m^2. Similar high values were found for Austria [Böck et al. (1988)], [Zechner et al (1996)], Switzerland [Czarnecki et al. (1986)], [HSK (1986)], Northern Italy [Minach et al. (1988)], Czechoslovakia [Klener et al. (1996)] and for Hungary [Andrasi et al. (1988)]. [De Cort et al. (1990)], and [Raes et al. (1989)] published data for other countries.

Contamination of Foodstuffs

In [UNSCEAR (1988)] data on the contamination in foodstuffs from many countries are compiled. Whereas the contamination of foodstuffs in Western Europe already in 1986 [GSF (1986)] and four years later in 1990 by [Müller (1992)] was considered to be rather small, excepted for mushrooms and game-meat, this could not be assumed likewise for the highly exposed regions of the former USSR. Thus the fear of the population in such regions was obvious.

Distrust and fear also made the population leave some regions being only slightly contaminated. Although Russian scientists stated, as confirmed by scientists of many countries, supervised by the IAEO in Vienna, that the contaminations of the foodstuffs were below the internationally used recommendations, this information found no trust by the population. To overcome distrust and fear by the population, it was agreed between the successor states of the USSR, the Russian Federation, Belorussia, the Ukraine, and Federal Republic of Germany, that German scientists should carry out measurements of soil and food contaminations and whole body measurements, beginning with contaminated regions south of Kaluga and west and north of Klintsy in the Russian Federation (see Fig. 9a). The findings (see Table 15 and 16) from the years 1991 to 1993 [Berg et al. (1991)], [Heinemann et al. (1991), (1993), (1994)], [Schneider (1991)], [Hille et al. (1996)] were made accessible to the population in each single case, accompanied also by warnings, if necessary, e.g., on intolerable amounts of radioactivity to limit the consumption of certain products, mostly from restricted zones.

In the region between Kaluga and Bryansk (see Fig. 9a) the ^{137}Cs concentrations in food remained below the limits valid for the Russian Federation, for instance in milk below 370 Bq/L. In the higher contaminated area of the district of Klintsy, however, the concentrations in milk exceeded the limit considerably (see Table 15), e.g., in the villages of Veprin and Unetcha, but also in Usherpye, Guta Koretzkya, and Rozny. The measured contaminations of soil corresponded to the activities recorded in the deposition maps. In Belorussia however, in 1993 no ^{137}Cs burdening of milk beyond the valid limit of 110 Bq/L was detected, since dairy farming in highly contaminated areas had been strictly forbidden.

The contamination of further foodstuffs is compiled in Table 16. Cereals, but also potatoes, and garden products were burdened without exceptions slightly in all regions. Fish from stagnant waters, meat, especially meat of game, and products from the woodlands, such as mushrooms or blueberries, often could not be graded harmless, since in rural areas mushrooms and berries from the forest were continuously consumed not only occasionally.

Table 15a: Dose rate, soil deposition and ^{137}Cs concentration in cow milk in different locations of Russia (Country Bryansk, District Klincy), based on [Berg et al (1991)], [Heinemann et al. (1991), (1993), (1994)], [Hille et al. (1996)].

Location	Dose Rate (µSv/h)			Soil (kBq/m²)			Milk (Bq/L)		
	1991	1992	1993	1991	1992	1993	1991	1992	1993
Beresovka	-	0.29	0.16	-	238	234	-	357	274
Blizna	-	0.25	0.18	-	176	130	-	461	707
Dushkino	0.25	-	-	229	-	-	104	-	-
Guta Korezkaya	0.53	0.53	0.50	962	625	443	666	535	868
Kivai	0.33	-	-	374	-	-	135	-	-
Klintsy	0.18	-	-	193	-	-	130	-	-
Lopatni	0.19	0.33	0.23	35	195	319	212	382	666
Mali Topa	0.22	-	-	239	-	-	90	-	-
Martyanovka	0.08	-	-	76	-	-	93	-	-
Olchovka	0.53	0.55	0.48	438	465	147	117	165	77
Pablichi	0.09	0.13	-	8	32	-	65	-	-
Pervoye Maya	0.22	-	-	110	-	-	212	-	-
Pestshanka	0.09	0.16	-	71	66	-	122	78	-
Rozny	0.36	0.42	0.48	540	340	424	561	388	161
Smotrovaya Buda	0.16	-	-	116	-	-	355	-	-
Tulu Kovshina	-	0.45	0.47	-	345	381	-	604	258
Turozna	0.43	0.41	-	350	273	-	107	139	-
Unecha	-	0.50	0.62	-	519	431	2062	1201	1101
Usherpye	0.55	0.78	1.03	566	409	1079	884	519	278
Velekiy Topal	0.21	-	-	253	-	-	43	-	-
Veprin	1.10	1.00	1.30	1001	1030	855	2128	1169	343

Table 15b: Dose rate, soil deposition and ^{137}Cs concentration in cow's milk in different locations of Belorussia (Country Gomel, Districts Vetka, Korma, Chechersk and Bragin), based on [Heinemann et al. (1993)], [Hille et al. (1996)].

Location	Dose Rate (µSv/h)	Soil (kBq/m²)	Milk (Bq/L)
	1993	1993	1993
Bartholomoyevka[a]	1.06	-	131
Chaltch	-	-	57
Raduga	0.31	497	-
Cherstin	0.90	454	-
Stolbun	0.48	223	-
Svetolovichi	0.44	458	-
Vetka	0.62	863	-
Gorodok	0.52	379	-
Korma	-	-	33
Volenzy	-	-	78
Lipa[a]	2.09	443	-
Otor	0.43	201	-
Sebrovitchi	1.54	809	-
Yaseni[b]	0.22	478	-
Savitchi[b]	3.94	3435	-
Soboli	0.32	284	-

[a] evacuated location,   [b] 30 km zone

Table 16: [137]Cs concentrations (range) in foodstuffs in the years 1991 and 1992 in the three Russian districts, in some regions of the Belorussia and Ukraine, based on [Hille et al. (1996)].

Location	Products		(Bq/kg)		
	Forest	Garden	Grain	Meat	Fish
Klintsy	<18 - 12320	<2.5 - 1270	79	4 -4590	23 - 117
Gordeyevka	89 - 18000	<4 - 514		<20 - 820	71 - 86000
Krasnaya Gora	119 - 16900	1.5 - 758	< 3 - 8	27 - 500	28 - 2925
Belorussia	<10 - 64000	<4.1 - 300	16 - 25	23 -7311	61 - 607
Ukraine	< 7 - 304	<1 - 10			

Whole Body Measurements of the Population

In Western Europe, especially in the Alpine region, by means of whole body measurements, the internal radiation exposure was controlled even directly after the arrival of the radioactive fallout clouds [Berg et al. (1986), (1987abc)], [Bogner et al. (1987)], [Schmier et al. (1988)], [Henrichs et al. (1992)]. In Fig. 11a the arithmetic means of the ^{137}Cs retention, standardized to the body weight, for children, women, and men from the environs of Munich are listed. The maximum value of about 21 Bq of ^{137}Cs and of 9 Bq of ^{134}Cs per kg of body weight was reached only one year after the accident. In the higher contaminated areas of the immediate foothills of the Alps, at far-reaching self-supply with foodstuffs produced locally, the values were higher by a factor of 6 to 8. In Fig. 11b the retention data of 12 pregnant

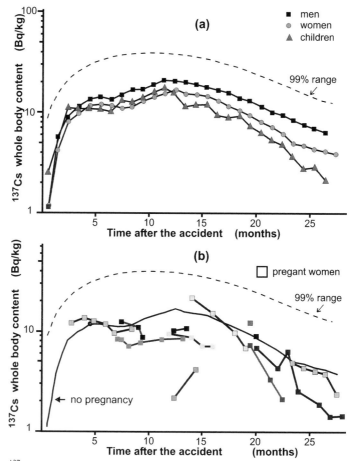

Figure 11: ^{137}Cs retention standardized to body weight for a reference group from Munich: (a) arithmetic means for adults and children, and (b) twelve pregnant women.

women compared to non-expectants are presented. The curves show that in the case of the pregnants, no higher Cs burden was measured, but that at the beginning of the second year after the accident the decrease in the body, combined with the diminished intake, was faster than in non-expectants.

Data on caesium whole-body content for persons from exposed regions such as Novosybkov, Unetcha, Veprin, and Kaluga in the Russian Federation were published by [Likhtarev et al. (1992)], [Perevoznikov et al. (1992), (1993), (1994)], and for Byelorussia by [Minenko et al. (1996)].

In July 1986 at Novosybkov, at the maximum about 1.7 kBq/kg (120 kBq totally) of caesium were found on average for males, and about half that for females [Merwin et al. (1993)]. In 1987 the activity levels dropped for adults to about 0.28 kBq/kg (20 kBq totally). In the extremely contaminated villages Unetcha and Veprin (see Table 15) a whole-body activity of about 4.7 kBq/kg was measured in 1986, which decreased until spring 1987 only to a third, in the months of summer, however, activity increased once more from consumption of local products, especially from the forest. In the region of Kaluga levels of about 100 Bq/kg were found in 1986, which decreased by a factor of about 5 until 1990.

During the measuring campaign by German scientists during the years 1991 to 1993 a large number of whole-body measurements were carried out by mobile whole-body counters [Hill et al. (1992)], [Hille et al. (1996)]. As many as 227427 inhabitants from different regions of the Russian Federation, but also 41758 persons from Belorussia and 14989 from the Ukraine were measured (see Table 17). For an easier radiological interpretation of the results, the measured persons were subgrouped into three categories:

Category I: less than 7 kBq total activity in adults or less than 4 kBq in children

Category II: between 7 and 25 kBq in adults or 4 to 15 kBq in children respectively

Category III: more than 25 kBq in adults or more than 15 kBq in children

The classification was chosen in such a way, that the category I corresponded to an effective dose of not more than 0.3 mSv/y, assuming a one-year presence of 7 kBq of incorporated caesium, category II to doses between 0.3 and 1 mSv/y, and category III at more than 1 mSv/y.

Table 17 shows that in the region of Bryansk and in Belorussia about 70-90 % of the measured persons could be classifed as the category I, only about 4-23 % as the category II, and 1-7 % as the category III, with exception of Rovno in the Ukraine, where nearly two thirds of all persons had incorporated more than 7 kBq. At the other areas, like Kaluga, Orel, and Tula, and in the Ukraine, all persons could be assigned to category I.

Table 17: [137]Cs content in the whole body in the years 1991, 1992 and 1993 in Russia, Belorussia and Ukraine, (a) adults, (c) children, based on [Hill et al. (1992)], [Hille et al. (1996)].

Year Location	Number	Person - Category		(%)
		I [a]<7 kBq [c]<4 KBq	**II** [a]7-25 kBq [c]4-15 kBq	**III** [a]>25 kBq [c]>15 kBq
Russia				
1991	163033	93.7	5.3	1.0
Bryansk	80336			
Gordeyevka	6901	70.7	25.4	3.9
Klintsy	46977	91.6	6.4	2.0
Krasnya Gora	7071	81.0	16.3	2.7
Novosybkov	5928	91.5	7.8	0.8
Zlynka	8315	80.3	17.8	1.9
Starodub	5144	99.6	0.4	0.0
Kaluga	53573			
Shvastovichi	9952	96.0	3.8	0.2
Ludinovo	25240	99.8	0.2	0.0
Shistra	9982	99.6	0.4	0.0
Ulyanovo	8399	96.9	3.1	0.0
Orel	1447			
Bolchov	1447	99.9	0.1	0.0
Tula	27677			
Arsenyevo	4953	100.0	0.0	0.0
Plavsk	4060	100.0	0.0	0.0
Uslovaya	18664	100.0	0.0	0.0
1992	49858	85.8	11.7	2.5
1993	14836	70.0	22.9	7.1
Belorussia				
1992	29229	95.2	4.4	0.4
1993	12556	83.8	12.4	3.8
Ukraine				
1992 (Fastov)	11373	100.0	0.0	0.0
1993	36126	84.5	11.9	3.7
Rovno	2773	38.6	34.9	26.5

2.3.1.7. *Reactor Accident at Sosnovy Bor in March 1992*

On March 24th, 1992, in Sosnovy Bor a grave incident occurred in a RBMK-reactor, in the course of which radionuclides were released. One of the 1700 fuel rods had burst. Noble gases, radioactive iodine, and a series of other fission products were released at small amounts. [Toivonen et al. (1992)] observed, besides the already mentioned radionuclides, additionally ^{95}Zr, ^{95}Nb, ^{99}Mo, ^{103,106}Ru, ^{140}Ba, ^{134}Cs, ^{137}Cs, ^{141,143,144}Ce, and ^{239}Np at a range of 1-10 mBq/m^3 in the air 100 km away at south Finland.

2.3.2. Contamination from Plane or Satellite Crashes Involving on Board Nuclear Material

Loss of aeroplanes equipped with nuclear weapons

In the past, a series of crashes of military aircrafts with nuclear weapons on board have occurred. The exact number of these accidents and their consequences are widely unknown. In two cases reported from USA, bombs were destroyed and plutonium was released locally. Fortunately, however, no nuclear fission reaction happened. Reports on such accidents from other countries are unavailable.

The crash of a B29-bomber of the USA on April 11th, 1950 was made public by [Greenpeace (1996b)]. The plane started from the Kirtland Air Force Base equipped with nuclear weapons. At the crash a bomb in fact was destroyed, however, releasing no fissionable material, since the cover around the fissile material remained tight.

On December 5th, 1965 on the US-aircraft carrier "Ticonderoga" a plane of the type A-4E with a thermonuclear bomb on board had an accident [Greenpeace (1996b)]. The plane sank with the crew and the bomb into the 4900 m deep water of the Pacific Ocean about 350 km east of Okinawa and was not recovered again.

Air crash nearby Palomares, Spain in 1966

Near the South Spanish village of Palomares, after the crash of an american B-52 bomber, components of the conventional explosives of the igniter of two hydrogen bombs exploded. At this explosion ^{239}Pu was released as insoluble plutonium oxide occurred with particle sizes 0.3 to 30 μm. The wind dispersed the particles, so that houses, vegetation, and the soil at Palomares were contaminated. In order to prevent inhabitants from evacuation, great efforts were exerted to inform the population and to control the incorporation. Some hundreds of persons were examined for incorporation without exceeding the limit [Brunner et al. (1995)]. Furthermore a decontamination and restoration program was started. At more highly contaminated places the top soil after moistening by water was removed mechanically down to a depth of 5 to 8 cm, altogether 280 m^3 were cleared away. Less contaminated areas were wet ploughed 25 to 30 cm deep. Surfaces only slightly contaminated were fixed and the vegetation first was composted in an improvised dump, finally, however, burnt in a dry stream bed during seaward wind. All the waste was collected and transported to the Savannah River plant in the USA. Over a period of several years the crops harvested were checked for contamination, however, no significant increases of radioactivity have been verified.

Plane crash near Thule, Greenland in 1968

In January 1968, the crew of a B-52 bomber, which had caught fire, was forced to plan an emergency landing at the air force base of Thule in North Greenland. However, the aeroplane crashed 7 miles west of the air base with four hydrogen bombs aboard [Eisenbud (1987)], [Brunner et al. (1995)], [Greenpeace (1996b)]. During the crash onto the ice, the chemical igniters in the nuclear weapons exploded [Aarkrog (1971c)] and fuel and fragments of the plane caught fire. An unknown amount of ^{239}Pu and ^{241}Am was released. The contamination was limited to the fragments of the plane and to the fire site of about 700x150 m^2. Since the ice was covered by 10 cm of snow and after the accident a further 2 cm of powdery snow fell, the contaminated materials froze at -40° C or were covered by the snow. The ice itself had been burst only at the point of the impact and froze up rapidly. At the impact place itself only minor contamination was discovered. The contaminated fragments were collected and packed into containers, as well as 12000 m^3 of snow, were transferred into fuel barrels, and all together was transported to the Savannah River plant. The contaminated ice was covered by a mixture of coal and sand in order to accelerate the melting and dilution process in the water at snow melting in springtime and summer.

In the following 11 years in the arctic biosphere, an extensive radioecological research program of Danish scientists supplied interesting information on the behaviour of plutonium, deposited at the sea, at the sea bottom, in sediments and Benthic organisms. The studies of [Aarkrog et al. (1984)] proved, however, that there should be no fear of endangering of the Eskimos living nearby or of the commercial fishing. No worthwhile amounts of plutonium were found, neither in plants, collected from the surface water, nor in fish, or marine mammalians.

Crashes of satellites with nuclear power sources

Several of the satellites developed by the USA, as well as by the USSR were equipped with nuclear power sources. Whereas the USA used for the power supply of her space program radionuclide generators, the so-called SNAP (**S**atellite **N**uclear **A**uxiliary **P**ower), the USSR favoured nuclear reactors.

The SNAP installations were generators, using the heat of decay of the α-emitter ^{238}Pu (half-life 86.4 a) at an order of 0.4 PBq for the production of power [Eisenbud (1987)]. At first it was assumed that after re-entry and burning up in the atmosphere, the release of ^{238}Pu from the batteries might contribute only a small additional part to the that already present from weapon tests. The realisation, however, that after the start of rockets not reaching the orbit, ^{238}Pu might be not widely scattered, but go down close to the starting point, led to a change of the design in that way, that the batteries were equipped with utmost secure wrappings. Between 1961 and 1984 the USA installed 34 radionuclide generators in 19 systems of space flight.

In 1970, the space capsule "Apollo 13", which had been damaged on its way to the moon, had aboard a SNAP-generator, being lost in the South-Pacific after re-entry into the atmosphere of the Earth. Since radioactivity could not be proven, it was assumed that the wrapping had remained undamaged. On April 21st, 1964, a navigation satellite did not reach its planned orbit after the start and re-entered the atmosphere at an altitude of 46 km above the Indian Ocean. The radioisotope unit of the type SNAP9-A contained about 0.63 PBq of ^{238}Pu [Krey (1967ab)]. In the upper layers of the atmosphere in fact ^{238}Pu remaining from former nuclear weapon tests was present, but four months after the crash at a height of 33 km an additional ^{238}Pu could be observed by sampling taken by means of balloons. At that time it was estimated that after a distribution period of two years, about 0.6 PBq might be deposited onto the surface of the Earth [Krey et al. (1970)].

While the USA sent only one reactor into space in the year 1965 [Eisenbud (1987)], the USSR shot more than 30 reactors in satellites into space, all circling the globe at altitudes of 700 to 800 km. The first reactor was on board COSMOS 198, which was started on December 27th, 1967. In the year 1973 there had been problems at the starting of another satellite which crashed into the Pacific Ocean close to the Japanese coast. On January 24th, 1978 the Soviet satellite COSMOS 954 crashed down [Nilsen et al. (1994)], [Eisenbud (1987)], [Krey et al. (1979)]. It had been equipped with a nuclear reactor containing 20 kg of highly enriched uranium as fuel. At the re-entry into the atmosphere radioactive fragments were scattered and a track 1000 km long was drawn, starting northeast of the Great Slave Lake reaching to Baker Lake close to the Hudson Bay [Steward et al. (1980)]. At this time, the core contained at a rough estimate 3.1 TBq of ^{90}Sr, 0.2 PBq of ^{131}I, and 3.2 TBq of ^{137}Cs. After intensive search by the Canadians, 65 kg of the fragments were recovered, some of them showing dose rates up to 5 Sv/h. It was estimated, however, that 75 % of the original material remained in the upper atmosphere. In air, water, and foodstuff samples no contaminations were observed.

Because of such accidents of satellites with a nuclear power supply, the IAEA developed a guide [IAEA (1996b)] in order to help authorities with emergency planning.

2.3.3.   Contamination by Ships with Nuclear Propulsion

Civilian ships

Worldwide, 13 ships were built with nuclear drive for civilian use. Nine of them were put in to operation in the former Soviet Union and only four in other countries, none of these four ships were operated commercially. Eight of them were Russian icebreakers, the "Lenin" 1959-1989, the "Arktika" 1975, the "Sibir" 1977, the "Rossiya" 1985, the "Taimyr" 1989, the "Sojuz Vaigach" 1990, and the "Jamal" 1993, and the container ship "Sevmorput" 1988, which all are used commercially with the exception of the "Lenin" which is off duty now. These ships were built to maintain shipping routes along the north coast of Siberia incase of 1.2 to 2 m thick ice blocking the fairway channel. The service base "Atomflot", 2 km east of Rosta, a suburb of Murmansk, belongs to the servicing infrastructure for those ships and furthermore five conventional ships: the "Imandra" 1981, the "Lotta" 1961, the "Serebryanka" 1975, the "Volodarsky" 1929 (1969), and the "Lepse" 1936 (1962). Their PWR reactors are run with uranium enriched to 30-40 % as fuel, which is exchanged every three to four years at the base "Atomflot" within a service period of about one and a half months. Only the third cooling circuit is in direct contact with the environment using sea water to condense steam and to cool. The container ship "Sevmorput" and the two icebreakers "Vaigach" and "Taymyr" each have one reactor, the other ships are equipped with two reactors.

In 1959 a serious reactor accident happened on the icebreaker "Lenin", which was equipped with three reactors of older design then. Problems with the cooling system had occurred on several occasions. In 1966 after 82000 nautical miles, maintenance work became necessary. On this occasion it was realized that a part of the fuel elements had been deformed, apparently from overheating during a period of defective cooling. Part of the fuel rods, 40 %, were transferred to the "Lepse" and stored there. When in 1967 difficulties recurred and also other damages at the reactor core arose once more, the "Lenin" was towed into the Kara Sea, the three reactors were dismantled, and sunk in the Tsivolky Bay east of Novaya Zemlya. Thereafter two new reactors were installed and the ship could be put in to operation again in 1970.

About one third of the burnt up fuel elements, unloaded from the ship reactors, which could not be processed in the reprocessing plant Mayak (see 2.1.1.4), were stored on the "Imandra" or "Lotta". The "Imandra" is equipped with a tank for the intermediate storage of contaminated water, which then can be discharged to the tanker "Serebryanka". Until 1986, radioactive contaminated cooling water was dumped into the Barent Sea. After a three- to four-year running time of the reactor, about 170 m^3 of liquid waste at an activity of 22 GBq resulted, one half from the primary circuit,

25 % from cleaning works at the reactor, and the rest from the storage of the burnt up fuel elements. In a special factory for waste management located at the Atomflot base this liquid waste was purified until a radioactivity concentration of less than 11 kBq/m^3 was achieved and thereafter it was drained off into the fjord of Murmansk. The total solid waste resulting from the reactors of the civilian-used ships amounts to a volume of 104 m^3 containing 7.7 TBq of radionuclides.

Up to 1980 the "Lepse" had been the service ship to store fuel element sets upon, she was replaced by the "Lotta" and "Imandra". She also was used to transport and discharge radioactive waste at Novaya Zemlya. Especially on this ship the deformed fuel elements were stored, originating from the reactor of the "Lenin". Since the deformed fuel elements jammed at insertion into containers, the workers used forging hammers to push them further downwards. A part of the fuel rods was damaged and therefore was not allowed to be delivered to the reprocessing plant. They remained on board, together with other fuel rods. In 1993 the inventory of radioactivity was estimated to be 28 PBq, 0.63 PBq of it being transuranic elements. To protect against the radiation, in 1980 in the sector where the damaged fuel rods are stored, a cement cover was installed, a kind of sarcophagus, which nowadays complicates the recovery of the damaged rods. An international advisory panel with participants from the CEC, the USA, Norway, and Russia came to the conclusion that the 624 spent fuel assemblies aboard the "Lepse" have been stored with insufficient safe-guards. The fuel assemblies were cleared away, the ship broken up, and the scraps disposed of as radioactive waste.

Military ships

A nuclear drive using a reactor as a heat source for the turbines can be built very compactly and consumes during operation neither oxygen nor fossil fuel, important especially for submarines. Thus no combustion products result that have to be disposed of, no fuel tanks are necessary and the replenishing of fuel is necessary only after a long period of operation, about every three to four years. The nuclear power has proved its worth in submarines, allowing longer periods, larger submerging depths, and thus larger ranges of action, than conventional submarines. A similar benefit for aircraft carriers, which mostly have to operate far away from their bases, possibly without support logistics for a long time.

This is the reason why the first nuclear-powered warship of the world was a submarine, the "Nautilus (SSN 571)", put into service by the USA in 1955. Within two years, she covered a distance of almost 63000 miles, more than half submerged, before the fuel elements of the reactor had to be exchanged for the first time. In 1958 she even was under way from the Pacific Ocean to the Atlantic below the ice of the north pole. The first nuclear powered surface vessel of the world was the American cruiser "USS Long Beach (GN9)", put into service in 1961, followed by the aircraft carrier "USS Enterprise (CVN65)" which during three years covered 207000 miles without further fuel supply.

Since this time a series of nuclear-powered ships have been built by the USA and also in other countries: by the USA 106 nuclear submarines, attack and ballistic missile submarines, a cruiser, and eight aircraft carriers, by the former USSR, 247 nuclear submarines and five nuclear-powered surface vessels, among them three battle cruisers; by Great Britain 18, by France 11, and by China six nuclear powered submarines. Meanwhile, already a series of ships were put off duty, among them no less than 138 Russian nuclear submarines. By servicing, fuel element exchange, and especially during breaking up of the ships, special problems arose in order to avoid contamination of the environment and to handle radioactive waste safely for interim storage or ultimate disposal. Additionally, a series of grave accidents happened with loss of human lives, ship sinking, and release of radioactive material.

On April 10th, 1963, at the first dive after a maintenance period of nine months, the American nuclear submarine "USS Thresher" sank 160 km east of Cape Cod with her crew of 129 persons. The sinking probably had been caused by a water leak in the lines of the engine-room, so that the ship was hampered in surfacing and now lies crushed by the water pressure into six parts at a depth of 2600 m. Samples for determining radioactivity in the sediments of this region have been found to have only low concentrations of ^{60}Co at 12 Bq/kg [Nilsen et al. (1996)].

On March 10th, 1968, a non-nuclear-powered Soviet submarine "(Project 929M)"sank in the Pacific Ocean 1200 km northwest of the island of Oahu, Hawaii, but with SS-N-5 missiles aboard, equipped with three nuclear warheads and possibly further two nuclear-armed torpedoes [Greenpeace (1996b)].

In the same year on May 22nd, 1968 the American nuclear submarine "SSK Scorpion" sank southwest of the Azores with her crew of 99 persons. She is lying at a depth of 3600 m, burst into two parts. Probably the accident was caused by an explosion of a torpedo on board. As endangering potential for the environment, additionally to the nuclear reactor, two torpedoes with nuclear warheads were aboard. Samples from the sediments of this marine region, however, showed only low contaminations [Nilsen et al. (1996)].

Two years later, on April 8th, 1970, the Soviet nuclear submarine "K8 (Project 627A)"got into distress at sea in the Bay of Biscay, 490 km northwest of Spain. Simultaneously, at two different places fires broke out, one of them in the central part of the boat. The boat surfaced, but the fires could not be extinguished. Since the reactor broke down and the auxiliary

Diesel engine could not be started, the ship remained without power supply. On April 10th, water began to intrude into compartments 7 and 8. A part of the crew was evacuated to the escort vessel. The next day, on April 11th, the ship sank with two nuclear reactors and two nuclear torpedoes to a depth of 4680 m. Fifty persons were killed at this occasion.

A further Soviet nuclear submarine, the "K219 (Project 667A)" got lost on October 6th, 1986. She sank north of Bermuda into the Atlantic Ocean with sixteen SS-N-5 missiles, equipped with two nuclear warheads and two reactors aboard. Gas escaped, produced by reaction of water with the missile fuel, destroying the rocket runway. A fire and a short circuit put the emergency system out of action, and one person lost her life while making the effort to retract the control rods into the reactor. The submarine surfaced and the second reactor was put into operation, which had been powered off at the time of the explosion. Nevertheless, the state of the boat worsened rapidly. After the second reactor stopped working, the greater part of the crew went ashore and when the ship started sinking, also the remaining members of the crew left. The cause of the explosion has remained uncertain.

On April 7th, 1989 another Soviet nuclear submarine, the "278 Komosolets (Project 685)" sank in the Norwegian Sea thence into the Atlantic Ocean. By means of its titanium cover this boat was able to submerge 1000 m deep. On its way home to the base Zapadnaya Litsa at the Peninsula of Kola, 180 km south of the Bear Island at a depth of 160 m, a fire broke out in the seventh compartment of the boat. It surfaced eleven minutes later and the fire could not be extinguished, but even spread further. By a short circuit the reactor was put out of operation and the boat remained without power supply. The compressed-air system, damaged by the fire, ran to empty and the boat lost stability, with the consequence of sinking, lying at a depth of 1685 m. Although a part of the crew could leave the ship in time, 41 persons were drowned. The exact reasons for the fire have remained unknown. Furthermore, in the Soviet navy a series of other grave accidents happened without loss of the ships, in which the reactors got out of control by damages in the adjustment or by loss of coolant. During these accidental happenings, considerable amounts of radioactive substances were released into the environment.

On August 12th, 2000, the Russian submarine "Kursk" sank to the bottom of the Barents Sea and the wreckage laid at a depth of 108 meters. After 14 months, the submarine was lifted to the surface and plodded to Roslyakovo docks near Murmansk. Until now, the exact causes of this accident have not been published. However, at the time of the accident, NATO ships detected two explosions. The prevailing version supposed that the vessel was damaged by an explosion of a defective torpedo. The

"Kursk" was part of Russia's Oscar II-class (Antey, type 949A) submarine fleet and was powered with nuclear reactors of a similar type (OK-650b) to those on board the submarine "Komsomolets" (OK-650b-3). At the time of the accident 118 men were aboard. Investigations of the wreckage revealed extensive destruction at the fore ship. During the rescue operation before lifting to the surface, Norwegian divers discovered, that all segments of the submarine were flooded. Measurements by the Norwegian Radiation Protection Authority (NRPA) and the Murmansk Meteorological Institute suggest that there was no discharge of radioactivity from the wreckage of the "Kursk" into the Barents Sea. The two reactors, equipped with an automatic shutdown and passive cooling system, are not discharging any radioactivity. The estimated inventory of the "Kursk" reactors should be 5.6 PBq of ^{90}Sr and 6.2 PBq of ^{137}Cs [Bellona (2003)].

A further potential endangering of the environment is represented by the bases of nuclear powered ships, where servicing, exchange of the fuel rods, or breaking up of the ships is carried out. This is a problem for today's Russia, which intends putting out of action 159 nuclear submarines by 2003. The most important bases are located on the Peninsula of Kola, e.g., the base Zapadnaya Litsa. At these bases 7000 m^3 of light- or medium-contaminated radioactive waste is stored at a total activity of 3.7 TBq, and further 8000 m^3 solid waste. The total activity amounts to about 37 TBq.

2.4.   Radioactive Waste

In storage of radioactive waste, as a matter of principle, one has to distinguish between an interim storage [IAEA (1998f)], limited in time, being inevitable for the treatment of the waste, and the ultimate disposal. The problems of management, interim storage, and ultimate disposal are extraordinarily complex and politically explosive, since nobody wants to accept such dumps near his residence. In contrast to other, highly poisonous waste, the half-lives of decay are known. However, some half-lives are of a duration which exceed our imagination. In the history of humankind, 1000 years represent an immense time, but even 24000 years, the half-life of ^{239}Pu, are in geological terms only a short time period. The development of human society and changes in the environment are uncertainties, with which we are saddling future generations. They give cause for violent, emotionally charged, political discussions on the ethical justification of the storage of radioactive waste, mostly involving nuclear power [Persson (1995)]. A great number of papers of the IAEA dealt with the subject "radioactive waste", since its safe storage is not only a national problem, but needs, considering the long periods of time, international control.

2.4.1. Classification of Radioactive Waste

Radioactive wastes are not only classified by their physical and chemical properties, but also whether they are of civilian or military origin. Furthermore they are subdivided, according to their activity, to the nature of their radionuclides, and to the chemical compounds they are bound to [IAEA (1997a)]. A further useful differentiation is between those wastes that contain only short-lived radionuclides, requiring less expenditure. Six different categories are to be distinguished:

- radioactive waste to be decontrolled,
- "low level" waste with exclusively short-lived radionuclides (mostly < 100 d, e.g., in Germany),
- low level waste with long-lived radionuclides,
- medium radioactive waste with exclusively short-lived radionuclides,
- medium radioactive waste with long-lived radionuclides,
- high level radioactive waste.

In the waste to be decontrolled, radionuclides are present only at concentrations below objectionable limits, so that they can be handled like conventional wastes. Low level wastes are interpreted as lightly contaminated wastes, originating from research- and development-laboratories, from hospitals, or from industry, e.g., contaminated paper, biological matter, or building materials. Exceptions are discarded radiation sources, usually belonging to the medium-radioactive wastes. Nuclear fuel from civilian-used reactors and/or solid residues from the reprocessing of the fuel rods, remainders from reprocessing of military applications are classed with the high-level wastes. A further category are the transuranic wastes, originating mostly from military application, and containing predominantly α-emitting nuclides [Eisenbud (1987)]. Because of the high and long lasting endangering potential of such wastes, in the USA, in Japan, and France research programs were started, in order to isolate these partially very long lived radionuclides and to transform them to shorter lived or stable isotopes.

There exist no internationally binding classifications. Therefore the upper limit for liquid low-level waste varies between $1.5 \cdot 10^7$ and $3.7 \cdot 10^9$ Bq/m^3 and the lower limit for high-level between $3.7 \cdot 10^9$ and $3.7 \cdot 10^{13}$ Bq/m^3 [Herrmann (1983)].

2.4.2. Treatment of Radioactive Waste

Prior to ultimate disposal of radioactive waste, it is useful to make available processes for the reduction of the volume of the waste [IAEA (1992ab), (1995), (1997b)]. The potentialities to recycle substances from the waste or to decontaminate the waste such that it may be considered low level waste, diminishes the total volume substantially. Combustion, compaction, evaporation, filtration, ion exchange, or precipitation are some of the fundamental concepts of the volume reduction. A further important criterion, the waste should be chemically stable on a long-term basis. Special procedures belong to it to impede the migration of radioactive material in the biosphere by immobilisation, which means by solidifying with cement, bitumen, or glazing [IAEA (1992a)], thus rendering difficult the release into the environment. The treated wastes are then packed additionally into 200 L high-grade steel barrels or heavy-walled steel containers [IAEA (1993a)]. At the disposal of highly radioactive waste, e.g., from reprocessing plants or as whole fuel rods, these procedures have to include sufficient care for the extraction of the decay heat, leading otherwise to temperatures of some hundred degrees Celsius, and assurance that no critical mass and thereby a chain reaction may be achieved. The consequences of an inadequate treatment of wastes were demonstrated for instance by the accidents at Kysthym and at the Lake Karachay (see 2.1.1.4).

Since the place of the waste origin, the waste management facilities and the waste dumps are not always located at the same site, the radioactive waste has to be transported. For a safe transport special containers have been designed [IAEA (1996c)].

2.4.3.   Ultimate Disposal of Radioactive Waste

Ultimate disposal is the final step in the treatment of waste. Which places are suitable for ultimate disposal, where they are established, and which solution the different countries prefer, are issues not yet threshed out politically. Ultimate disposal nevertheless should fulfil the following important criteria:

– to exclude a spreading of radioactive substances from waste to the biosphere, at least for periods within which the activity of the radionuclides has not decayed to a negligible extent,
– to be maintenance-free,
– to guarantee security from unauthorized access and intervention,
– to guarantee sufficient protection against external exposure to radiation originating from the stored up waste.

The behaviour of radioactive waste in the sea was described in detail in the reports of [NAS-NCR (1957a), (1959)], [IAEA (1983b)] and in large depths of the sea of [IAEA (1986a), (1988a)]. Simultaneously attempts were made to find possibilities for the safe storage of radioactive waste, e.g., in geological formations in layers near the surface as well as in deeper layers [IAEA (1980b), (1981ab), (1982ab), (1983acde), (1984ab), (1985abc), (1986b), (1989ab), (1990), (1993b)]. The publications of [IAEA (1985a), (1997b)] comprise international guidelines for the establishment of storage areas for light and medium radioactive waste in shallow grounds and in rock caverns and case studies of such areas already established and in use [IAEA (1985c)].

The preferential concept of storage is, to store waste near the surface (low-level) or in deep geological layers (high-level) without the option of retrieval and not relying upon long-term monitoring or servicing. Security should be guaranteed by conditioning on the one hand (physic-chemical form, solidifying, packing) and on the other hand by isolation from the biosphere by one or several, natural or engineered barriers around the stored waste. Barriers may be for instance: a package container with walls constructed in such a way that the release of radioactive substances is considerably slowed, or rock with filling material with high absorption properties. Salt formations deep in the geological underground have already been recommended by [NAS-NRC (1957b)], [Gear (1975)] as ultimate disposal for radioactive wastes. Salt stocks are very stable and have no water-bearing strata inside. They recommend themselves by sealing clefts and hollows due to their plastic properties and deformability. The high heat conductivity of stone salt ensures a fast carrying off of heat of embedded high-level waste. An alternative are rock formations, which themselves are to a high degree free of water, are not in contact with ground water-bearing strata, are highly packed, and stay tight even at the entry of decay heat.

For decades it was routine practice to store radionuclides in dumps close to the surface or in shallow water. Already in 1946 low-level waste from research and development centres were disposed of into the Atlantic, Pacific, or the Gulf of Mexico, mixed with cement and packed into steel barrels. After investigations in sea regions where radioactive waste was disposed, a committee of marine scientists of the National Research Council came to the conclusion that comparatively large amounts of low-level waste may be deposited with relative safety in shallow coastal waters [NAS-NRC (1959)]. However, this practice was stopped in 1970. By that time, the USA had disposed of about 35 PBq into the sea and the European countries about 26 PBq. Compared to that, by nuclear weapon tests alone, in the Pacific 700 times more activity was deposited, namely 42000 PBq of ^{3}H, 70 PBq of ^{14}C, 420 EBq of ^{137}Cs, 263 PBq of ^{90}Sr ,and 6.3 PBq of ^{239}Pu [Eisenbud (1981)]. In 1993, in the white book of the office of the Russian president, all the submerging and dumping into the Barents and Kara Sea, carried out in the past, was listed in detail. According to this list, in the Arctic waters six nuclear reactors together with their spent fuel and a further ten without fuel, all from nuclear submarines, as well as a shielding assembly of an icebreaker and other solid wastes have been submerged. The points of submergence are located particularly in the flat fjords of Novaya Zemlya in the Kara Sea at a depth of 12 to 135 m and in the Novaya Zemlya Trough at a depth of 380 m. Liquid low-level wastes were dumped into the open Barents and Kara Sea. In 1993, an international group of experts estimated the inventory of high-level radioactive waste altogether to be 37 PBq with an additional 4.7 PBq in

1994. The group of experts established that the contamination in the main Kara Sea is still low, compared to other oceans [Sjölblom et al. (1998)], [IAEA (1998g)]. Only those radionuclides were found, that either result from global nuclear weapon tests, from draining of reprocessing plants in Western Europe, or from the Chernobyl accident. Augmented levels of activity in sediments were detected only in the immediate vicinity of the low-level waste containers, but not outside the fjords or in the open Kara Sea.

In many countries ultimate disposal in formations close to the surface and in geological formations is favoured today, and also is being applied in practice [Gömmel (1997)]. At the beginning of the 1980s the deposition into the sea was stopped after a voluntary memorandum by many countries. Dumps in crystalline rock (e.g., granite) for ultimate disposal are favoured and were considered by Argentina, Great Britain, France, Finland, India, Japan, Canada, Russia, Sweden, Switzerland, Spain, South Africa, USA, and China, followed by deposition in salt formations in Germany, the Netherlands, Russia, Spain, USA, and in clay by Argentina, Belgium, Great Britain, France, Italy, and Spain. In Europe the following ultimate disposal areas exist:

– close to surface ultimate disposals at Centre de l'Aube and Centre de la Manche in France, Drigg and Dounraey in Great Britain, at Ignalina in Lithuania, El Cabril in Spain, and at Dukovay in the Czech Republic,
– depth storage areas in Germany at Morsleben, Olkiluoto in Finland, and at Forsmark in Sweden.

2.5.   Contamination by Particle Accelerators

In accelerators radioactive substances can be produced in targets (see Chap. 2, 2.2). Due to loss of the beam or insufficient shielding of the accelerator, however, radionuclides may be generated outside the beam tube too, for instance in the shielding soil, in the ground water, or in the surrounding air, as well as from the dust particles contained therein [Hoyer (1968)], [Gabriel (1970)], [Thomas (1970)], [Borak et al. (1972)], [Stapleton et al. (1972)], [Lucci et al. (1973)], [Phillips et al. (1986)].

The extent and type of radionuclides produced and which of them may endanger the environment, depends on the design of the accelerator, and on its operating conditions. If the ground, which at subterranean-installed accelerators also serves for shielding, communicates with the ground water, there may be a danger that by leaching, radionuclides enter the drinking water [Borak et al. (1972)], [Bull et al. (1997)], [Baker et al. (1997)]. Dust activated in the air may be forwarded by the wind and can burden people living in the neighbourhood by inhalation. The majority of the radionuclides generated are extremely short-lived, so that they disintegrate quickly after switching off the accelerator and decay fully already during the further transport. Longer-lived nuclides are produced only in minor quantities outside the controlled area. These include, dependent on the composition of the ground, the following isotopes: ^{3}H, ^{7}Be, ^{22,24}Na, ^{32}P, ^{35}S, ^{47}Sc, ^{45}Ca, ^{55}Fe, ^{54}Mn, ^{51}Cr, ^{56,57,58,60}Co, and ^{65}Zn.

2.6.   Contamination by Medical and Industrial Applications of Radionuclides

Besides the application of radium and thorium a major usage of radionuclides in medicine and engineering [NCRP 56 (1977)], [NCRP 95 (1987)] did not begin until, after the discovery of nuclear fission, suitable radionuclides were available cost-effectively.

In the first phase in medicine, the radionuclides ^{131}I, ^{51}Cr, ^{203}Hg, ^{57}Co were used for the diagnosis of thyroid, liver, and kidney diseases. Since, at that time, no precautions were taken against the release of radionuclides into normal canalization, in hospitals with large numbers of examinations considerable amounts entered the sewage plants and rivers. Today, these radionuclides for scintigraphy or positron emission tomography are replaced to a great extent by very short-lived radionuclides, like for instance ^{99m}Tc (half-life 6 h), or ^{18}F (1.8 h) (see further in Chap. 2, Table 5). The activities of the radionuclides used with radioimmunoassays can, by highly developed measuring techniques, be administered in such low doses that they do not present a danger of major contamination of the environment. With respect to therapy, however, very high activities of ^{226}Ra, ^{137}Cs, ^{60}Co, and ^{131}I are sometime applied. In the case of radium, caesium, and cobalt this concerns enclosed radiation sources, which nearly exclusively are used in tumour therapy. Iodine is exceptional and is administered to patients in unclosed form as ^{131}I at the treatment of thyroid diseases in doses of 37 MBq up to 7.4 GBq. As the major part is excreted by urine, faeces, and respiratory air it may thus lead to a contamination of the environment. Due to the short half-life of radioiodine, architectural precautions, as well as filter and sewage installations, and precautions in the case of the premature death of the patient, general pollution can be avoided.

The plutonium batteries (SNAP), developed in USA, were not only used in space flight, but also were implanted in humans as a power supply for pacemakers until the end of the 1980s [NCRP 95 (1987)]. About 2500 of those batteries were used, each of them containing about 250 mg (150 GBq) of ^{238}Pu. The analogous thermal generators, developed in the former Soviet Union, ran by radioisotopes (**r**adioisotope **t**hermal **g**enerator = RTG), and served as a power source for installations in regions with missing infrastructure. These RTG's contained ^{90}Sr activities of 1850 TBq to 9620 TBq, which at the beginning yielded electric outputs of up to 80 W. They were used above all in lighthouses along the Arctic coast, 132 of them out of 2000 at the island of Kola and Novaya Zemlya being thus equipped. The total activity was estimated to be between 225 PBq and 1270 PBq [Nilsen et al. (1994)]. In December 2001 three woodsman found the cores of two unshielded nuclear batteries in a mountainous region in Georgia [IAEA (2002)]. The men suffered serious injuries from overexposure (dizziness, nausea, radiation burns) and were hospitalised in Tbilisi.

Smoke detectors are installed in large numbers in nearly all public buildings, hotels, offices, and factories and already have saved many of human lives. They consist of an ionization chamber and a radionuclide that provides a continuous flow of current. This current is reduced by smoke and from the current drop an alarm signal is derived. The first smoke alarm installations, containing radium (188 kBq-3.7 MBq ^{226}Ra), were introduced in 1951, but subsequently replaced by the cheaper ^{241}Am. About 12 million smoke detectors were sold as of the mid-1980s with a total activity of about 0.31 TBq of ^{241}Am as dioxide [Eisenbud (1987)]. Not only in the case of luminous paints, but also in the case of seeds (see 1.3.2), at the control of seams and at material testing, radioactive nuclides (^{3}H, ^{147}Pm, ^{192}Ir) are used. In the case of enclosed sources, which are employed in therapy units and at material testing and contain large amounts of activity, three grave accidents are know to have happened by slovenliness.

In December 1983 at Ciudad Juarez in Mexico a ^{60}Co source of 16.7 TBq was to be put out of service. The source was improperly disposed of and ended up in a scrap yard [UNSCEAR (1988), (1993)]. Probably the source, containing 6000 pellets, broke asunder, dispersing them partly during the transport. Subsequently several streets and a few hundred houses were contaminated. By use as scrap material in some iron produced factories, thousands of tons of contaminated steel were sold in Mexico and to the USA. About a thousand people were substantially irradiated by this slovenliness. A very similar accident occurred in Istanbul, Turkey, in 1998/99 when two ^{60}Co source devices were sold as scrap metal [IAEA (2000a)]. While opening the shielding equipment, several persons were strongly exposed with clinical signs of the acute radiation syndrome.

In the year 1984 in Morocco near Mohammedia during an examination of a seam welding a ^{192}Ir source was lost from its shielding container. The source dropped to the bottom, was found by a passer-by, and was taken home. There was no contamination, but he and his whole family, eight persons in total, died from overexposure.

In September 1987 in Goiânia in Brazil a ^{137}Cs source of 50.9 TBq was removed from a therapy device and dismantled by junk dealers [IAEA (1988b)]. The opening of the source led to local contamination of inhabited areas. As a result, 129 persons were exposed [UNSCEAR (1993)] and four of them died. For decontamination seven houses had to be pulled down, and a large quantity of contaminated ground had to be carried off, in all producing 3100 m^3 of radioactive wastes. Several more accidents happened in Lilo, Georgia (USA) [IAEA (2000b)], and in Yanango, Peru, in 1999 [IAEA (2000c)].

2.7. Danger of Contamination by Criminal Dealing in Radioactive Material

Since the beginning of the 1990s illegal dealing in radioactive substances has strongly increased. From 1984 to 1991, in Germany alone, more than 150 cases were registered [Baumann et al. (1995)], [Becker et al. (1995)]. Apart from the increasing danger that radioactive substances, as well as highly enriched uranium and plutonium, might fall into the hands of terrorists, the risk increases that by improper transport, package, and disposal radioactive material might be released into the environment, leading to contamination and irradiation of the general population.

In March 1992 in Germany at Augsburg, 1.1 kg of uranium pellets enriched from 2.2 % to 3.2 %, and four months later, 320 g of highly pure uranium oxide were confiscated. In December 1992, at Holzkirchen near Munich, a ^{137}Cs source at an activity of 120 MBq was seized, in August 1993 in Helsinki, 2 MBq of ^{252}Cf, in March 1994, uranium-pellets and uranium-powder at an ^{235}U-enrichment of 2 %, 6 %, and 88 %, and in July 1994, 900 g of uranium-pellets enriched to 3.3 to 4.4 % of ^{235}U. The limit was reached, when at the Munich airport 390 g of weapon graded plutonium at a percentage of 87 % of ^{239}Pu (0.8 TBq) were confiscated. In October 1994, three thieves stole a metal container from a radioactive waste repository near Tammiku in Estonia, containing a radioactive source [IAEA (1998h)], which probably was sold on the black market.

The increase of nuclear criminality certainly has been forced by the decline of the Soviet Union, since after breaking asunder, in the individual successor states radioactive material has been insufficiently and unreliably supervised because of financial difficulties. By careful analysis of the isotopic composition (finger print) the origin of the nuclear fuels could be established with high probability [Koch (1995)] and the results indicated a provenance from the former Eastern block countries.

Chapter 4

APPLICATION OF ENVIRONMENTAL RADIONUCLIDES IN RADIOCHRONOLOGY

Jan Košler
Charles University, Faculty of Science, Department of Geochemistry, Mineralogy and Mineral Resources. CZ – 12843 Prague 2, Czech Republic

Jan Šilar
Charles University, Faculty of Science, Department of Hydrology, Engineering Geology and Applied Geophysics. CZ – 12843 Prague 2, Czech Republic

Emil Jelínek
Charles University, Faculty of Science, Department of Geochemistry, Mineralogy and Mineral Resources. CZ – 12843 Prague 2, Czech Republic

Since the discovery of radioactivity by Henri Becquerel in 1896, naturally occurring radionuclides (Chapt. 2) have greatly contributed to our ability to measure the ages of terrestrial and extra-terrestrial rocks and minerals, groundwater reservoirs, and biological samples, such as bones or made-made objects. Radiochronology utilizes the radioactive decay of terrestrial nuclides to their progeny atoms, as well as the decay of nuclides produced by cosmic radiation, to derive the ages of natural and man-made objects and the time and duration of processes in nature. Since the publication of the first radiometric age determination by Ernest Rutherford in 1904, radiochronology has grown into a separate discipline of science that is integrated into isotope geology, cosmochemistry, hydrogeology, archaeology and the environmental sciences.

Radiometric dating brought new insight into the dating methods used in geology and archaeology. Both the relative and absolute methods of dating are used in these sciences. The different methods and concepts of time are discussed in Part 4.1.

R. Tykva and D. Berg (eds.), Man-Made and Natural Radioactivity
in Environmental Pollution and Radiochronology, 147-271.
© 2004 *Kluwer Academic Publishers. Printed in the Netherlands.*

1. THE GENERAL CONCEPT OF RADIOCHRONOLOGY

Jan Košler

The basic equation for radiogenic dating is derived from Eq. 12, Chap. 1 with $\lambda_B=0$:

$$N_B = N_A \left(e^{\lambda_A \cdot t} - 1\right) \tag{1}$$

The radiogenic growth of a stable progeny isotope D^* ($\lambda_D=0$) by decay of its parent radionuclide P with decay constant λ_P in a closed system is described as:

$$N_{D^*} = N_P \cdot \left(e^{\lambda_P \cdot t} - 1\right) \tag{2}$$

This equation is valid for a normal (past→present→future) time scale. However, an inverse time coordinate with time running from the present to the past is often used in radiogenic dating. This definition of the time axis is reflected in the decay equation as

$$N_{D^*} = N_P \cdot \left(e^{\lambda_P \cdot (t-\tau)} - 1\right) \tag{3}$$

where $\tau=0$ means the present, t corresponds to the time elapsed since the closure of an isotopic system (decay time), and N_{D^*} and N_P refer to the number of progeny and parent atoms currently present in the sample, respectively.

If the sample at the time of isotopic closure contained a certain proportion of the progeny atoms $N_{D_{init}}$, Eq. 3 must be modified to

$$N_{D^*} = N_{D_{init}} + N_P \cdot \left(e^{\lambda_P \cdot t} - 1\right) \tag{4}$$

It is common practice that Eq. 4 is normalised to the number of atoms of a stable isotope of the same element in the sample, D_{stab}:

$$\frac{N_{D^*}}{N_{D_{stab}}} = \frac{N_{D_{init}}}{N_{D_{stab}}} + \frac{N_P}{N_{D_{stab}}} \cdot \left(e^{\lambda_P \cdot t} - 1\right) \tag{5}$$

The rationale of this normalisation is that it is practical in mass spectrometry to measure the amount of an isotope in the sample relative to the amount of another isotope in the same sample, i.e., as an isotopic ratio, rather than as the absolute number of atoms present. When plotted in $N_{D^*}/N_{D_{stab}}$ vs. $N_P/N_{D_{stab}}$ coordinates, Eq. 5 represents a straight line where $N_{D0}/N_{D_{stab}}$ corresponds to the ordinate-intercept and the term $(e^{\lambda_P \cdot t}-1)$ is the slope. The age t of a sample at the present time is expressed as

$$t = \frac{1}{\lambda_p} \cdot \ln(1 + k) \tag{6}$$

where the slope k is:

$$k = \frac{\dfrac{N_{D^*}}{N_{D_{stab}}} - \dfrac{N_{D_{init}}}{N_{D_{stab}}}}{\dfrac{N_P}{N_{D_{stab}}}}$$

Accordingly, the compositions of a suite of cogenetic samples, i.e., samples of the same age which also have identical initial isotopic compositions and which differ in their parent/progeny element ratios, should lie on a straight line. However, as the isotopic ratios and the element concentrations are determined experimentally in a mass spectrometer, an experimental error is always associated with each data point. The most commonly used method of fitting a straight line to a set of experimental data points is least-square regression, i.e., a method which minimises the squares of the data deviations from a straight line. Experimental errors in Eq. 5 are associated with both coordinates and therefore a two-error regression, which employs weights that are inversely proportional to the squares of the deviations, is usually employed [McIntyre et al. (1966)], [York (1969)], [Brooks et al. (1972)]. In practice, the most common way of assigning errors to data points is to use a blanket error (calculated from the long-term reproducibility of standards) for the x-coordinate, i.e., for the parent/progeny ratio, and the experimental error – calculated from mass spectrometric measurement – for the y-coordinate, i.e., for the daughter isotopic ratio. The errors in the x and y coordinates are usually independent. However, in some radiochronological methods, such as common Pb and U(Th)-Pb dating, the errors exhibit a high degree of correlation (e.g., for N_{207Pb}/N_{204Pb} vs. N_{206Pb}/N_{204Pb} most of the error is due to the low precision of the measurement of the small N_{204Pb} value, which is common to both ratios). Isotopic data with correlated errors must be treated using special statistical techniques [York (1969)], [Rodick et. al. (1987)].

When the data deviations from a straight line are less than, or equal to, the analytical uncertainties, the term isochron has been proposed for the regression line fitted to the isotopic data (see e.g., 7.2, Fig. 22). Numerous statistical methods are used for the calculation of isochrons [Vugrinovich (1981)], [Castorina et al. (1989)], [Kent et al. (1990)], [Kullerud (1991)]. If the scatter of the data exceeds the expected analytical uncertainties, the term errorchron should be used instead [Brooks et al. (1972)]. The excess scatter is usually attributed to a post-crystallization disturbance, initial isotopic inhomogeneity or improper sampling. Various criteria have been proposed for discrimination between isochrons and errorchrons; probably the most commonly used is the value of the mean squared weighted deviation (MSWD) [Wendt et al. (1991)]. This value is calculated as the sum of the squared residuals divided by the number of degrees of freedom, i.e., the number of data points minus two. The MSWD value can be attributed to every regression calculated for more than two data points. If the average scatter of the data equals the predicted analytical scatter, then MSWD=1. This criterion can be used to assess the quality of the regression fit provided that the estimation of the analytical errors is realistic.

2.   RADIOCARBON

Jan Šilar

All materials containing carbon of atmospheric origin with a measurable concentration of radiocarbon can be dated or used for environmental investigation, i.e., for example, wood, charcoal, textiles, flesh, bones, foodstuffs, carbonate sediments, mortar, shells, groundwater, seawater, and other materials.

2.1.   The Origin of Radiocarbon

Radiocarbon is a cosmogenic radionuclide (see Chap. 2, 1.2.2 and 3.4.3). It originates primarily due to the effect of neutrons, produced in the atmosphere by cosmic radiation, on the atoms of nitrogen according to the reaction:

$$^{14}N \ (n,p) \ ^{14}C$$

[Libby (1955)] calculated that 2.6 ^{14}C atoms per second are formed per square centimetre of the Earth's surface (see Chap. 2, Table 3). Taking into account the equilibrium between the ^{14}C produced in the atmosphere and that which decays, he developed the geochronometric dating method known as radiocarbon dating. Libby published the principles of radiocarbon dating in 1955 [Libby (1955)].

It has been generally accepted that, within several minutes or, at most, several hours, virtually all the ^{14}C atoms will have undergone oxidation to form carbon dioxide. However, the exact mechanism(s) involved are not well understood. Natural ^{14}C makes its initial entry to the terrestrial carbon cycle as $^{14}CO_2$ [Taylor (1987)].

In addition to the cosmogenic production of radiocarbon in the atmosphere, *in situ* production has been considered [Zito et al. (1980)], [Taylor (1987)], [Florkowski (1992)], [Lal (1992)]. The approximate radiocarbon production rates on the Earth were summarized by Lal (1992) in Table 1.

The cosmogenic production of ^{14}C in the atmosphere and the processes controlling its subsequent incorporation into the global carbon cycle are considered to be the main factors involved in radiocarbon dating. It has been concluded that, even under the most favourable circumstances, the effect of the *in situ* production of ^{14}C would have been negligible, even for high altitude locations, certainly during the Holocene period and probably for the last 50000 years [Harkness et al. (1974)], [Radnell et al. (1979)], [Cain and Suess (1976)]. For materials with indicated ages in excess of 50000-60000 years, that may contain higher amounts of nitrogen than are found in wood (e.g., bone), the effect might conceivably be noticeable [Taylor (1987)].

Table 1: Approximate radiocarbon production rates on the Earth in a column of 1 m^2 cross section based on [a][Lal (1992)], [b][Lingenfelter (1963)], [c]assuming a mean radiogenic neutron flux of 0.463 neutrons $m^{-2}s^{-1}$.

	^{14}C atoms $(m^{-2}\cdot s^{-1})$
Global average cosmic-ray production in the Earth's atmosphere:	$\approx$ [b]20000
Integrated in situ cosmogenic production rate in a rock exposed at:	
a) 5 km (above sea level)	30
b) 2 km (above sea level)	5
[a]Integrated in situ cosmogenic production rate in the oceans	1
[c]Integrated radiogenic production rate in the crust	1

2.2. Natural Radiocarbon Cycle

Radiocarbon in the atmosphere is oxidized to $^{14}CO_2$, which is assimilated by plants during photosynthesis. Animals consume plants and subsequently radiocarbon reaches all living organic matter in the biosphere (Chap. 2, 1.2.2 and 3.4.3). Due to the disintegration of organic matter in the soil, radiocarbon reaches the soil air and dissolves in soil water, groundwater and in other parts of the hydrosphere. It reaches the shells of terrestrial, fresh-water and marine molluscs and other calcium-carbonate-containing parts of organisms. It also becomes a part of calcium-carbonate sediments precipitated from water. Consequently, radiocarbon, together with the stable carbon isotopes, is incorporated into the atmosphere, biosphere, hydrosphere and lithosphere. The processes and interrelations between the reservoirs are shown in Fig. 1.

All the carbon in the Earth's crust can be considered as being divided into two reservoirs, one dynamic and the other dormant [Broecker et al. (1959)]. The dynamic reservoir consists of the atmosphere, the oceans and other water bodies, and organic matter either living or undergoing decay. The dormant reservoir consists mainly of deposits of coal and oil, limestone and dolomite beds, and the organic fraction of shales. Within the dynamic reservoir, various mixing processes allow the cosmic ray-produced radiocarbon to be distributed to all parts, replacing the radiocarbon lost by radioactive decay.

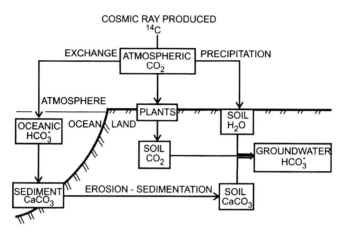

Figure 1:   Processes involving ^{14}C and interrelations between atmosphere, upper lithosphere and hydrosphere, based on [Fontes (1983)].

On the other hand, the dormant reservoir, having had no access to fresh radiocarbon for millions of years, has lost through decay any radiocarbon that was once present.

The parts of the dynamic reservoir have characteristic exchange rates ranging from an order of a few years for the atmosphere (and most of the biosphere) up to 1000 years for parts of the ocean [Taylor (1987)].

Due to different exchange rates, radiocarbon is present in reservoirs in different initial concentrations in modern organic as well as inorganic materials. These have to be taken into account when evaluating the results of radiocarbon dating methods. This factor affects the dating of samples of marine origin as well as the dating of samples which contain carbon of non-atmospheric origin that is free of ^{14}C, e.g., dating of groundwater or Quaternary carbonate fresh-water sediments. Thus, when dating natural materials by using radiocarbon, the effects of the particular carbon cycle processes in the environment and the reservoir have to be considered (see 2.5-2.6, 2.11-2.12 and 4.3).

Detailed analyses of the radiocarbon cycle were published by [Libby (1955)], [Craig (1953), (1954)], [Münnich (1963)], [Suess (1965), (1986)] and in monographs on radiocarbon dating by [Geyh (1971)], [Gupta et al. (1985)] and [Taylor (1987)]. [Bolin et al. (1979)] and [Taylor (1992)] presented a thorough analysis of the carbon cycle. [Levin et al. (2000)] concluded that the radiocarbon lifetime is perfectly suited for dating of carbon pools interacting with the atmospheric the CO_2 reservoir on the time scale of several hundred to several thousand years. [Nydal (2000)] summarized the results of the use of radiocarbon as a tracer in the study of the global carbon cycle, and particularly the exchange of CO_2 between the atmosphere and the ocean.

2.3. Stable Isotope of Carbon ^{13}C and its Behaviour

Stable isotopes and radioisotopes of the same element react to geochemical processes in the same way and to physical processes in a similar physical manner. These, however, are influenced by fractionation. This is the change in the proportion of the particular isotopes of that element in matter during the processes to which it is exposed. The isotopic composition of various components of the biosphere, hydrosphere, and lithosphere is specific and is shaped by environmental processes, especially those linked with a change in the state and/or chemical composition, e.g., evaporation and condensation of water, dissolving and condensation of carbonates, photosynthesis, etc.

Among the stable isotopes of carbon, ^{12}C accounts for 98.9 % and ^{13}C for 1.1 % of all the carbon on the Earth [Craig (1953)]. The abundance of ^{14}C is negligible. The stable isotope ^{13}C reacts in a similar manner as the other carbon isotopes to the processes during the natural carbon cycle to which the organic as well as inorganic matter is exposed. Thus, due to its similar behaviour, its concentration is used to clarify those processes that could influence the results of radiocarbon dating.

The concentration of ^{13}C is expressed with respect to the concentration of a standard sample of calcium carbonate of marine limestone whose isotopic composition is sufficiently stable. The relative difference from the standard sample is called the $\delta^{13}C$ value:

$$\delta^{13}C = 1000 \cdot \frac{R - R_{st}}{R_{st}} [\text{‰}] \qquad (7)$$

where R is the ratio of the mass of ^{13}C to ^{12}C in the investigated sample and R_{st} is the ratio in the standard sample as measured in a mass spectrometer.

The primary world standard for ^{13}C is the calcium carbonate of the Peedee Belemnite limestone of Upper Cretaceous age in South Carolina (PDB standard) [Craig (1953)].

Fractionation of carbon isotopes occurs during biochemical and geochemical reactions and during transition from the biosphere to the hydrosphere or the lithosphere and vice versa. Hence, the carbon in organic and inorganic matter, which has different origins, has characteristic isotopic compositions, and the $\delta^{13}C$ value can clarify the geochemical processes. Geochemical processes could influence the ^{14}C concentration and the results of the radiocarbon dating methods. [Deines (1980)] summarized the processes shaping the isotopic composition of reduced organic carbon in plant materials, humic materials, organic matter of sediments and hydrocarbons.

Living organisms preferentially absorb carbon ^{12}C, so that the carbon in carbon dioxide originating from the disintegration of organic matter has less ^{13}C than that originating from carbonates in limestones of inorganic origin.

Inorganic marine limestones have very constant $^{13}C/^{12}C$ ratios. Some of them are related to the PDB world standard and are used as reference standards for determining the $\delta^{13}C$ value (=0 ‰). On the other hand, organic matter, e.g., plant tissue and matter of biogenic origin, such as CO_2 in soil atmosphere, usually have lower $\delta^{13}C$ values, close to -25 ‰. The carbon in organic matter is relatively enriched in the isotope ^{12}C. Groundwater containing carbon of biogenic and inorganic origin has $\delta^{13}C$ values between 0 and -25 ‰. In normal fresh groundwater these values are between -10 and -18 ‰ [Brinkmann et al. (1960)], [Vogel et al. (1963)]. A thorough review of the behavior of ^{13}C during geochemical processes, during the water-rock interaction, and also in relation to forming of mollusc shells was published by [Burchardt et al. (1980)].

CO$_2$ of inorganic or even magmatic origin in groundwater is manifested by higher ("less negative") or even positive $\delta^{13}C$ values closer to 0. Thus, carbon in bicarbonates in groundwater affected by CO_2 of magmatic origin is isotopically heavier than that in fresh groundwater with atmospheric and soil-produced CO_2 [Šilar (1976)]. Consequently, tufa (travertine) precipitated from thermal water containing magmatic CO_2 has higher $\delta^{13}C$ values in comparison with deposits precipitated from fresh groundwater [Demovič et al. (1972)]. The isotope fractionation between bicarbonate and crystalline carbonate leads to about 2.0 ‰ enrichment in ^{13}C in the carbonate. Consequently, the precipitated calcite will have a somewhat higher ^{13}C content than the dissolved inorganic carbon in water [Burchardt et al. (1980)].

Normalization of ^{14}C by ^{13}C

Both isotopes, ^{13}C and ^{14}C, are exposed to fractionation. They react similarly and the $\delta^{13}C$ values can be used to determine the ^{14}C fractionation in order to eliminate its effects in ^{14}C dating. This is carried out by normalizing, i.e., by adjusting the ^{14}C input values using the $\delta^{13}C$ value of the measured samples. Normalization by $\delta^{13}C$ decreases the systematic errors in ^{14}C dating originating either due to isotopic fractionation, which is conditioned genetically, or due to the chemical processing of samples [Geyh (1971)].

The fractionation of ^{14}C is about double that of ^{13}C [Craig (1954)]. Because wood was used as the standard in many laboratories before the need for normalization was generally accepted, it has been agreed to normalize all samples to be dated by a factor of -25 ‰ [Olsson (1991)].

It is desirable to report the $\delta^{13}C$ values relative to the PDB standard. By convention, the ^{13}C isotopic fractionation in all samples, irrespective of environment, is taken into account by normalizing with respect to the PDB value of -25 ‰. This is the postulated mean value for terrestrial wood. The normalized ^{14}C sample activity A_{sn} relates to the measured sample activity A_s as follows fom the equation [Stuiver et al. (1977)]:

$$A_{sn} = A_s \cdot [1 - \frac{2 \cdot (25 + \delta^{13}C)}{1000}] \qquad (8)$$

where $\delta^{13}C$ is measured or estimated in per mill relative to PDB.

The normalization takes into account the processes to which the carbon of the dated sample was exposed when being incorporated into the sample material and it reflects the analogous behaviour of ^{14}C and ^{13}C.

There is considerable evidence that the ^{14}C variations in land plants, plant-derived matter and animals are due to the photosynthetic cycle used to fix carbon. These differences are related to three photosynthetic cycles: The Calvin-Benson cycle (C), the Hatch-Slack cycle (HS), and the Crassulacean acid metabolism cycle (CAM). Whether or not a $\delta^{13}C$ correction is required, it is necessary to establish which photosynthetic cycles are utilized by the plants [Sheppard (1975)].

Most plants in the temperate regions utilize the Calvin-Benson cycle [Lerman (1972)] and have $\delta^{13}C$ values of about -27 ‰ and the resulting age corrections are low. The $\delta^{13}C$ values and resulting corrections are given for other materials in Table 2. The $\delta^{13}C$ normalization corrections have low values for dating of wood and charcoal (the most frequent materials in archaeological dating). Thus, such corrections are sometimes considered unnecessary because the corrections are smaller than their respective errors [Sheppard (1975)]. However, when dating Hatch-Slack and CAM plant material and bones, the $\delta^{13}C$ corrections are recommended.

Table 2: Correction of radiocarbon ages for isotopic fractionation $\delta^{13}C$: Calvin-Benson cycle (C), the Hatch-Slack cycle (HC), Crassulacean acid metabolism cycle (CAM) [Lerman (1972)]: [a] Nordic and temperate peats, humus and soils, [b] Wheat, oats, barley, rice, rye, etc., and other related grasses, [c] Maize, sorghum, millet, panic grass, etc., and related grasses, [d] Cactus, agave, pineapple, Tillandsia, etc.

Plants	Photosynthetic Pathway	$\delta^{13}C$ (‰)	Age Correction (y)
Wood and wood charcoal	C	- 25 ± 5	0 ± 80
Tree leaves	C	- 27 ± 5	- 30 ± 80
Peat, humus, soil[a]	C	- 27 ± 7	- 30 ± 110
Grains[b]	C	- 23 ± 4	+ 30 ± 60
Leaves, straw of grasses & sedges (totora (cat-tail))[b]	C	- 27 ± 4	+ 30 ± 60
Grains[c]	HS	- 10 ± 3	+ 240 ± 50
Leaves, straw of grasses (e.g., papyrus)	HS	- 130 ± 4	+ 200 ± 50
Succulents[d]	CAM	- 17 ± 8	+ 130 ± 120

The $\delta^{13}C$ values in the organic matter of lake sediments are also significant in studies of past climate and environmental changes in the continents. They reflect (together with $\delta^{18}O$) the changes in the temperature in the environment. Fig. 2 depicts age corrections in relation to $\delta^{13}C$ in nature.

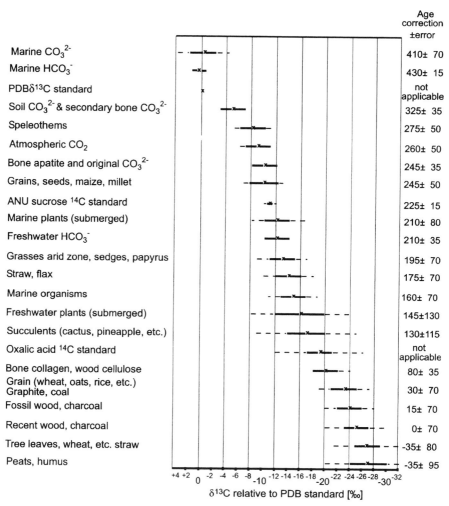

Figure 2: ^{13}C variations relative to Peedee Belemnite limestone standard (PDB) and the age corrections in different samples, based on [Stuiver and Polach (1977)].

2.4.   Open and Closed Systems and Dynamic Equilibrium of ^{14}C

Due to simultaneous origin and disintegration, a dynamic equilibrium of radiocarbon has been established in the atmosphere and also, more or less, in open systems that are in contact with the atmosphere (i.e., in living organisms). In closed systems, in which the recharging of radionuclides has been terminated (e.g., in dead organic matter), the radionuclide content decreases because of continuous disintegration. As the radionuclide content decreases, the specific activity of matter within a closed system decreases.

2.5.   The Concept of the Radiocarbon Dating Method

If the supply of radiocarbon from the environment into carbon-containing matter is interrupted (e.g., to a living organism by its death), then the elapsed time since that moment can be calculated from the ^{14}C concentration at the present time, the estimated concentration at the moment of interruption, and the half-life of radiocarbon. This is called the radiocarbon age of that matter.

Radiocarbon dating is based on the following assumptions given by [Libby (1955)]:

– If the cosmic radiation has remained at its present intensity for 20000 or 30000 years, and if the carbon reservoir has not changed appreciably over that time, then, at the present time, there exists a complete balance between the rate of disintegration of radiocarbon atoms and the rate of assimilation of new radiocarbon atoms for all the material in the life cycle.

– When death occurs, however, the assimilation process is abruptly halted, and only the disintegration process remains.

– The materials contain the original carbon atoms present at the time death occurred.

If all of these assumptions were valid it would follow that, under natural conditions:

– The specific radiocarbon activities of all living organic matter which is in contact with atmosphere would be equal,

– The original radiocarbon activity of organic matter at the time of its death could be replaced by the radiocarbon activity of living organic matter, i.e., by the activity of a modern standard sample.

[Taylor (1987)] has summarized the assumptions of radiocarbon dating as follows:

– The concentration of ^{14}C in each carbon reservoir has remained essentially constant over the ^{14}C time scale.

- There has been complete and rapid mixing of ^{14}C throughout the various carbon reservoirs on a global basis.
- The carbon isotope ratios in samples have not been altered except by ^{14}C decay since these sample materials ceased to be an active part of one of the carbon reservoirs (e.g. on the death of an organism).
- The half-life of ^{14}C is accurately known.
- The natural levels of ^{14}C can be measured to appropriate levels of accuracy and precision.

The natural radiocarbon cycle was affected by the testing of nuclear weapons after 1950, which introduced large amounts of anthropogenic radiocarbon into the environment and subsequently placed limitations on the radiocarbon dating method. Samples that originated after 1950 are considered to be "modern" and their radiocarbon ages are related to the year 1950 in terms of "years B.P.", (before the present), where "the present" is to be understood to be the year 1950.

2.6.   The Radiocarbon Age

Due to the decay of radiocarbon, the original initial specific activity A_0 (see Fig. 3 and Chap. 1, Eq. 4) decreases to A after time t_r according to

$$A = A_0 \cdot e^{-\lambda \cdot t_r} \qquad (9)$$

where $\lambda = ln\ 2/T_{1/2}$ is the radioactive decay constant.

If A_0, A and the half-life $T_{1/2}$ of radiocarbon is known, it follows that the time t_r, which has elapsed since the interruption of the recharge of radiocarbon, equals

$$t_r = \frac{T_{1/2}}{ln\ 2} \cdot ln\frac{A_0}{A} = 8033 \cdot ln\left(\frac{A_0}{A}\right) \qquad (10)$$

when the Libby ^{14}C half-life of 5570 years is used.

This t_r value is the radiocarbon age of the sample. Samples as old as approximately 50000 years can be dated. For radiocarbon dating purposes, "by international agreement, conventional ^{14}C age is calculated using the Libby half-life of 5570±30 years" [Kra (1986)]. Radiocarbon age calculated using the 5730±40 year half-life [Godwin (1962)] is 1.03 times higher than that calculated using the Libby half-life.

The radiocarbon dating method consists of comparing the initial specific ^{14}C activity A_0 of the dated sample and its present specific activity A (Eq. 10). The non-measurable A_0 is replaced by that of a modern standard sample (see 2.6.2). Therefore, the measurements necessary for dating include measurement of the activity of the standard sample, of the investigated sample and, in addition, measurement of the activity of the background of the measuring device, i.e., of a non-active or "dead" sample (see Chap. 5, 1.2). If all three measurements are performed in the same device under the same conditions, it is not necessary to calculate the specific activities in absolute units. Instead, counts per unit of time (usually counts per minute, cpm) can be used for evaluation (see Chap. 5, 1.1).

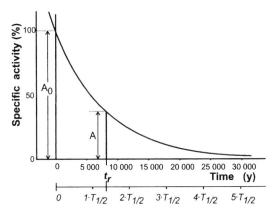

Figure 3: Decay of radiocarbon: the initial ^{14}C specific activity A_0, the final ^{14}C specific activity A and the radiocarbon age t_r of the sample. A_0 and A are related to the modern standard activity and are given as a percent of modern carbon (pmc).

2.6.1. Initial Radiocarbon Activity of the Sample

The initial radiocarbon activity of a sample resulted from the dynamic equilibrium between the supply of radiocarbon from the atmosphere to the sample and its decay. This is the activity the sample had at the moment it changed from an open system to a closed system. This occurred in the case of an organic sample at the time of its death, in the case of groundwater, at the time of its recharge and, in the case of Quaternary calcium carbonate deposits, at the time of their precipitation from water.

However, the production of radiocarbon in the atmosphere in the past was not constant. [Libby (1955)] admits the possibility of variations in cosmic radiation in the past, contrary to the original assumption that the ^{14}C formation rate has been constant. These variations were recognized by Suess (1965). They have resulted in fluctuating initial activity in living organic matter over the millennia and have influenced the results of radiocarbon dating.

When dating samples containing inorganic carbon, a lower initial radiocarbon activity has to be adopted compared to that used for samples of organic carbon, and the resulting radiocarbon ages have to be corrected (see 2.7 and 4.4).

The radiocarbon concentration in the shells and conches of molluscs is influenced by its concentration in the bicarbonates in the water in which the molluscs live. This results in apparent ages being calculated that are several hundred years too old (the "reservoir effect", see 2.2).

The initial concentrations of radiocarbon in organic as well as inorganic samples and their resulting radiocarbon ages may also be influenced by primary contamination by allochthonous carbon, e.g., by CO_2 of magmatic origin or derived from geochemical processes (see 4.3). This influence must be taken into account when evaluating the results of dating. Primary contamination by allochthonous grains of calcium carbonate should be checked when dating mortar. This, however, should be distinguished from secondary contamination of samples by the roots of plants, improper packing of samples, admixture of foreign material, etc., which is reflected in the final radiocarbon concentration and which can often be avoided (see 2.11).

2.6.2.   Standard Modern Radiocarbon Activity

Modern radiocarbon activity derived from a conventionally accepted standard sample is used as a substitute for the initial radiocarbon activity of the organic matter to be dated. The errors inherent in this assumption should be compensated for as follows:

- The fluctuation in the ^{14}C activity of the atmosphere is corrected by dendrochronological calibration of the resulting ^{14}C ages.
- The reservoir effect is accounted for by a correction to the ^{14}C age.
- The geochemical processes influencing the dating of groundwater and Quaternary calcium carbonate are taken into account by empirically or analytically determining the initial activity from the modern standard activity.

The international accepted value of the standard modern radiocarbon activity is 95% of the activity of the NBS (National Bureau of Standards) oxalic acid in Anno Domini (A.D.) 1950 [Olsson (1970)]. It corresponds to wood grown in A.D. 1890 not yet contaminated with CO_2 whose activity was corrected for ^{14}C decay to correspond to A.D. 1950 [Gupta et al. (1985)]. The NBS Oxalic acid (Ox) standard activity multiplied by 0.95 is accepted as the primary reference standard for modern radiocarbon activity. Unfortunately, Ox is no longer available; therefore, the NBS New Oxalic acid (NOx) standard must be introduced as an international standard. The Australian National University (ANU) Sucrose standard is an internationally accepted secondary standard. The Chinese Charred Sucrose (Ch-Suc) standard is a national secondary standard. Primary and secondary ^{14}C reference standard correction ratios are given in Table 3.

The Ox standard is normalized to $\delta^{13}C=-19\%o$ with respect to the PDB limestone standard, the NOx and the ANU-Suc standards are normalized to $\delta^{13}C=-25\%o$ and the Ch-Suc standard is not normalized. These secondary standards are related to the primary reference standard of modern radiocarbon activity by means of the pertinent correction ratios.

Table 3: Correction ratios of primary (Ox, $\delta^{13}C=-19‰$) and secondary ^{14}C reference standards (NOx, $\delta^{13}C=-25‰$, ANU, $\delta^{13}C=-25‰$ and Ch-Suc - not normalized, $\delta^{13}C=0‰$).

Reference standard	Correction ratio	References
NOx/(0.95·Ox)	1.3407 ± 0.001	[Mann (1983)]
ANU-Suc/(0.95·Ox)	1.5081 ± 0.002	[Currie et al (1980)]
Ch-Suc/(0.95·Ox)	1.3620 ± 0.003	[Qui Xou Hua et al.(1983)], [Polach (1989)]

In their calculations, most laboratories use an initial activity value A_{ON} that is 95 % of the measured net oxalic acid activity (counting rate) A_{Ox} today, normalized for ^{13}C fractionation according to [Stuiver et al. (1977)]:

$$A_{ON} = 0.95 \cdot A_{Ox} \cdot \left(1 - \frac{2 \cdot \left(19 + \delta^{13}C\right)}{1000} \right) \qquad (11)$$

2.6.3. Final ^{14}C Activity of the Sample

The ^{14}C activity A of the sample is measured using one of the techniques mentioned in Chap. 5, 1. The value of the activity of the sample should be normalized according to its $\delta^{13}C$.

The $\delta^{13}C$ correction should be carried out if a $\delta^{13}C$ value is available for the sample. Otherwise, the resulting ^{14}C ages can be corrected according to Table 2. The $\delta^{13}C$ corrections are indispensable when dating materials containing inorganic carbon, such as calcium carbonate sediments, shells and groundwater.

2.7. Calculation of the Radiocarbon Age

The radiocarbon age t_r can be calculated according to Eq. 10. The assumed initial activity A_0 corresponds to A_{ON} (Eq. 11), i.e., to the modern radiocarbon standard If an inorganic sample (e.g., groundwater or a Quaternary carbonate sample) has an initial activity lower than A_0 because of the dissolving of ^{14}C-free calcium carbonate in soil and rocks, then a correction factor $C<1$ can be used to adjust Eq. 10:

$$t_r = \frac{T_{1/2}}{\ln 2} \cdot \ln \frac{C \cdot A_0}{A} = 8033 \cdot \ln \frac{C \cdot A_{ON}}{A} \qquad (12)$$

C expresses the proportion of the "true" initial ^{14}C activity of the sample as a percentage of the modern ^{14}C activity.

2.8.   Precision of the Radiocarbon Age

By convention, the standard deviation σ_{tr} (Chap. 5, 1.7) of a radiocarbon age includes only counting errors related to the stochastic nature of radioactive decay. As given by the general formula (Eq. 10 and 11), the calculation of the radiocarbon age can be expressed using the standard deviation [Libby (1955)]

$$\sigma_{t_r} = \frac{T_{1/2}}{\ln 2} \cdot \sqrt{\left(\frac{\sigma_A}{A}\right)^2 + \left(\frac{\sigma_{A0}}{A_0}\right)^2} \qquad (13)$$

where σ_A and σ_{A0} are the standard deviations of activities A and A_0, respectively.

A and A_0 are the activities and result from the measurement of the count rates of the investigated and standard samples, respectively, from which the background B of the counter has been subtracted. Thus, σ_A and σ_{A0} include σ_B.

σ_{tr} means that there is a 68 % chance that the radiocarbon age will be within the range $[t-\sigma_{tr}]$ and $[t+\sigma_{tr}]$. It should be emphasized that σ_{tr} is not an outside limit for the radiocarbon age. Use of the 95 % ($2\sigma_{tr}$) confidence limits [Sheppard (1975)] (see Chap. 5, Table 9) is a statistically better approach to the reporting of radiocarbon age errors.

2.9.   Limits of Radiocarbon Dating

In a particular radiocarbon dating laboratory, the limits of dating are determined by the capability of the measuring device to distinguish the ^{14}C sample activity from the activity of the background and from the modern standard activity.

The maximum detectable age is determined from the minimum detectable ^{14}C activity of the sample, which can be distinguished statistically (Chap. 5, 1.7) from the mean activity of the background. Using a 2σ detection criterion corresponding to the 95 % statistical confidence level, the maximum detectable age in years, t_{max}, is [Gupta et al. (1985)]

$$t_{max} = 8033 \cdot \ln\left[\sqrt{\frac{T}{8}}\right] + 8033 \cdot \ln\left[\frac{A_{ON}}{\sqrt{B}}\right] \qquad (14)$$

where T is the counting time in minutes assumed to be equal for each sample, background B (cpm), and modern carbon standard A_{ON} (net cpm).

The minimum detectable age in years is determined from the detectable ^{14}C activity of the sample, which can be distinguished statistically from the modern reference standard. Using a 2σ detection criterion corresponding to the 95 % statistical confidence level, this is given by:

$$t_{min} = -8033 \cdot ln\left[1 - 2\sqrt{\frac{2}{A_{ON} \cdot T}} \right] \qquad (15)$$

or, using 68 % confidence level (1·σ deviation) [Gupta et al. (1985)]:

$$t_{min} = -8033 \cdot ln\left[1 - \sqrt{\frac{2}{A_{ON} \cdot T}} \right] \qquad (16)$$

In general, with conventional decay counting, the practical limitations imposed by the sample sizes generally available from archaeological sources and the problem of removing contamination reduce the maximum range in the vast majority of samples to about 50000 years. The minimum detectable radiocarbon age is about 250 years at the 95% confidence level [Sheppard (1975)].

2.10. Basic Conditions Affecting Radiocarbon Dating

The quality of the results of radiocarbon dating in general depends on the validity of the assumptions (see 2.5). It also depends on the precision and statistics of the measurements, i.e., on the measuring techniques used, and on the circumstances that influence the statistical evaluation.

The radiocarbon age may not be identical with the true age of the sample. It merely expresses a time relationship between the initial activity A_0 and the present activity A. Both, A_0 and A, may have been influenced by different natural processes, for instance isotopic effects or exchange of carbon between the "dead" sample and the environment at any time (the ^{14}C building rate not constant, this is not an ideal closed system). Thus, radiocarbon dating should not be considered to be absolute. In contrast to radiocarbon dating, the dendrochronological or varve dating method really counts the number of sideric years according to yearly events reflected in wood or in glaciofluvial sediments, respectively.

If the true initial activity of the sample is less than the activity of the modern carbon standard A_{ON}, then a greater age results from Eq. 12 and vice versa.

The fluctuation of the solar activity results in initial A_0 values that are different from A_{ON}. De Vries (1958) proved the fluctuation of ^{14}C concentration on a global scale. This aspect has been studied in detail and it is now possible to correct radiocarbon ages by calibration (see 2.13).

$A_0<A_{ON}$ may be a result of dilution of the cosmogenic radiocarbon in the atmospheric CO_2 by CO_2 of volcanic origin [Chatters et al. (1969)] or postvolcanic origin [Bruns et al. (1980)]; e.g., living grass around CO_2 vents in a mineral-spring region in western Bohemia has exhibited a ^{14}C activity corresponding to 11.7 % of the modern carbon standard and an apparent ^{14}C age of about 17000 years before the present (B.P.) [Šilar (1976)].

In samples of inorganic substances (e.g., samples of Quaternary carbonate sediments or groundwater), the initial ^{14}C can be diluted due to the admixing of non-active ("dead") carbon from dissolved calcite from soils and rocks. If the dated substance contains a secondary admixture of non-active carbon, then its final activity *A* will have decreased and its radiocarbon age will be greater than the true age. On the other hand, the radiocarbon age will be less if a secondary admixture of modern carbon has increased the present activity *A* of the sample. Both cases can occur due to secondary contamination, which can greatly influence the results of dating. Radiocarbon groundwater ages can be influenced by mixing of groundwaters of different origins.

2.11.   Sample Collection, Pre-treatment and Storing

The method of dating with ^{14}C has extensive use in archaeology, Quaternary geology, hydrology, oceanography, environmental studies, and technology (see 3). Due to the different types of samples like wood, charcoal, textiles, flesh, bones, foodstuffs, carbonate, mortar, shells, groundwater, seawater and different laboratory procedures, the collector should be familiar with the requirements of the laboratory which will analyse the samples, and he should also be aware of the purpose of the dating.

The ^{14}C dating procedure starts with collecting the sample in the field. Its amount must be large enough to be processed and measured in the measuring device of the particular laboratory. It has to be collected and stored in a manner that prevents contamination. It must also fulfil other requirements in order to provide the expected geochronological information.

Errors in ^{14}C dating due to contamination were first reported and analysed by Johnson (1955) who has shown that it is essential to exercise great care in cataloguing and storing samples for measurement. Samples, which have been stored for a long time in museums, may be contaminated due to the activity of microorganisms, to dust, to other contaminants or to mixing of different samples. The potential for contamination by root hairs, moulds and similar factors should be recognized by the collector in the field and the samples should be subjected to specialized cleaning and testing in the laboratory.

The way in which samples are collected, handled and reported can greatly influence the results of dating and interpretation. Some basic principles should be observed to avoid errors.

2.11.1. Sampling Record

From the time of sampling, every sample should be accompanied by a description indicating the:
  – Collector's sample number
  – Date of collecting

- Location of the sampling site (topographic description and coordinates)
- Type of material
- Name of the collector
- Name and address of the submitter
- Date of submitting to the ^{14}C dating laboratory
- Sample size, weight and description
- Archaeological and/or geological description of the sample
- Archaeological and/or geological description of the sampling site, and of the stratigraphic position of the sample, accompanied by a drawing and/or photograph
- In the case of groundwater samples, basic hydrogeological, geochemical and technical information on the borehole or spring
- Remarks on potential contamination by roots, by humic acids or by alteration of the mineralogical calcium-carbonate species
- Estimated age
- How the sample was collected and treated
- Whether more samples are available, if required.

After submission to the ^{14}C laboratory, the sample is given a laboratory sample number which accompanies it during the entire laboratory treatment.

After the ^{14}C measurement and calculation of the ^{14}C age, the result of dating is given a code (identical with that of the laboratory) and measurement number. The code and measurement number should be used in the dating reports, should be added to the sampling record and should be referred to when publishing the results of dating.

2.11.2. Representativeness of Samples

Samples should be representative of the strata or events to be dated. When selecting samples, the true meaning of the date that will be obtained and how this may be tied to the problem being investigated should be considered.

Wood samples: The carbon in wood was fixed at the time the plant grew and each tree ring is of different age. To determine the time of death of the tree, the outermost rings as well as small twigs and branches are most suitable for dating. The resulting date of death of the tree may not be identical with the time the wood was used for construction, artwork, as fuel, or the time the tree trunk was deposited in sediments or stranded as driftwood. Tree trunks found in alluvial deposits in river plains may be older than the layer in which they were found because they may have been redeposited and buried later [Růžičková et al. (1993)], [Jílek et al. (1995)]. Even wood of autochthonous trunks found *in situ* and standing upright may be older than the organic matter of layers at the same level within which the trunks were buried.

Charcoal samples: It should be considered whether contemporaneous or older wood was used as fuel. Old wood and driftwood of unknown origin could have been used where wood is scarce. Charcoal found at fireplaces along a coast may have originated from old driftwood brought from remote places. The ^{14}C age of charcoal from a fire pit where peat was used does not date the fire, but is older.

Shell and calcium-carbonate samples: If the shells were obtained from living animals (e.g., collected as food), their ^{14}C age will correspond to the time of collection. Marine shells show somewhat higher ^{14}C ages than contemporaneous land plant materials due to the reservoir effect (see 2.2, 2.6 and 4.3). Fresh water shells may show high ^{14}C ages if non-active carbon from rocks becomes a part of bicarbonates dissolved in water. Shells collected from beach litter to be used for jewellery may be older than the culture that used them.

The mineral composition of shells and other marine carbonates (e.g., corals) varies from pure aragonite to pure calcite, with many species contributing a mixture of the mineral species. Recrystallization of fossil carbonate material into young material may occur readily. Recrystallized material is often only calcite, thereby affording a measure of recognition [Gupta et al. (1985)]. The reprecipitated carbonate exhibits a calcite structure with a different isotopic composition, as influenced by exchange with bicarbonates in the immediate environment [Taylor (1987)]. This may be reflected in the powdery surface of the shell. Consequently, shells with fresh shiny surface should be preferred when collecting samples for dating Quaternary deposits. When in doubt, the potential degree of recrystallization of the shell calcium carbonate should be determined during the subsequent laboratory pre-treatment by x-ray diffraction and infrared spectroscopy and by light and scanning electron microscopy [Vita-Finzi (1980)] in order to remove the secondary calcite.

Any processes and circumstances that may have changed the position of the sample, its composition and its relationship to the dated event should be taken into account.

Fragments of wood and charcoal may have been redeposited due to soil creep or human activities. Wooden artefacts may have been manufactured of old wood. The ^{14}C age of the tissue of mummy wrappings and the style of the embalming techniques of an Egyptian mummy may not correspond to the historical style and artistic pattern [Šilar (1979)] (see 4.2). The origin and age at the time of use must be considered when dating ancient writing materials such as paper, parchment and textiles [Burleigh et al. (1983)]. These may have been manufactured of old wood or of old textiles or may have been reused.

Such phenomena should be considered and thus the samples should be collected by trained archaeologists, historians and geologists able to analyse the sample origin. Records accompanied by sketches and photographs should be made at the time of field sampling in order to identify the position of the samples in relationship to the stratigraphy of the site and to the history and events studied.

Short-lived organic matter (e.g., grain, grass, thin branches in fire-places, textiles, and flesh) has a higher information value than long-lived wood, e.g. thick beams used for construction purposes.

Perhaps the most important generalization is as follows [Sine (1964)]: We most often try to date an event by dating materials known to be associated physically with that event. It is only by the most careful field work that we can be sure that the material was formed contemporaneously with the event, and not merely used in conjunction with the event, having existed for many years previously.

2.11.3. Sampling Considering the Reservoir Effect

The reservoir effect (see 2.2 and 4.3) should be considered when dating the calcium-carbonate of mollusc shells and Quaternary tufa deposits. Thus, modern reference samples of the same material should be collected at the same time. Because of atom bomb-produced radiocarbon, samples collected before 1950 provide the best information on the reservoir effect [Mangerud et al. (1975)]. The geological position of the sampling site and the circumstances that could potentially affect the ^{14}C age of samples (e.g., the origin of water from which a tufa deposit precipitates) should be recorded.

2.11.4. Sampling of Groundwater

Only well-equipped boreholes with reliable logs and well-documented springs should be used for sampling to identify aquifers and to avoid the mixing of waters. An aquifer is a body of rock that is sufficiently permeable to conduct groundwater. The chemical composition of the water and the hydrogeology of the site should be known. A record should be kept on the sampling procedure and the circumstances that could have influenced the ^{14}C concentration in the bicarbonates of the groundwater.

The most frequently used technique for extracting carbon is to precipitate bicarbonates and carbonates in the field in the form of $BaCO_3$ by adding a solution of about 300 grams of $BaCl_2$ in 1000 ml of water to a volume of about 60 to 100 litres of water which has previously been alkalized to pH>8. Flocculants are added to facilitate the precipitation. Special plastic containers with a funnel-shaped bottom can be obtained from IAEA for this purpose. A 1000 ml plastic bottle is screwed onto the opening at the funnel bottom to collect the $BaCO_3$ precipitate. Clean drums, carboys and other large containers are also suitable. For more detailed information on how to extract carbon from groundwater, see the IAEA instruction [Sine (1981)].

Another technique for extracting carbon from water is to trap carbonates and bicarbonates on a strongly basic ion exchanger as developed by Crosby et al. (1965). Water isolated from the atmosphere flows through two sampling cylinders. The first contains a weakly basic ion exchanger to trap the ions of the strong acids. The second contains a strongly basic ion exchanger to trap the bicarbonates. To sample calcium bicarbonate-type water, the method was simplified by filling both cylinders with a strongly basic ion exchanger, as it is unnecessary to eliminate the ions of strong acids [Šilar (1976)]. The ion-exchanger sampling technique has the advantage of decreasing the risk of contamination by atmospheric CO_2 but a high enough hydraulic head must be available (e.g., using a pump) to ensure that the sampled water flows through the sampling cylinders.

In the case of water with very low dissolved solids or with a high concentration of sulphates, the bicarbonates and carbonates can be extracted by acidification and trapping the evolved CO_2 in NaOH solution [Morgenstern et al. (1990)].

Samples of groundwater for all isotopic analyses of stable isotopes as well as radionuclides should be taken at the same time and together with additional samples for chemical analyses. Field measurements of the basic data required for the evaluation of the results (pH, temperature, conductivity, etc.) should be carried out simultaneously.

2.11.5. Sources of Carbon and of ^{14}C

The sources of carbon and of ^{14}C in the sample should be known and potential anomalies in the initial ^{14}C concentration due to primary contamination, e.g., by magmatic CO_2, should be taken into account. The same applies to samples of calcium carbonate precipitated from groundwater containing dissolved magmatic CO_2. Such situations can be clarified by analysing the sampling site geologically, by taking unaffected reference samples from beyond the potential reach of magmatic CO_2 vents and by analysing the ^{13}C composition of the sampled material.

2.11.6. Sampling and Contamination

When collecting samples of a layer of organic sediments like peat, bog with wood, etc., secondary contamination by roots of living plants and by humic acids from the upper layers must be expected. Thus, different fractions of the sampled material should be separated mechanically during sampling and the contaminating matter (rootlets, insects, etc.) removed while the sample is fresh. The chemical fractions (humic acids and cellulose) are separated later during laboratory processing.

When collecting carbonate samples (tufa, travertine or mortar), the potential contamination with allochthonous non-active calcium carbonate should be examined by petrological analysis in order to select a suitable pretreatment technique.

2.11.7. Transport of Samples

Secondary contamination of samples during transportation should be avoided. Samples should not be wrapped in paper or other organic tissue but should be placed in double plastic bags with a label in between, in plastic bottles or in metallic foil.

2.11.8. Storing of Samples

Samples should be properly stored to avoid secondary contamination by microorganisms. Aerobic bacterial activity results in lower radiocarbon age while anaerobic activity does not affect the determination. Saprophytic moulds growing on a sample do not affect the dating, as they live on the sample's substance [Geyh (1971)], [Gupta et al. (1985)]. Contamination by algae should be prevented as this introduces carbon from atmospheric CO_2 into the sample. This danger can be reduced by drying, but this makes further processing of the sample difficult.

Storing samples under their natural humidity in a cool dark place facilitates further processing and the separation of humic acids.

Under no circumstances should samples be treated by any organic preservatives. An example where this has occurred are the bones of a hippopotamus recovered from a river terrace at Leeds in England and preserved during the last century. These bones were generally assigned to the last interglacial period, circa 120000 years ago, but have persistently yielded a ^{14}C age of 30000 years and this has raised doubts about the local glacial chronology. Finally, painstaking research revealed that, in all probability, the bones were soaked in gelatine to aid their preservation. The genuine antiquity of the Leeds deposit was eventually confirmed by the proven absence of a measurable quantity of ^{14}C in an unvarnished molar [Harkness (1981)].

2.11.9. Necessary Amount of Sample

The amount of sample required depends on the ^{14}C measuring techniques used. In general, for proportional gas counting or liquid scintillation counting, about 2 to 5 grams of equivalent carbon are required, while for direct accelerator mass spectrometry (AMS) counting, samples of the order of milligrams order are sufficient. Informative values are given in Table 4.

For ^{14}C analyses of water, for instance, about 2 grams of carbon are necessary. The laboratory of the International Atomic Energy Agency in Vienna (IAEA) requires 2.5 grams. This amount, corresponding to about 12.5 grams of bicarbonates, can be extracted from the volume of water [Sine (1981)]

$$V = 12500/[C]$$

where V is the volume of water in litres $[C]$ is the concentration of carbonates and bicarbonates in water in parts per million (ppm).

Table 4: Minimum amount of material (dry weight) to be collected for routine dating by radiometry, after [Gupta et al. (1985)].

Description of suitable materials	Dry weight (g)		Comments
	Minimum	**Desirable**	
Charcoal	2	10 - 20	Rich, black flakes
	5 - 10	50 - 200	Sandy, brown
Shell	20	100	Hard shiny surface
	30	150 - 200	Powdery or pitted
Grass, leaves	5	35 - 50	
Paper, cloth	3	25 - 30	
Peat, Gyttja	30	80 - 120	Dark brown
	60	150 - 200	Light grey-brown
Materials for special projects in collaboration with the radiocarbon laboratory:			
Bone - apatite	500	1000	Up to 20000 yrs B.P.
- carbonate	200	600	Seldom valid date
- collagen	300	800 - 1500	Up to 10000 yrs B.P.
- charred	500	1500 - 3000	Up to 36000 yrs B.P.
Soil - organic matter		2000 - 5000	Humic acid, humin
- carbonate		100 - 500	Pedogenic, nodules

2.11.10. Pre-treatment of Samples

The purpose of pre-treatment of samples is to isolate the fraction of the carbon that is least likely to have been altered [Gupta et al. (1985)], i.e., to exclude the effects of any post-depositional alteration or contamination by foreign carbon. Sample pre-treatment strategies have been summarized by [Taylor (1987)]. Individual laboratories usually use their own standard procedures, which are published as their own manuals or in the journal "Radiocarbon".

Mechanical Pre-treatment

Mechanical pre-treatment is a significant phase in the dating procedures because, if a contaminant has not been removed from the sample during this phase, it cannot be eliminated in the subsequent phases.

Mechanical pre-treatment consists of removing any contaminating particles, such as roots, soil and sand, and separating those parts of the sample which should be dated separately. In the case of shells and bones, the external layer that has been exposed to the environment has to be scratched off. Removal of potential contaminants, which could not be removed in the field, must be completed before homogenization of the sample. The procedures were described, e.g., by [Gupta et al. (1985)]. A simple technique for mechanical pre-treatment of mortar to separate lime for dating from aggregate was developed by [Folk et al. (1979)].

When dating materials which have grown in layers (e.g., the wood of tree rings, layers of sediments or calcium carbonate from dripstones), separating the individual layers increases the precision and authenticity of the dating as it reduces the mutual mixing and contamination of materials with different ages. This is significant when dating marine or lake sediments in which the trend and rate of sedimentation are being determined. Layers of crystalline dripstone in caves should be collected from stalagmites where individual layers can be separated for chemical treatment [Geyh (1971)]. Secondary layers of calcium carbonate or secondary calcium-carbonate cement in an oolithic limestone sample can influence the ^{14}C activity of the bulk sample and hence affect the resulting age [Šilar (1980)]. Thus they should be separated.

For homogenization, wood is disintegrated using a saw or chisel, if necessary, and charcoal and carbonate samples are crushed.

Chemical Pre-treatment

Chemical pre-treatment differs depending on the type of sample. Detailed descriptions of the chemical pre-treatment of samples as practiced in the different ^{14}C laboratories are published in the journal "Radiocarbon".

Pre-treatment of wood and charcoal generally consists of boiling in HCl to remove carbonates and washing until a neutral reaction is obtained, dissolving humic acids with NaOH solution and precipitating with HCl, washing in distilled water until neutral and then drying.

Dating of different organic fractions may result in different radiocarbon ages. Special procedures for pre-treatment and separating different fractions of organic sediments and wood have been presented by [Gupta et al. (1985)] and [Olsson (1986), (1991), (1992)].

For calcium-carbonate samples, no special chemical pre-treatment is necessary. The calcium carbonate is converted to CO_2 by acidification, and is further treated in a standard way, depending on the particular laboratory. When dating speleothems (dripstone), tufa and oolithic sediments, different ^{14}C activities and ^{14}C ages of the individual microscopic layers can be expected, due either to the original layering or to secondary recrystallization and consequent contamination. Thus, different calcium-carbonate fractions of the sample should be chemically treated separately, if possible. Contamination with allochthonous grains of ^{14}C-free calcium carbonate must be expected in tufa and travertine sediments in streams. The kind of chemical pre-treatment should be considered in advance when collecting samples and processing them mechanically.

$BaCO_3$ samples for groundwater dating as well as bicarbonates trapped on an ion exchanger are converted to CO_2 by acidification and then processed in a standard way depending on the particular ^{14}C laboratory. For CO_2 proportional gas counting, particular attention has to be paid to the removal of electro-negative impurities, which sometimes occur in thermal and mineral water, in water contaminated by fertilizers or industrial wastes, and in boreholes or in mines where water may be contaminated due to blasting. One part of the $BaCO_3$ precipitate or ion exchanger should be separated and stored for $\delta^{13}C$ analysis.

Radiocarbon dating of bones is a special problem of radiochronometry, mainly because of the chemical composition of the bones and the geochemical processes to which they are exposed. Bone material is present in many sites where other organics are not present or, if present, have a questionable relationship with an actual or purported cultural expression. It is also an important source of data for a wide range of studies [Taylor (1992)].

Contemporary bone comprises about 80 % inorganic matter, about 18 % collagen, and about 2.5 % other proteins and fats. The inorganic component is basically composed of crystals of calcium phosphate with the structure of hydroxy-apatite, containing small portions of carbonate ions incorporated into the crystal structure [Gupta et al. (1985)]. Because of the possibility of continuous exchange between the carbonate in the bone and that in the surrounding environment, the inorganic fraction of bone often exhibits ^{14}C dates that are inconsistent with the organic fractions. Because of this, the aim in laboratory treatment of bone samples is to dissolve away the inorganic carbon with dilute HCl and date the collagen residue [Haynes (1967)]. If the bone has been in contact with humus materials, humic acid can contaminate the collagen and cause erroneous dates. Satisfactory dates can be obtained if the humic acid is removed by NaOH treatment after the collagen has been separated chemically from the inorganic fraction by the use of HCl [Sheppard (1975)].

The aim of pre-treatment of bones is to separate the inorganic calcium phosphate from collagen and the other organic compounds so as to be able to analyse both fractions separately. Because of the exchange of carbonate in bones with the surrounding environment, the dating of inorganic carbon in bones is accompanied by the risk of contamination and hence dating of the organic fraction is preferred. The pre-treatment techniques for bones and methods of separation of the inorganic and organic fraction and of subsequent separation of particular organic fractions as well as the problems of evaluation of the ^{14}C ages of bones have been discussed and summarized by Gupta and Polach (1985) and by Taylor (1987), (1992).

Collagen-based pre-treatment methods for bone yield inconsistent results for samples with low protein preservation, as frequently found in bones from the semiarid zones of Australia and North America. Gillespie (1989) has proposed the use of non-collageneous proteins, particularly blood proteins, for accelerator mass spectrometry (AMS) ^{14}C dating because of their better preservation. The amino-acid profiles of collagen and non-collageneous proteins suggest that such differential preservation may be due to physico-chemical differences, and help to explain the poor results from dating low-collagen bones.

With the advent of AMS radiocarbon dating of very small samples, a much greater opportunity now exists for research into specialized materials. Investigations of the proteins of bone and teeth for archaeological purposes suggest that much more information might be obtained by appropriate study of individual proteins in these tissues [Sobel et al. (1995)]. Even tooth enamel is a valuable archaeological resource that, however is only abundant in megafaunal remains. The carbonate fraction of tooth enamel is used for ^{14}C dating using AMS, but the procedure requires removal of additional exogenous carbonate [Hedges et al. (1995)].

Radiocarbon dating of bones has often been regarded with mistrust and scepticism because of the ambiguous results obtained for various fractions of the samples. Taylor (1987) describes an example of a carbonate fraction that was significantly older than the organic component of an artefact that played an important role in discussions concerning the timing of the initial occupation of the Western Hemisphere. An age of 27000+3000/-2000 ^{14}C years B.P. was initially assigned to an apparent tool made from a caribou tibia from an Old Crow Basin locality in the Yukon Territory, Canada [Irving et al. (1973)]. Subsequent ^{14}C analysis by AMS techniques determined that the ^{14}C age of the organic fraction of this bone was 1350±150 ^{14}C years B.P. [Nelson et al. (1986b)].

After pre-treatment, all kinds of samples can be treated according to the measuring technique used (Chap. 5, 1.5).

2.12. Factors Affecting Radiocarbon Dating

Systematic discrepancies between the early ^{14}C dating results and the historical age of Egyptological samples [Libby (1955)] have indicated fluctuations in the ^{14}C production in the atmosphere and in the initial ^{14}C activity of organic samples in the past. [De Vries (1958)] proved this phenomenon, later called the "de Vries effect" on a global scale for the past 300 years.

Dendrochronological calibration can compensate the discrepancy between the ^{14}C and the real age of a sample. The calibration curve has been compiled by correlating the dendrochronological age of wood, determined according to tree rings, with their pertinent ^{14}C age (see 2.13).

Organic materials from the late nineteen-century have been found to be deficient with respect to the expected ^{14}C activity up to several percent ("Suess effect"). During the industrial revolution, the specific ^{14}C activity in the atmosphere decreased due to the emission of ^{14}C-free CO_2 from fossil fuel combustion [Suess (1955)]. The end of the 19th century, therefore, was used to define the "zero age". At this point the wood grown was substituted by synthetic modern carbon standards (see 2.6.2).

At the beginning of the testing of nuclear weapon (Chap. 3, 2), ^{14}C activity increased considerably (see Fig. 4), and almost doubled until 1963, compared to the middle of the 19th century [Nydal (1964), (2000)], [Nydal et al. (1979)]. This increase is called the "nuclear bomb effect". Today it is not possible to check the relationship between the standard modern radiocarbon activity (95 % activity of the NBS oxalic acid) and that of the sample in the past. Both effects inhibit this check by collecting living specimens. Thus, to establish the initial ^{14}C activity in the various reservoirs with respect to the NBS oxalic acid standard, one must collect independently dated samples or samples deposited pre-1950, preferably pre-1890 [Gupta et al. (1985)].

The increased ^{14}C concentration in atmospheric CO_2 still persists. Due to the "nuclear bomb effect", radiocarbon in post-1950 environmental samples can be used rather as an environmental tracer.

The level shows seasonal variations and is slowly declining [Tenu et al. (1998)]. An analysis of changes of the radiocarbon concentration in the atmosphere and biosphere in relation to the global carbon cycle was given by [Levin et al. (2000)].

The final activity A of a sample must be authentic and not influenced by any factors other than radioactive decay (see Eq. 12). Any additional contamination of a sample by an admixture of allochthonous carbon after it has been converted to a closed system changes its ^{14}C activity and thus its radiocarbon age. Errors due to such contamination are high, especially for old samples contaminated by modern carbon. This is evident from the fact that contamination of a non-active (i.e., "infinitely" old) sample by 50 % modern carbon makes it only 5730 years old.

The influence of contamination on radiocarbon dating has been analysed by [Olsson (1972), (1979), 1980), (1986), (1991), (1992)] and [Olsson et al. (1992)] (see Fig. 5) and is also discussed in monographs on radiocarbon dating [Geyh (1971)], [Gupta et al. (1985)], [Taylor (1987)].

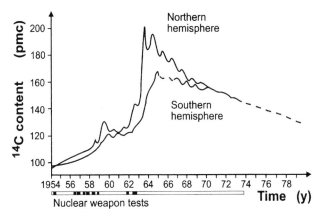

Figure 4: The "nuclear bomb effect": radiocarbon content of surface air normalized to modern carbon standard (pmc), based on [Vogel et al. (1975)], [Levin et al. (1980)], [Rauert (1980)].

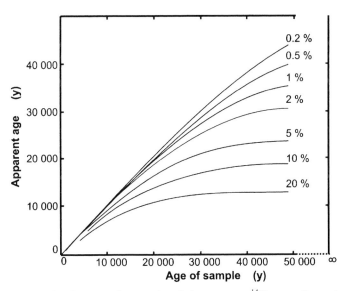

Figure 5: Relationship between the actual and the apparent ¹⁴C ages of samples contaminated with varying percentages of modern carbon, after [Olsson et al (1992)].

2.13. Calibration of Radiocarbon Age

The ^{14}C- and dendrochronological age of sequoia (*Sequoia gigantea*) and of bristlecone pine trees (*Pinus aristata*, now *Pinus longaeva*) between Anno Domini (A.D.) 1855 back to 1100 Before Christ (B.C.) were correlated by [Suess (1965)] who recognized:
- variations of ^{14}C on the time scale of the order of 100 years, which show correlation with average sunspot numbers, and
- those with a time constant of more than 1000 years, which involve the total ^{14}C inventory. The reasons of the cyclical ^{14}C variations were analysed e.g. by [Damon (1992)] and by [Sonett (1992)].

Correlation curves between the ^{14}C and dendrochronological ages have been elaborated from long series of samples of the high-precision ^{14}C calibration measurements by some ^{14}C laboratories (see [Becker (1992)]). [Michael et al. (1972)], [Damon et al. (1972)], and [Klein et al. (1980)] compiled calibration tables. These have made it possible to deduce the corrected calendar-year age by calibration of the ^{14}C age as far back as 8000 years B.P.

Unlike the ^{14}C age, which is expressed in terms of years B.P. (see 2.5), the calibrated ages are related to the standard ("Christian") calendar as B.C. and A.D.

Later, parallel determinations of the ^{14}C and ionium dates on a stalagmite from a cave in South Africa have provided evidence of variations in the ^{14}C content in the atmosphere beyond the range of the California tree-ring sequence to as far back as 40000 B.P. The calcium carbonate crystals of a 2.8 m high stalagmite, still growing, were used for both thorium and ^{14}C dating [Vogel (1983)]. Twenty paired ^{14}C and U/Th dates covering most of the past 50000 years were obtained from stalagmites in caves in South Africa [Vogel et al. (1997)].

For practical use, the calibration data of different laboratories for different time intervals were summarized in the Calibration Issue of Radiocarbon [Stuiver (1986)]. The calibration data were extended as far back as 13300 calibrated years B.P. [Stuiver et al. (1986)]. Soon, the calibration was additionally extended by incorporating further refinements and a new calibration data set covering nearly 22000 calibrated years, which corresponds to about 18400 ^{14}C years. The extending of ^{14}C age calibration beyond 10000 ^{14}C years was possible due to U/Th and ^{14}C dating of corals contemporaneous with the last glaciation off the south coast of Barbados [Bard et al. (1993)]. Computer programs for calibration were developed [Stuiver et al. (1986c)], [Van Der Plicht (1993)] and revised [Stuiver et al. (1993b)]. [Wohlfarth et al. (1995)] argue, however, that the use of the pre-Holocene part of the calibration program is premature and inadvisable.

Calibration of ^{14}C age data has become a routine tool for correcting ^{14}C age values in archaeological and Quaternary research. It makes the potential error due to the ^{14}C half-life redundant.

Due to the "wiggles" on the calibration curve, a ^{14}C age in some time intervals can yield ambiguous results as it corresponds to more than one real age and, on the basis of the date alone, it is impossible to determine to which age [Stuiver et al. (1966)]. When including the pertinent statistical errors in the band on the calibration curve, it can happen that a long calendar-year interval corresponds to a single ^{14}C year (see Fig. 6). [Ferguson et al. (1966)] have suggested a possible solution of this problem consisting in a comparison of a whole series of radiocarbon dates, whose relative ages are known, with a calibration curve [Neustupný (1970)]. The ascending or descending limbs of the calibration curve may be recognized and identified according to the changing of radiocarbon age values in the series whose general trend of relative age is known.

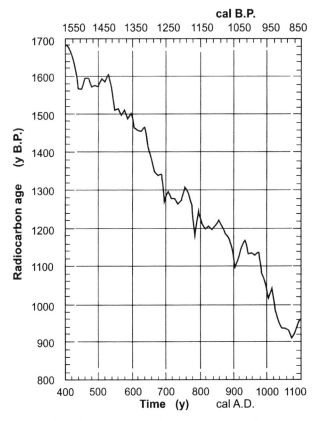

Figure 6: Section of a calibration curve, after [Stuiver et al. (1993a)].

2.14. Reporting of Radiocarbon Ages

Due to the large number of active radiocarbon dating laboratories, it became necessary to standardize the evaluation and reporting of the radiocarbon dates. Conventions were adopted during some of the International Radiocarbon Dating Conferences, which are organized regularly by one of the major ^{14}C laboratories [Cook et al. (1995)].

The following recommendations/resolutions were adopted by the Twelfth International Radiocarbon Dating Conference in Trondheim in 1986 [Mook (1986)]:

– Conventional ^{14}C age – based on $T_{1/2}$=5568 years and NBS oxalic acid activity – is to be reported in years B.P. where 0 B.P. is the year 1950.

– Dendrochronologically calibrated ages are to be reported as calibrated (cal) A.D. or cal B.C., or, if required, cal B.P.

– Historical non-^{14}C age values are generally given in A.D. or B.C. The use of A.D./B.C. by archaeologists in connection with ^{14}C ages is discouraged. It should be noted, however, that, in the past, extensive use has been made of A.D./B.C. dates obtained by subtracting 1950 years from conventional age B.P.

– Calibration curves for ^{14}C should contain real time on the horizontal axis (progressing time to the right) and conventional ^{14}C age on the vertical axis (increasing age in the upward direction). The abscissa scale thus indicates cal B.C. and cal A.D. (or, if required, cal B.P.), the ordinate scale indicates years B.P.

– The standard deviation of the calibration curve (σ_c) is to be incorporated in the standard deviation of the ^{14}C measurement (σ) by applying $(\sigma^2+\sigma_c^2)^{1/2}$ before the calibration is carried out.

– The calibration curves by Stuiver and Pearson for the period 500 B.C. to the present, and by Pearson and Stuiver for the period 2500 B.C. to 500 B.C., have been sufficiently double-checked to serve as the officially recommended calibration curves for these periods.

– The calibration graph used should contain a reference to the curve's author(s) in the upper right-hand corner (e.g., on computer printouts). Computer printouts should identify the authors of the curve for each calibrated date.

– ^{14}C ages of samples from the Southern Hemisphere should be diminished by 30 years before applying existing calibration curves.

– Other useful hints for collecting, submitting, recording, and reporting radiocarbon samples were published by [Kra (1986)].

When reporting radiocarbon ages and the respective errors, the practice of rounding off data according to Table 5 is used. The convention of rounded-off data, however, decreases their informative value for statistical processing.

Table 5: Rounding off of age and errors in reporting, after [Polach et al. (1983b)], [Gupta et al. (1985)].

Age (y)	Rounded off	to Error
0 - 1000	nearest 10	nearest 5 up
1000 - 10000	nearest 10	nearest 10 up
10000 - 25000	nearest 50	nearest 10 up
> 25000	nearest 100	nearest 50 up
under exceptional circumstances consider		
< 1000	nearest 5	nearest 5
> 25000	nearest 50	nearest 50

3. TRITIUM
Jan Šilar

3.1. Tritium in the Atmosphere and Hydrosphere

Environmental tritium (^{3}H) is a cosmogenic radionuclide, which is produced primarily by the reaction of neutrons (see Chap. 2, 1.2.2 and 3.4.2), originating in cosmic radiation, interacting with nitrogen ^{14}N:

$$^{14}N \ (n, ^3H) \ ^{12}C$$

and decays according to

$$^3H = {}^3He + ß^-$$

The cosmogenic tritium produced in the atmosphere becomes part of the water molecules of the hydrosphere and of the hydrological cycle. The half-life of tritium is considered to be 12.35 years [Moser et al. (1980)] or 12.43 years [Payne (1983)]. Its practical application for dating in hydrology, however, is limited due to its relatively short half-life and the high amounts of tritium introduced into the atmosphere during the tests of thermonuclear weapons after 1952.

The specific activity of tritium in water is usually expressed in tritium units (T.U.): 1 T.U. corresponds to the proportion of 1 atom of ^{3}H to 10^{18} atoms of ^{1}H. This corresponds to 0.118 disintegrations per second in 1 litre of water, i.e., to a specific activity of 0.118 Bq·kg^{-1}. Another term, the "tritium ratio" (T.R.), is preferred to tritium units by some authors as the term "unit" should not be applied to a number ratio [IAEA (1983f)].

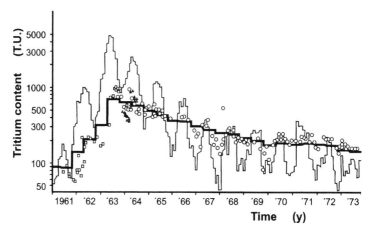

Figure 7:   The nuclear bomb effect: Measured ^{3}H contents in the Rhine River at three different stations. Model values of semi-annual ^{3}H contents (thick line) and monthly ^{3}H
contents in precipitation (thin line) approximately valid for the Frankfort area after
[Weiss et al. (1975)], [Rauert (1980)].

Prior to 1952 under natural conditions, the concentration of tritium in precipitation was between 5 and 20 T.U. (0.59-2.36 Bq·kg^{-1}), depending on the
geographic location [Payne (1983)]. It was about 6 T.U. in central Europe
(0.708 Bq·kg^{-1}) [Roether (1967)].

After 1952, due to the thermonuclear explosions, the concentration of tritium increased abruptly by about three orders of magnitude [Payne (1983)] in
the Northern Hemisphere, with maximum concentrations of about 6000 T.U.
(708 Bq·kg^{-1}) in 1963 (see Fig. 7). Since about 1965, the concentration of ^{3}H
in natural water has been declining. In some regions, however, it has been increasing due to the operation of nuclear facilities.

[Thatcher et al. (1965)], [Suess (1969)], [Nuti (1991)] have made the
following observations from tritium measurements in precipitation:

- During the year, the concentration of ^{3}H changes and is higher in the late
  spring and in the summer than in the winter (Fig. 7). ^{3}H increases in the
  atmosphere due to the so-called "spring injection" from the stratosphere
  into the troposphere.

- ^{3}H increases with the geographic latitude, as it is mostly introduced into
  the troposphere in polar regions where the tropopause is nearer to the
  earth's surface.

- Due to the fact that the thermonuclear explosions took place in the
  Northern Hemisphere, the tritium content is much higher in the Northern
  Hemisphere than in the Southern Hemisphere. The exchange between
  the two hemispheres is low because they have quite separate air mass

circulation (Chap. 2, 3.1.1). In addition, the tritium content of the oceans is very low, (a few T.U. at the surface) and there is much more oceanic vapour in the Southern hemisphere.

Under natural conditions, tritium not affected by nuclear weapon tests could be used for dating of groundwater in a similar manner as radiocarbon. It acts as a conservative tracer of the part of the water molecule exposed to each process of the hydrological cycle. Unlike the radiocarbon dating method, the usual geochemical processes of the water-rock interaction do not influence tritium dating (see 4.4). The time range of tritium dating, however, is limited, due to its short half-life.

Before the nuclear weapon tests, the original yearly average ^{3}H- concentration in the atmosphere was 6 T.U. (0.708 Bq·kg^{-1}). Theoretically, water which originated before the tests could be dated for up to 50 years. In practice, however, this is impossible due to the nuclear explosions and the high admixture of tritium in precipitation after 1952/53. Even a very low admixture of precipitation in groundwater increases the tritium concentration in the old water.

Generally, tritium is not suitable for groundwater dating [Hebert (1996)], because the input function, i.e., the initial tritium concentration in groundwater, varies with time and because the mixing processes of different water components already start in the unsaturated zone.

3.2. Tritium as an Environmental Tracer

Due to the circumstances mentioned above, tritium has lost its significance for conventional radiometric dating of groundwater. It has become, however, a useful environmental tracer. It can be used for dating when analysing the time shift of its peak concentration in ground or soil water behind its peak in precipitation, taking into account its residual concentration due to its decay. Moreover, its presence in groundwater in concentrations higher than in the present precipitation indicates the origin of the groundwater after 1952. It can also be used in dating products of the food industry (e.g., wine and spirits, see 4.6).

3.3. Sampling of Water for Tritium Analyses

The most laboratories require samples of 500 or 1000 ml for tritium analyses. Glass bottles are preferred for sampling. The bottles have to be dry. Contamination by precipitation and long storage in plastic bottles must be avoided.

4.   RADIOCARBON AND TRITIUM DATING IN SCIENCE AND TECHNOLOGY

Jan Šilar

The importance of radiocarbon and tritium dating methods soon became apparent in practical applications following the development of efficient measuring techniques. Because of its half-life, radiocarbon became an important tool for dating, especially in archaeology and in Quaternary geology. In combination with tritium and certain stable isotopes, it has been used for studying processes in the hydrological cycle. Both radionuclides are also used in analysing environmental problems and in technology.

Radiocarbon dating brought new insights into geological and archaeological dating in which mainly stratigraphic methods of relative dating had formerly been used. Dilette Polach [Polach (1980)] published a bibliography of papers on radiocarbon dating including works on applications in science and technology before 1968. Later developments were outlined in the volume edited by [Taylor (1992)].

4.1.   Different Concepts of Chronology: Relative and Absolute Dating

The prehistoric past and evolution of the Earth including man constitute one of the principal fields of interest of geology and archaeology. In archaeology, geology and other time-related sciences, referring of an event to a time scale and determining of the duration of a certain period is significant for interpretation of the prehistoric past. Such arranging of events in their proper sequence of time is called "chronology" [Glossary of Geology (1980)]. In geology, the term "geochronology" is also used. In this respect, archaeologists and geologists chronologically order events or happenings in their proper sequence in time.

The classical geological determining of time has been based on two constitutive laws:
  – The law of superposition: in any sequence of sedimentary strata or of extrusive igneous rocks that has not been overturned, the youngest stratum is at the top and the oldest at the base.
  – The law of faunal assemblages: similar assemblages of fossil organisms indicate similar geological ages for the rocks [Glossary of Geology (1980)]. Each stratum represents a period of time, which corresponds to the period necessary for its deposition.

The mentioned principles form a base for relative dating, which has long been the main task of chronology. This kind of interpretation of the past is called "relative chronology", or "relative dating". It is used in classifying the relative ages of strata. "Chronostratigraphy" is the branch of stratigraphy that deals with the ages of strata and their time relations [Glossary of Geology (1980)].

From time to time, efforts have been made to express time in archaeology and geology in absolute conventional time units, such as years. These efforts made it possible to express the duration of processes and the length of time-spans in years B.C. or A.D. In general, these techniques are based on natural phenomena that undergo progressive changes at uniform rates throughout time. By knowing the rate of change and the amount of change, the number of years that have elapsed since the process of change began can be computed. Working back from the present, the investigator can ascertain the date on which the process was initiated [Michels (1973)]. Some authors have improperly called the methods developed for this purpose, absolute chronology or absolute dating. Better terms, such as chronometry, chronometric dating, or if referred to geological events, geochronometry have been proposed.

In geology, dating is conceived as age determination of naturally occurring substances or relics by any of a variety of methods based on the amount of change, happening at a constant measurable rate, in a component. The changes may be chemical, or induced or spontaneous nuclear, and may take place over a period of time [Glossary of Geology (1980)].

The majority of chronometric methods is based either on determining the changes in concentration of a radionuclide in the sample due to its decay or on measuring the concentration of a radionuclide in the sample and/or of its decay products. These methods are called "radiometric dating" [Glossary of Geology (1980)].

The changes can also be of geological or biological nature such as, e.g., the yearly alternation of varves (layers of laminated thin-bedded glaciolacustrine sediments) is the base of varve chronology developed in Sweden by De Geer in 1912 [Hohl (1981)]. By counting the varves and parallelization from outcrop to outcrop, conclusions can be drawn on the duration of some processes in years. This became the first geochronological method used in Quaternary geological research. In a similar manner, counting of tree rings and relating them to each other in suitable samples can yield a time series that is significant for dating events in the past, especially in relation to weather conditions and climate. These methods of dating are considered "absolute" as they are based on direct counting of natural events based on passing of sideric years.

Radiocarbon dating, in spite of attributing historical and geological events to the conventional time scale of the Christian calendar, is not an absolute dating method (unlike varve chronology and dendrochronology). The radiocarbon calendar is not linear [Switsur (1986)] and doubts arise as to whether the ^{14}C dates agree with the observed stratigraphy [Kinness et al. (1986)]. Thus radiocarbon dates must be corrected by calibration (see 2.13).

The standard unit of time used in chronometric dating is the calendar year. Measurements of time intervals between an event and the present are expressed in the number of years that have elapsed between the two temporal points. Therefore, chronometric dating allows us to do two different but related things: We can quantify time, and we can date events by reference to the Christian calendar [Michels (1973)].

Unlike stratigraphy, chronometric dating does not deal with lithostratigraphic or biostratigraphic units but with real time expressed in years. Due to the physical nature of radiometric and other techniques used in chronometric dating, its results are connected with errors and uncertainties, which, however, may be quantified in terms of statistical deviations or degrees of probability. Another source of errors lies in geochemical and other natural processes such as leaching, migration, fluctuating production of cosmogenic isotopes, mixing of different solutions and groundwater components, etc. These factors change the prerequisites under which the chronometric methods were developed. The latter errors cannot be expressed by means of a statistical deviation, but must be borne in mind when assigning an "absolute" time scale, i.e., the Christian calendar, to historical or geological events. Chronometric data recorded in the "absolute" time scale do not constitute a task *"per se"*. Nonetheless, they do have substantial value in their links with the conventional stratigraphic or archaeological time scales which were elaborated using working methods of relative dating, often with a very high accuracy. Disregarding the statistical deviations and factual errors in the geochemical premises of the "absolute" dating in comparison with the obviously precise results of the "relative" dating led to misunderstandings and underestimation of the efficiency of chronometric dating methods. However, it became evident that chronometric dating would be useful in geological and archaeological research and some of the chronometric methods became routine working tools.

In the connections analysed, radiocarbon dating is a radiometric chronometric method, which, through dendrochronological calibration, is linked with the conventional absolute time scale of the Christian calendar.

4.2.   Radiocarbon in Archaeology

Since the beginnings of radiocarbon dating, its application has been directed to archaeology [Libby (1955)]. The limit for radiocarbon dating is about 50000 years B.P., (see 2.5). Grootes et al. (1975), however, reported a limit of 75000 years when using isotopic enrichment of ^{14}C. It then covers the whole period from the present back to the Paleolithic age. The basic principles of the archaeological approach and the evaluation of the results of radiocarbon dating are given, e.g., by [Michels (1973)], [Polach (1981)], [Taylor (1987)], [Polach et al. (1981)], and in [Taylor et al. (1992)].

Until recently, archaeologists had to attribute particular events to a proper sequence without knowing their links in space. They explored stratified deposits to discover chronological sequences and to compare them at different sites. The introduction of the ^{14}C dating in the 1950s coincided with a shift in the focus of the research away from the excavation of highly stratified, single sites. Area investigations typically involved excavations of a wide variety of site types distributed within and between major prehistoric periods. Concomitant with the shift from single-site to area investigation, prehistorians adopted a new paradigm for examining variability in the prehistoric record [Henry (1992)]. Without ^{14}C dating, such investigations would not have been able to fit prehistoric occupations, which often lack stratigraphic links, into common chronological sequences.

^{14}C dating profoundly changed the concept of cultural relationships within North Africa, and between North Africa and other areas. Most of this area is extreme desert, and erosion has destroyed much of the original context of many archaeological sites. Those that are preserved tend to occur in widely separated patches, and there are rarely opportunities for direct stratigraphic or other correlations between these patches [Wendorf (1992)]. Due to the shift from single-site to area investigations, ^{14}C dating has had perhaps the greatest influence on shaping theoretical developments within Near Eastern prehistory. When dating seeds attributable to agriculture using accelerator mass spectrometry (AMS) (Chap. 5, 2), the beginning of agriculture in the Middle East could be placed at approximately 10000 B.P. [Henry (1992)].

Sometimes, radiocarbon dating had to surmount doubts. For example, until recently, it had been generally accepted that the radiocarbon dates for the destruction levels at Akrotiri resulting from the Minoan eruption of Thera on the Aegean Island of Santorin "did not make sense", interference by carbon dioxide emanations from the volcano, deficient in ^{14}C, sometimes being suggested as the cause. The scatter was said to be unacceptably large, but above all, after calibration, the majority of the dates were at least a century earlier than expected on the basis of accepted Minoan chronology, established by archaeological linkages to the Egyptian astronomically anchored calendar. However, reassessment of linkages and using accurate high-precision calibration curves made the calibrated dates more recent by about half a century. The topic is of more than local interest because of the tendency for opponents of radiocarbon to use these dates as "yet another example of failure" and of the challenge to science-based techniques to provide an unambiguous result for this major catastrophe [Aitken et al. (1988)].

Sometimes the ^{14}C age of an archaeological object does not correspond to the art style of other objects with which the dated one was considered to be associated. This seeming contradiction does not necessarily prove errors in dating, but in some cases it indicates that the objects had been exposed to some later handling. Thus, for example, in the collection of Egyptian mummies of the Náprstek Ethnographic Museum in Prague, the mummy wrapping P 624c, consisting of a well preserved flax tissue, showed dendrochronologically corrected ages of 1028±144 and 928±124 yrs B.C. [Šilar (1979)]. According to the embalming techniques, the age of the respective mummy P 624b was of the Third intermediate period (1087-664 yrs B.C.). According to the Egyptological dating of the coffin, it should be of the Late till Greek periods (400-30 yrs B.C.) [Strouhal et al. (1976)]. The corrected radiocarbon ages are rather in agreement with the age derived from the embalming techniques than with that of the coffin. Therefore, the possibility that an older mummy had been displaced in a newer coffin cannot be excluded.

Archaeological samples of known historical age can be used as reference samples, if available in sufficient quantity. For example, stalks of the reed (*Phragmites australis (Cav.) Trin. ex Steud.*), which were excavated during the restoration of the zikkurat (earth pyramid) at Aqar Quf west of Baghdad, were used as reference samples for ^{14}C dating of the Egyptological samples described above. The reed comes from straw bands, which were used to reinforce the adobe structure between the layers of bricks [Šilar (1979)].

^{14}C dating contributed to determining the time of origin of the well-known prehistoric paintings in caves in North-Eastern Spain and in South-Western France. The radiocarbon dates for the charcoal used to draw stylistically similar bisons are 14000±400 yrs B.P. in the Spanish caves of Altamira, 12990±200 yrs B.P. in El Castillo, and 12890±160 yrs B.P. for a bison of a different style in the French Pyrenean cave of Niaux. So far, dates have been derived from the style or dated remains left by prehistoric visitors and could be biased by prolonged occupation or visits unrelated to painting activity. The results of ^{14}C dating show that painting dates derived from human activities should be used with caution [Valladas et al. (1992)].

Direct dating by radiocarbon techniques of pictographs has not yet been possible, mainly because of the difficulty of separating inorganic carbon from the organic material in the pigment. [Russ et al. (1990)] report on a new technique that allows this separation to be effected by a low-temperature, low-pressure oxygen plasma to selectively oxidize the organic component; this can then be analysed using standard ^{14}C methods. The technique was applied to a portion of a pictograph from the Lower Pecos region of Southwest Texas. The date of 3865±100 yrs B.P. is consistent with that expected on the basis of archaeological inference.

The introduction of accelerator mass spectrometry (AMS) techniques (Chapt. 5, 2) made it possible to date very small samples and to use radiocarbon dating to solve problems in a manner that had not been possible previously. For example, the AMS technique was used for dating of wrought iron [Igaki et al. (1994)], [Nakamura et al. (1995)] in evolving CO_2 from iron samples by heating and converting the CO_2 into graphite targets for AMS dating. It was also used to date pottery by dating the smoke-derived carbon separated from potsherds [Delqué Količ (1995)]. By dating blood residues on prehistoric stone tools, the time of their use was found directly, rather than by stratigraphic or other archaeological inferential techniques [Nelson et al. (1986a)]. [14]C dating contributed to clarifying the origin of significant ancient relics. Thus, for example, very small samples from the shroud of Turin, a linen cloth on which appear the imprints of the front and back of a crucified man and believed by many to be a true relic of the Passion of Christ, have been dated using AMS in laboratories at Arizona, Oxford, and Zurich. As controls, three samples whose ages had been determined independently were also dated. The results provide conclusive evidence that the linen of the Shroud of Turin is medieval [Damon et al. (1989)]. The procedures dictated by the Turin ecclesiastical authorities to accomplish the dating and the scientific approach to this task have been described by [Gove (1989)]. The flax from which the linen of the Turin shroud was woven was harvested in A.D. 1325 with an uncertainty of ±33 years [Damon et al. (1989)], [Gove (1992)]. [Dale (1987)] suggests that, whether relic or fake, the Shroud of Turin may well be recognized as one of the masterpieces of Christian art.

In other cases, ^{14}C dating offers explication of challenging historical problems. For instance, evaluating a climatic anomaly from ^{14}C measurements of tree rings together with evaluating biblical records and ^{14}C dating of wood found in caves at the Dead Sea were used by [Aardsma (1995)] to explain an inconsistency in biblical chronology. Biblical and secular data seem to tell entirely different stories about Near Eastern history prior to the first millennium B.C. Most modern critical scholars in Bible-related fields regard this as proof of the nonhistorical nature of the premonarchical biblical narrative. However, the incongruity between biblical and secular data seems also to be explainable by the postulate that exactly 1000 years were accidentally dropped from traditional biblical chronology just prior to the first millennium B.C. [Aardsma (1995)] evaluates this postulate relative to extra-biblical data for the exodus, Joseph's famine and Lot's observation that the Jordan valley was "well watered" in his day. Biblical chronology, when corrected by the restoration of the lost 1000 yrs, promises to reduce present uncertainties in secular historical chronologies in the 3rd and 4th millennia B.C. by at least one order of magnitude. It seems reasonable to suggest that the accidental removal of 1000 yrs from 1 Kings 6:1 by some ancient scribe is the true source of the many apparent incongruities between the biblical narrative and secular data that fuel the postulate of no historicity at present.

It is still not generally accepted that ^{14}C dating has any usefulness for archaeologists. There are two types of problems that archaeologists would like to solve with the help of ^{14}C dating: dating of specific events such as the destruction horizon of a site and dating if the duration of archaeological phenomena such as the occupation of a settlement or a cemetery, or the duration of successive cultural groups in a geographically defined space [Ottaway (1986)]. Feedback between archaeologists supplying the samples and using radiocarbon dating and physicists carrying out ^{14}C dating measurements is necessary in order to determine the time relationships for a meaningful archaeological interpretation of the results.

The contribution of radiocarbon dating to archaeology was summarized by [Taylor (2000)] and by [O Bar-Yosef (2000)].

4.3.   Radiocarbon in Geology

Radiocarbon dating covers the time range of the Holocene and Upper Pleistocene back to the Pleniglacial during the Würm period in the Alpine glaciation system or the Weichselian (Vistulanian) period in the Nordic glaciation system. Both periods started about 70000 years B.P. Samples of organic materials as well as inorganic (calcium carbonate) materials have been successfully dated.

Radiocarbon dating provided a link between the conventional stratigraphy of the Quaternary and the calendar time scale. In combination with micropaleontological and paleobotanical studies, and with stable-isotope analyses of Quaternary samples, it has contributed to the knowledge of temperature changes during the Upper Pliocene and Holocene.

AMS dating proved very effective in the dating of marine and lake sediments where only small samples were available from the drilling cores and where separate dating of thin layers AMS was essential for determining the rate of sedimentation and, in combination with paleontological analyses, to clarify paleoclimatic events. For example, the rate of sedimentation could be determined by AMS dating of $CaCO_3$ of foraminifera from the pelagic sediments of the Atlantic Ocean between 3.0 and 5.9 cm per 1000 years [Thomson et al. (1995)] and from shelf-sediment cores from the Danish coast of the North Sea from 1 to 70 m per 1000 years [Heier-Nielsen et al. (1995)].

Originally, dating of samples containing calcium-carbonate was accepted with mistrust. Libby (1955) deduced that material that is likely to be supported in its growth in any significant part by carbon from limestone is, of course, useless. Since that time, however, ^{14}C dating of carbonate lake sediments, shells, mortar and groundwater has advanced and now encompasses a wide range of radiocarbon dating topics. The processes involved in carbonate geochemistry make ^{14}C dating of these inorganic carbon materials more complex and difficult compared to organic carbon and they must be taken into account.

When dating carbonate sediments or the shells of molluscs, some special circumstances should be kept in mind:

The initial radiocarbon activity of calcium carbonate in marine shells is influenced by the reservoir effect (see 2.2) of the ocean, i.e., by the slow exchange between atmospheric carbon dioxide and the ocean, which results in a lower radiocarbon activity in marine samples. Consequently, the shells of living marine molluscs before the year 1950 appear to have ages of up to several hundred years. Because of the ocean circulation pattern, these apparent ages differ regionally. Thus, for a cold marine environment, [Mangerud et al. (1975)] found that the mean apparent radiocarbon age of marine shells, collected alive before the start of atom bomb testing, and also before the main input of dead carbon derived from fossil fuels, is 440 years for the coast of Norway, 510 years for Spitsbergen, and 750 years for Ellesmere Island, Arctic Canada. For temperate and tropical oceans, [Druffel et al. (1983)] found that the apparent radiocarbon age of surface waters, as measured in banded corals in Florida and Belize for the preanthropogenic period, ranged from 280 to 520 years B.P. while the apparent radiocarbon age of subsurface water was about 670 years B.P. The concentration of bomb-produced radiocarbon was much higher in corals from temperate regions than in corals from tropical regions. On the basis of ^{14}C measurements in corals grown between A.D. 1800 and 1974, [Druffel et al. (1978)] reported bomb ^{14}C to be present as early as 1959 in Gulf Stream surface waters. On the basis of the existence of a time delay between the atmospheric and oceanic ^{14}C maxima, they deduced that several years are required to approach a steady state (Chap. 2, 3.4.3). The final activity of the sample can be influenced by any allochthonous admixture of calcium carbonate, e.g., by aeolic contributions deposited in marine carbonate sediments [Olsson (1972)], by erosion of the calcium carbonate particles and redeposition on the bottom of sedimentary basins [Shepard (1967)], by substitution of the original aragonite of the shells by calcite through recrystallization [Vita-Finzi (1980)], [Yates (1986)] or by the presence of non-active carbonate nuclei of ooids and, at the same time, by the presence of the more active cement of oolithic sediments [Šilar (1980)]. For dating Quaternary tufa deposits, the initial ^{14}C activity must be known; this was studied on modern and recent fresh-water tufa deposits [Srdoč et al. (1983)] and in fine-grained detrital lake sediments [Geyh et al. (1970)], [Geyh (1971)]. [Hebert et al. (1983)] also discussed the aspects of ^{14}C dating of tufa sediments and speleothems (dripstone), the initial ^{14}C activity, sampling, sample processing and evaluation of results.

The dating of organic as well as inorganic carbon in lake sediments, and dating of speleothems, shells, peat, and groundwater and correlating it with paleontological dating and stable isotope analyses contributed considerably to solving problems of Quaternary geology, paleoclimatology and hydrology [Geyh et al. (1970)], [Monge Soares (1993)], [Bath (1983)], [Pazdur et al. (1994)], [Maloney et al. (1995)]. The problems of accuracy and precision of radiocarbon dating of lake sediments were thoroughly analysed by [Olsson (1991)].

Radiocarbon dating of the organic matter deposited in fluvial sediments makes it possible to obtain information on the evolution of a river and its floodplain. For example, [Jílek et al. (1995)] could established the stratigraphy and the evolution of the floodplain by radiocarbon dating of wood in the Holocene floodplain in the upper and middle reaches of the Labe River (Elbe) in central Bohemia (Czech Republic) in combination with geomorphologic, sedimentological and mineralogical investigation of sediments. The results imply that several abrupt changes in temperature and precipitation occurred during the Holocene. These changes led to intervals of hydrological disequilibrium, which caused the formation of two Holocene terraces and a contemporaneous floodplain. During the Holocene, there were four periods during which large tree trunks were deposited in the fluvial sediments, indicating periods of extensive flooding.

In karstified areas, dating of speleothems (dripstone) was used to investigate geological evolution and climatic events. ^{14}C age values of speleothems were correlated with glaciation events and changes of climate in the Upper Pleistocene. Paleokarst forms, located in raised beach deposits in Southwest England and South Wales, known as "pipes", were dated by ^{14}C and also by thermoluminiscence, to correlate them with the evolution of the beaches [Pazdur et al. (1995)]. Similarly, ^{14}C dating of stalagmites from caves in the Bahamas was used to determine low sea level stands about 22000 years B.P. [Spalding et al. (1972)]. Formation of tufa deposits in different regions could be dated, compared and related to the paleoclimatic evolution in the Quaternary [Srdoč et al. (1989)], [Horvatinčić et al. (1989)]. Simultaneous dating of organic carbon from wood and inorganic carbon from tufa was used both to clarify the processes by which tufa was formed [Srdoč et al. (1980)], [Jílek et al. (1987)], [Hiller et al. (1991)]. The radiocarbon age was correlated with paleontological dating and paleontological climatic evidence [Šilar et al. (1988)] in order to verify the reliability of ^{14}C dating of freshwater calcium carbonate.

Climatic changes could be determined also in paleohydrological records by means of ^{18}O and ^{2}H in groundwater. In the hydrosphere the concentrations of the stable isotopes ^{18}O and ^{2}H are expressed as $\delta^{18}O$ and $\delta^{2}H$ in relation to the world standard of mean ocean water (SMOW), similarly to $\delta^{13}C$

(see 2.3, Eq. 7). The fractionation processes during evaporation and condensation depend on the temperature and influenced considerably $\delta^{18}O$ and δ^2H. Precipitation and groundwater originating in a cold climate are marked by highly negative values of $\delta^{18}O$ and δ^2H. Conversely, samples from a warm climatic zone show less negative values. For example, Zojer (1992) has shown (see Fig. 8a) that the $\delta^{18}O$, δ^2H and ^{14}C concentrations of the samples from the Styrian Basin (Austria), a periglacial region, are not constant. In this region, deep groundwater exists with infiltration going back to the last glacial period. About 30000 years ago (low ^{14}C concentration), the climate worsened at the end of the interstadial phase, later reaching a temperature minimum at 20000 to 15000 years ago, followed by a rapid increase in the temperature. Ferronsky et al. (1992) drew similar conclusions for the Syr Darya artesian basin in Kazakhstan and for the Syrian Desert correlating ^{14}C groundwater age values with $\delta^{18}O$ in groundwater (Fig. 8b,c).

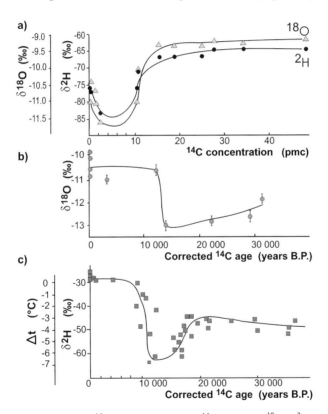

Figure 8: Correlation of the ^{14}C concentration or ^{14}C age to $\delta^{18}O$, δ^2H and change in temperature Δt in samples from: a) Styrian Basin (Austria) after [Zojer (1992)], b) Syr Darya artesian basin after [Ferronsky et al. (1992)], c) Syrian Desert, after [Ferronsky et al. (1992)].

A comparable phenomenon was observed in the Cretaceous basin of Třeboň in the Czech Republic. In the boreholes at three sites (see 4.4.4, Fig. 12) reaching to different depths with casing open at the lower sections (A, B, C, D, and E), the $\delta^{18}O$ values decrease with increasing ^{14}C groundwater age and indicate lower temperature at the time of groundwater recharging during the Pleistocene (lower part of Fig. 12). The ^{14}C groundwater age increases with the depth of the confined aquifers [Šilar and Šilar (1995)]. Deep aquifers in the Lorraine, in the Aquitaine and in the Paris Basins [Dray et al. (1998)] indicate a climatic change in the Late Glacial on the basis of the groundwater isotope data obtained.

The isotope data ($\delta^{18}O$, $\delta^{2}H$, ^{14}C concentration, $\delta^{13}C$), as well as noble gases and hydrochemical data, were used for paleoreconstruction of recharging of the groundwater system in the aquifers of the Chad Basin. The overall results are consistent with a major recharging episode in the Chad Basin during the Late Pleistocene and the quite rapid flow velocities are inferred from the ^{14}C gradient over the 180 km section. It is considered that an outlet controlled this flow at times of high rainfall, but all effective recharging ceased quite suddenly at the end of the Late Glacial Maximum. There is no evidence for renewed groundwater circulation during the Holocene. This event contains some evidence of some climatic amelioration towards the end of the Late Pleistocene recharging episode [Edmunds et al. (1998)].

Isotope techniques became a significant tool in investigating the humid and dry paleoclimatic episodes elsewhere in the arid regions in Africa, in Tunisia [Zouari et al. (1998)] and in the Grand Erg Oriental in Algeria [Guendouz (1998)].

The chronology of the groundwater recharging in the coastal aquifers could be determined in the Caen region (Northern coast of France). The chemical composition and stable isotope contents indicate that two successive mixings between fresh and sea water occurred in the confined part of the aquifer during the Late Quaternary. Recharging by marine water apparently took place during the Flandrian transgression (Holocene). The ^{14}C activities of the water are constant, independent of the saline fraction. The calculated ^{14}C ages are 4800-7500 years B.P. [Barbecot et al. (1998)]. A similar problem of marine-water intrusion into a coastal aquifer was investigated in the Orissa State in India. The groundwater salinity in aquifers occurring at intermediate depth (50-100 m) is largely due to the Flandrian transgression during the Holocene. This is based on the ^{14}C age values of groundwater coupled with sedimentological and paleontological studies [Kulkarni et al. (1998)].

In soil science, ^{14}C has been used to study the processes of pedogenesis [Geyh et al. (1998)]. The radiocarbon activity of the organic fractions in soil is extremely variable and the usefulness of using such an activity to infer age is quite limited, except under special conditions. The main difficulty with soil ^{14}C dating is to understand the processes involved in soil genesis, particularly the initial ^{14}C activity in the sources of the organic materials in the various fractions [Geyh et al. (1983)]. The procedures applied in individual laboratories have been published [Olsson (1979), (1980), 1986)], [Scharpenseel (1979)], [Sheppard et al. (1979)], [Kigoshi et al. (1980)], [Gupta et al. (1985)]. Separate dating of the fractions is indispensable for the correct interpretation of the soil genesis [Gilet-Blein et al. (1980)], [Geyh et al. (1983)]. [Harkness et al. (1996)] published examples and topical problems of ^{14}C application in soil science in a special volume of Radiocarbon "^{14}C and Soil Dynamics: Special Section".

4.4.   Radiocarbon and Tritium in Hydrology

[Münnich (1957)] adapted radiocarbon techniques for dating groundwater based on the fact that the carbon dioxide in the atmosphere and in soil containing radiocarbon is dissolved in water, reacts with the carbonates in the soil and in the rocks and is incorporated in bicarbonates, which are dissolved in groundwater.

The aim of groundwater dating is to clarify the time of its origin and to contribute to knowledge of the groundwater regimen by determining the groundwater "age". Radiocarbon groundwater age is the time that elapsed since the groundwater lost its supply of radiocarbon, i.e., since it infiltrated under the surface and through the soil. Strictly speaking, the term must be limited to individual water molecules. However, groundwater in general is a mixture of water particles of different origin with a different isotopic composition and its radiometric (^{14}C or ^{3}H) age only reflects the final isotopic composition of the mixture in the analysed sample. Thus, the term groundwater age should be considered a fictional value which rather reflects the mean groundwater residence time in an aquifer. The proper meaning of the hydrodynamic terms in this respect is defined by [Yurtsever (1983)].

Radiocarbon and tritium dating together with analyses of some other nuclides and of stable isotopes became the basis of what is called "isotope hydrology". Isotope methods have become a powerful tool for investigating the hydrological cycle. A comprehensive review of the application of isotopes in hydrogeology as well as in hydrology was published by [Clark et al. (1997)].

Water in the hydrological cycle is exposed to physical and geochemical processes and to isotopic fractionation. Thus, the concentrations of environmental isotopes in different parts of the hydrosphere changes in time and space. A characteristic geographic zonality and time-related fluctuations originate. The amount of radionuclides in closed hydrological systems decreases with time so that groundwater can be dated and time can be introduced into considerations about the hydrological cycle. Moreover, the proportion of the stable isotopes of oxygen and hydrogen reflects the temperature-dependent processes of evaporation and condensation.

Radionuclide dating by means of radiocarbon and tritium, together with analyses of stable isotopes, has revealed that the groundwater occurring in numerous hydrogeological structures does not belong to the present hydrological cycle, but to an older one or to cycles that started in the geological past under different climatic conditions and under different conditions for groundwater recharging [Sonntag et al. (1980)], [Šilar (1989)], [Verhagen et al. (1991)], [Geyh et al. (1997)].

Investigation of the hydrological cycle and of its fast exchange with respect to the geological time scale between partial reservoirs is successful if strong qualitative differences in the isotope configuration exist within the area investigated, resulting in significant differences in age between the recharge and discharge areas [Hebert 1996)].

The origin of groundwater can be linked with the geological history and with the conventional stratigraphic time scale and thus radionuclide dating can contribute to paleoclimatic studies. Moreover, it can contribute to the solution of ecological problems related to the hydrological cycle, to the evaluation of water resources and, especially, to groundwater protection.

Due to climatic fluctuations towards the end of the Pleistocene and the subsequent retreat of the ice caps and glaciers from large areas in the Northern hemisphere, large quantities of water from the molten ice joined the continental hydrological cycle, increasing the volume of water in the oceans and raising sea levels. These global events at the beginning of the Holocene intensified the hydrological cycle and changed the ecological conditions and the groundwater regimen on the continents. Radiocarbon dating of groundwater, of Quaternary sediments and of fossils, followed by geological and hydrological evaluation, provide a useful tool for clarifying these processes. The results of the study of past and current environmental changes by means of isotope techniques have been published in special volumes of Proceedings [IAEA (1993c), (1998i)]. A special volume dealing with global climate change and with the ^{14}C cycle in the oceans was published in Radiocarbon [Beck et al. (1996)].

Radiocarbon dating of groundwater is more complex than that of organic samples due to geochemical reactions between rocks and water and due to the mixing of the different components of groundwater in the lithosphere. On the other hand, groundwater dating by means of tritium is influenced by its short half-life. While radiocarbon dating is suitable for groundwater investigations in deeper aquifers with long groundwater residence times, tritium measurements can be used to prove recent or even modern groundwater origin, usually in unconfined aquifers with fast and shallow water circulation.

Table 6: Theoretically available environmental isotopes for groundwater dating, based on [Fontes (1983)], [A][Różański et al. (1979)], [B][Payne (1983)], [C][Loosli et al. (1979)], [D][Zito et al. (1980)]

Isotopes	Half-life (y)	Origin	Measurement	Time range (y)	Qualities	Limitations
^{85}Kr	10.8	[A]cosmic rays nuclear reactors and power plants	low-level β counting	since 1960	comparison with ^{3}H	heavy separation process, long counting time
^{3}H	[B]12.43	cosmic rays thermonuclear tests	low-level β counting	since 1952	ideal behaviour commonly suitable	modelisation, short time range
^{3}H-^{3}He	[B]12.43	cosmic rays thermonuclear tests reactors	low-level β counting mass spectrometry	since 1952 possibly 100 y time range	direct determination of turnover time	highly sensitive mass spectrometry, crustal production of ^{3}He
^{32}Si	≈100	cosmic rays nuclear tests	low-level β counting	≈1000 B.P.	few, link between ^{3}H, ^{39}Ar, ^{14}C	A_0 not known, sample 10 m^3; long counting time
^{39}Ar	[C]269	cosmic rays crustal	low-level β counting	≈2000 B.P.	no chemical interactions, comparison with ^{14}C	heavy separation process, sample 10 m^3, long counting time
^{14}C	5730	[D]cosmic rays thermonuclear tests crustal	low-level β counting accelerator	≈3·10^4 7·10^4	commonly suitable	complicated chemical and isotopic system
^{81}Kr	2.1·10^5	cosmic rays	low-level β counting	5·10^5	no chemical interactions, comparison with ^{14}C	perspective tool
^{36}Cl	3.06·10^5	cosmic rays nuclear tests	accelerator mass spectrometry	2·10^6	few chemical interactions, ideal for fossil waters	A_0 not known, access to accelerator
^{234}U-^{238}U	2.5·10^5	decay of uranides chain	low-level α-spectrometry	5·10^5	time range of very old waters	A_0 not known and variable chemical interactions

The presence of tritium is an indication of modern water in aquifers, indicating fast and shallow groundwater circulation and a risk of contamination.

Some seemingly contradictory results involving a low concentration of radiocarbon and a higher concentration of tritium can be ascribed to mixing of old with young components of groundwater (see 4.5). Both radiocarbon and tritium dating have become almost routine procedures.

The concentrations of both radiocarbon and tritium in the environment were influenced by the explosions of nuclear weapons after 1952 (see Figs. 4 and 7). Bomb-produced ^3H and ^{14}C became effective environmental tracers indicating groundwater that is of shallow and recent origin.

The power of groundwater dating increases with the simultaneous use of stable-isotope methods, particularly those analysing ^2H, ^{18}O, and ^{13}C. Thus, isotope-hydrology methods are usually applied jointly.

Dating by means of other radionuclides, e.g., ^{39}Ar, ^{81}Kr, ^{32}Si, ^{36}Cl and ^{129}I, has been developed and tested (see Table 6). The range of groundwater residence times and the groundwater age that can be investigated using a radionuclide depend mainly on its half-life and extend from between 0.2 to 5 times the half-life [Fröhlich (1992)].The radionuclides ^{81}Kr, ^{36}Cl and ^{129}I are suitable for dating groundwater whose age exceeds the dating limit of radiocarbon (ca. 50000 years), because of their long half-lives [Fröhlich (1992)]. The long half-lives of these substances make it difficult to measure the activities of these nuclides using a radiometric methods and they should be determined by AMS [Elmore et al. (1987)]. The use of the $^{234}U/^{238}U$ ratio as an age dating parameter for dating groundwater is also still an open question [Osmond (1980)]. The *in situ* production of the radionuclides used in hydrology and for groundwater dating has been discussed by [Florkowski (1992)] and [Andrews et al. (1992)].

The use of these methods has not become widespread in hydrological practice because of the necessity of processing large volumes of water when extracting the nuclides and the special analytical techniques required.

4.4.1.   Evaluation of Isotope Groundwater Data

The term "^{14}C groundwater age" must be considered to be a fictional value resulting from calculation according to Eq. 10. The initial activity A_0 of the sample may be influenced by interaction of the water and the rock and the final activity A may be affected by hydrodynamic processes within the aquifers. Both geochemical and hydrodynamic processes must be taken into account when evaluating the results of groundwater dating. Ideas about such processes are expressed in the form of geochemical and hydrodynamic models. An overview of ^{14}C analysis in the study of groundwater was presented by [Geyh (2000)].

4.4.2. Geochemical Models

Because of the ^{14}C-free carbon from the calcium carbonate in the soil and aquifers, the carbon of the bicarbonates in solution shows a lower activity in comparison with modern organic carbon. Due to the joint action of the recent and fossil carbon during the groundwater recharge, the initial ^{14}C concentration is between 100 % and 50 % of the modern activity. If 100 % modern is used as initial activity for dating according to Eq. 10, then the resulting groundwater age may be too high [Geyh (1980a)].

The major part of the radiocarbon reaches the groundwater from the soil air. While the normal CO_2 content of the atmosphere averages 0.03 % vol., CO_2 concentrations in the soil atmosphere are several orders of magnitude higher, up to 7 % vol., as indicated by [Miotke (1974)]. The ^{14}C concentration in groundwater and input values for ^{14}C groundwater dating are influenced by geochemical factors affecting the carbonate equilibrium [Matthess (1973)], [Pačes (1976)] and [Wigley (1976)].

It is a difficult problem in groundwater dating to evaluate the dilution of soil-produced carbon dioxide, which accounts for 100 % of modern radiocarbon, by radiocarbon-free carbon in dissolved carbonates, to determine the initial radiocarbon concentration in recharged groundwater. For this purpose, the geochemical processes forming the chemical composition of groundwater have been presented as geochemical models and summarized in the literature on isotope hydrology and groundwater dating by [Fontes et al. (1979)], [Mook (1980)], [Fontes (1983,1985)], [Geyh (1992)], [Verhagen et al. (1991)] and [Clark et al. (1997)]. The geochemical models take into account the fact that bicarbonates in groundwater incorporate both radiocarbon from the atmosphere and biosphere, and radiocarbon-free carbon from the lithosphere. They express the processes and factors influencing the carbon cycle in groundwater and quantify the corrected radiocarbon groundwater age.

In addition to the stoichiometric values of the carbon compounds and pH values, the models use the stable isotope ^{13}C as an indicator of these processes, because of the similar behaviour of ^{13}C and ^{14}C isotopes in the carbon cycle. Hence, the concentration of ^{13}C expressed as $\delta^{13}C$ (see 2.3) is one of the main input values of the more or less complex formulas of these models.

[Ingerson and Pearson (1964)] proposed a correction factor P for the initial activity A_0 which is actually identical with the factor C in Eq. 12:

$$P = \frac{\delta^{13}C_{sm} - \delta^{13}C_{ls}}{\delta^{13}C_{pl} - \delta^{13}C_{ls}} \qquad (17)$$

where $\delta^{13}C_{sm}$, $\delta^{13}C_{ls}$, and $\delta^{13}C_{pl}$ are the $\delta^{13}C$ values (see Eq. 7) in the sample, in the limestone, and in the plant-derived components, respectively. Assuming $\delta^{13}C_{ls}=0$ ‰, and $\delta^{13}C_{pl}=-25$ ‰, it follows that

$$P = \frac{\delta^{13}C_{sm}}{-25} \tag{18}$$

In investigating normal fresh groundwater resources, empirical values of the initial ^{14}C activity are often used for routine dating. This approach does not correspond to any of the geochemical models but is often used, as it does not result in unrealistic age values. An initial value of radiocarbon activity equal to 85 % of the modern activity, which is a common radiocarbon activity of shallow groundwater in humid and arid climatic regions, is often adopted [Vogel (1967, 1970)]. [Geyh (1972)] determined the initial ^{14}C activity for different lithological formations:

- in a karst region in Swabia it is between 60 and 75 % of the modern activity,
- in a loess region in Swabia it is 85 % of the modern activity,
- in the crystalline rocks in N.E. Brazil, it is between 90 and 100 % of the modern activity.

Similar values can also be used elsewhere in analogous geological formations for routine groundwater dating, but local hydrogeological factors and the consequences of potential errors must be considered in advance.

In groundwater systems with a short residence time, the effect of geochemical processes on the dilution of radiocarbon by inactive carbon of the rocks can be estimated if the modern contemporaneous age of groundwater can be considered as granted and related to the real modern ^{14}C activity of living organic matter. E.g., in the Bohemian Karst in the Czech Republic, radiocarbon, tritium and ^{13}C concentrations were studied in karst springs. The initial ^{14}C activity in ground water was calculated by relating the ^{14}C activity of real modern groundwater (as proved by high tritium as well as radiocarbon activities) to the ^{14}C activity of modern plant material represented by beach (*fagus*) leaves collected periodically in the area. In springs showing modern groundwater age, the ^{14}C activities corresponding to the initial ^{14}C groundwater activities were in the range between 79.42±1.51 pmc and 87.1±1.1 pmc [Šilar et al. (1999)].

Another approach in dating groundwater uses dissolved organic carbon (DOC) ground-water constituents instead of dissolved inorganic carbon (DIC) constituents. The method and technique have been discussed e.g., by a number of authors [Wassenaar et al. (1989,1992)], [Geyh et al. (1991)], [Long et al. (1992)], [Geyer et al. (1993)], [Geyer et al.(1994)], [Artinger et al. (1995)], and [Artinger et al. (1998)]. Different techniques of isolation and measurements of DOC are described by [Geyh et al. (1991)] and [Long et al. (1992)].

Groundwater dating by DOC has not yet become a routine method but, among other results, it clarifies some aspects of dating by DIC. [Geyh (1992)] deduces that the rather constant differences between the ^{14}C ages of the DIC and DOC components provide further evidence that hydrochemical reactions and isotope fractionation do not change the ^{14}C contents after groundwater recharging.

4.4.3. Hydrodynamic Models

Hydrodynamic models are concepts of groundwater-flow systems. These models express the relationship between the input and output of the radionuclide and the mixing of the individual groundwater components in a simplified mathematical form. They should be adequate for the hydrogeological system within the geological structure. They have been summarized, e.g., by [Fröhlich et al. (1977a)], [Yurtsever (1983)], [IAEA (1986c)], [Verhagen et al. (1991)], and by [IAEA (1994)]. The following models can be considered as basic (Fig. 9). For hydrogeological purposes, the piston flow model seems to be most appropriate for confined undisturbed aquifers, the exponential model for unconfined reservoirs and karst watersheds, and the injection model for confined aquifers with intrusion of young components.

Piston-Flow Model

The piston-flow model [Geyh (1980b)], [Yurtsever (1983)] considers a parallel flow of water particles, without mixing from the recharge area to the sampling point. The output concentration of the radionuclide is influenced only by its radioactive decay. The groundwater residence time is identical with the groundwater transition time through the hydrological system, with the age of the groundwater particles and with the radiometric age of the groundwater sample.

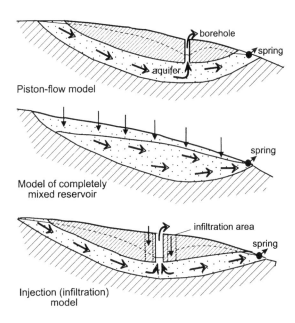

Figure 9: Some basic hydrodynamic models of groundwater flow.

The piston flow model is applicable in confined aquifers with low ground-water flow velocities. The groundwater flow pattern, velocities and hydraulic parameters of the aquifer can be derived from the age distribution in the area if steady state of the flow conditions can be assumed [Hanshaw et al. (1965)].

Model of a Completely Mixed Reservoir (Exponential model)

As a first approximation to the characteristics of a hydrological system, the so-called exponential model is used [Eriksson (1983)], [Siegenthaler et al. (1970)], also [Oeschger et al. (1974b)], [Fröhlich et al. (1977a)], [Geyh (1980b)], [Verhagen et al. (1991)]. Total mixing of all water components from different parts of the recharge area is assumed within the system which is considered to be a reservoir. The output concentration of the radionuclide at the outflow of the system equals the mean concentration in the system. A steady state of the recharge and discharge is assumed. The relative amount of individual water components of different origin with a certain residence time decreases exponentially with increasing residence time in the aquifer. The radiometric dating of completely mixed water yields an apparent radiometric age, which results from the final output concentration of the radionuclide.

For groundwater age values of up to a few thousand years, such as those that occur in radiocarbon dating in unconfined aquifers, the difference between the radiocarbon age and the mean residence time is not significant (and hence there is no great difference in using the piston flow or exponential model) [Geyh (1971)], [Oeschger et al. (1974b)]. With increasing ^{14}C groundwater age, this difference increases, as does the risk of erroneous conclusions.

The Injection Model

The injection model (also called the infiltration model) was proposed by [Fröhlich et al. (1977a)]. Significantly high tritium concentrations sometimes occur in confined aquifers, with radiocarbon groundwater age values of up to several thousand years. This phenomenon indicates admixing of contemporaneous groundwater with old groundwater in the aquifer. This frequently occurs in geological structures with multiple aquifers with different piezometric levels separated by aquicludes. The admixing of the contemporaneous (modern) tritium-containing water component occurs as an injection from the surface or from overlying aquifers to the lower aquifer through a fractured aquiclude or through leaky improper casing if the piezometric level in the deeper aquifer declines due to pumping. E.g., admixing of modern water with a confined aquifer due to its drainage by pumping from boreholes around a uranium mine was observed in Northern Bohemia [Mareš et al. (1978)]. The ratio of the two groundwater components can be calculated from the mass balance of tritium (see 4.5).

The Dispersion Model

The actual flow and mixing of natural as well as artificial tracers in hydrogeological systems is influenced by dispersion during the groundwater flow. For such cases, the use of dispersion models is recommended [Fröhlich et al. (1977a)]. Examples of these have been presented by [Geyh (1980b)], [Yurtsever (1983)], and [IAEA (1986c)] amongst others. They may be useful when evaluating tracer experiments [Klotz et al. (1980)]. In the hydrological application of environmental isotopes, however, the input of which is variable in space and time, their evaluation is very complex and has little practical use.

4.4.4. Radiocarbon and Tritium in Evaluating Groundwater Resources

Environmental radiocarbon and tritium have been used for investigation and dating of groundwater throughout the World. Numerous examples from different climatic regions with different geological settings were presented in the Proceedings of the symposia organized mainly by the International Atomic Energy Agency [IAEA (1963, 1967, 1970, 1974, 1978, 1984c, 1987, 1992c, 1996d, 1998i)]. Reports focused on arid regions were published by the [IAEA (1980a)], by [Verhagen et al. (1991)], and by [Adar et al. (1995)].

Radiocarbon dating can clarify the time of origin of groundwater and correlate it with geological and climatic events in the past.

A study in the Czech Republic and Slovakia provides an example. Fig. 10 depicts the frequency distribution of ^{14}C age values of 177 groundwater samples from the Bohemian Massif and from the Carpathian System illustrated by means of a histogram. In the histogram, each radiocarbon age including its standard statistical deviation 1σ is illustrated by a Gaussian curve with limits at unit area, which is the same for each represented radiocarbon age, in order to assign it with an adequate statistical weight. More accurate results with a low standard statistical deviation are represented by high and narrow Gaussian curves, while less accurate results are represented by lower and broader curves. The histogram is constructed by graphic summation of the Gaussian curves along the horizontal coordinate axis. Modern radiocarbon ages are represented as zero with the pertinent standard statistical deviation. Their number is given in the head lines as their peak reaches beyond the figure.

The climatic episodes of the Quaternary (Holocene, Pleistocene) are indicated below the x-coordinate. It is evident that a large part of the studied groundwater samples is not linked with the present short-term hydrological cycle, but rather with a long-term Quaternary cycle or cycles. This changes the opinion on the renewability of groundwater as a natural resource, which has to be considered in terms of the Quaternary time scale in thousands of years rather than in years. Moreover, the longer radiocarbon age and residence time indicate higher resistance of groundwater against pollution from the surface.

A relationship was observed amongst the radiocarbon groundwater age, the hydrogeological structures and the geomorphologic dissecting of aquifers; the oldest water was preserved in the deeper insulated parts of the basins, whereas faster circulation of younger groundwater prevails at their margins and/or in their dissected and uplifted parts.

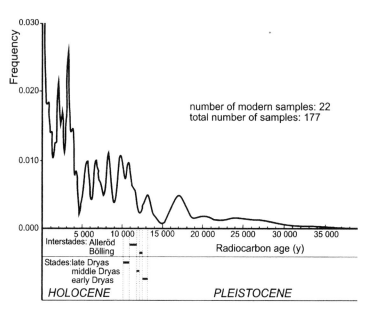

Figure 10: Frequency distribution of radiocarbon groundwater age values in the Bohemian Massif and the Carpathian System, after [Šilar (1990)].

Groundwater dating is even more significant in arid regions where groundwater recharge no longer exists and where groundwater resources originated in the geological past under different climatic conditions. Extracting groundwater resources in such regions results in their depletion and actually means groundwater mining similar to mining of non-renewable mineral resources. Groundwater development in such cases requires a special approach and economical considerations that are different from those in temperate humid regions. Fig. 11 depicts diagrams of distribution of radiocarbon groundwater age values in the desert and coastal regions in Egypt.

In different hydrogeological structures, even in a relatively small territory, the isotope composition and residence times of groundwater show a great diversity, which may be used for identification of the groundwater, its origin, components, and flow systems as is apparent from Figs. 12, 13 and 15.

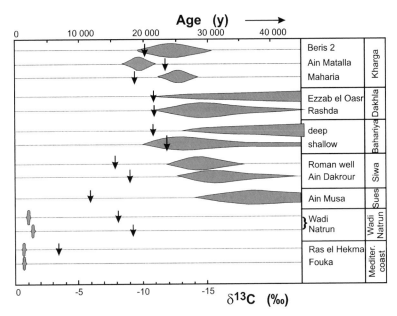

Figure 11: Distribution of radiocarbon "age values" of dissolved inorganic carbon in ground-waters from Egypt, after [Münnich et al. (1962)], $\downarrow = \delta^{13}C$.

Investigations by means of radiocarbon may be influenced by processes controlling the carbonate chemistry in groundwater. For example, during the study of groundwater resources in the Pinar del Rio Southern Coastal Plain in Western Cuba, analyses of radiocarbon and tritium together with stable isotopes (^{13}C, ^{18}O and ^{2}H) in rainfall, springs and wells were used. They determine the relationship between fresh water and marine water in the karstified limestone of Miocene age [Arellano et al. (1989)]. The intrusion of marine water into the karstified coastal aquifer results in oversaturation of the solution with respect to calcite, in low ^{14}C activity and in apparently high radiocarbon groundwater age. This is not in agreement with the available preliminary information related to the isotopic composition ($\delta^{18}O$, $\delta^{2}H$) measured in rainfall, springs and wells or with ^{3}H measurements [Arellano et al. (1987)]. The subsurface karst cavities and fissures influence the low ^{14}C activity. In this case, it is affected by diffusivity in the fissured medium with a porous matrix.

Systematic sampling followed by analyses to detect the concentrations of ^{2}H, ^{3}H, ^{13}C, ^{14}C, and ^{18}O were carried out in wells, rivers and small streams in the piedmont area and in the unsaturated zone in the detrital Tertiary of Madrid [López-Vera et al.(1996)]. A model could be established showing how the aquifer behaves. Its water resources could be estimated and new strategies for its exploitation taken into account. According to the results of the new

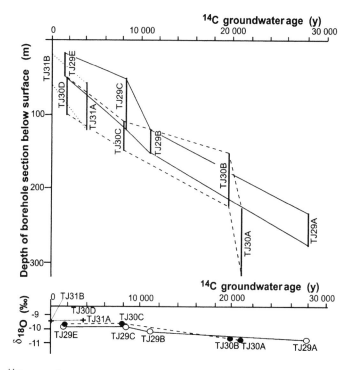

Figure 12: ¹⁴C groundwater age in confined aquifers in the southern part of the Cretaceous Třeboň basin (Czech Republic). Low temperature during the Pleistocene is indicated by the $\delta^{18}O$ values, which decrease with increasing ¹⁴C groundwater age and depth of the aquifers, after [Šilar et al. (1995)].

model, the recharging of the detrital Tertiary aquifer takes place simultaneously by different preferential means (zones of increased permeability) laterally through the mountainous margins and by infiltration in the stream beds of rivers flowing from the mountains. The precipitation in the surrounding mountains and the water in the rivers exhibit $\delta^{18}O$ and $\delta^{2}H$ values coinciding with the values of the groundwater in the wells at the valley bottoms. The samples in the unsaturated zone of the watershed divide areas, corresponding to the precipitation water, exhibited proportionate $\delta^{18}O$, $\delta^{2}H$ and deuterium excess values corresponding to those of water exposed to evaporation, which had no connection with the water of the saturated aquifer. The zones of discharge are the stream beds of small streams and rivers. Estimates of the radiometric ages of groundwater were revised. The major part of samples with radiocarbon age between 2000 and 6000 years B.P. exhibit tritium values of 2 to 5 T.U. These values have been reinterpreted as mixtures of waters. In some samples with ¹⁴C ages of up to 20000 years, the apparent aging was probably due to contamination of the sample by a fraction of calcium carbonate grains.

Radiocarbon and tritium, in combination with stable isotope analyses, may clarify the groundwater flow systems in correlation with the piezometric conditions.

Fig. 12 depicts the ^{14}C groundwater age increasing with the depth of the aquifer and δ^{18}O values indicating lower paleotemperatures at the time of the groundwater recharging of the lower aquifers.

The concentration of ^{14}C may be very specific under different hydrogeological conditions. Combination of the ^{3}H concentration and the values of $\delta^{13}C$ and $\delta^{18}O$ yields additional information about the different groundwater flow systems and regimes (Fig. 13).

The basin of Police nad Metují (a) consists of Cretaceous sandstone aquifers and marly aquicludes, which are as much as 400 m thick and are dissected by faults. Its margins are uplifted to a tableland and its surface is dissected by erosion. The groundwater circulation is fast, intensive and influenced by a high proportion of modern water.

The shallow northern part of the Třeboň basin (b) consists of Cretaceous and Tertiary sandstones and claystones as much as 110 m thick with a confined aquifer below a layer of claystone. A low ^{3}H activity and ^{14}C close to the modern standard indicate recent origin of the groundwater. The $\delta^{13}C$ value seems to be influenced by the organic carbon of peat deposits.

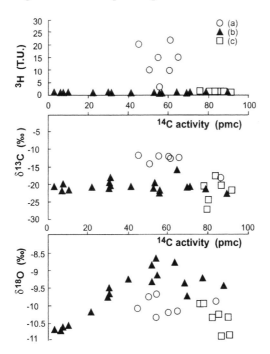

Figure 13: Activities of ^{14}C plotted versus ^{3}H, $\delta^{13}C$ and $\delta^{18}O$ in three basins (a)-(c) in the Bohemian Massif, after [Šilar et al. (1995)].

The deep southern part of the Třeboň basin (c) is a tectonic graben as much as 400 m thick, which is included in Fig. 12. It is filled with soft sandstones and claystones, with several confined aquifers. The lack of ^{3}H and ^{14}C indicates a high residence time. In some boreholes, groundwater of Pleistocene origin was encountered. The δ^{18}O indicates the origin of fossil groundwater components under cold climatic conditions. The groundwater circulation is slow and protected from surface interferences under natural piezometric conditions.

The isotope analysis reflects the geological structure and the geomorphology of the basins and improves the understanding of groundwater flow.

4.5.   Radiocarbon and Tritium and Environmental Problems

Radiocarbon was used to study and monitor air pollution due to combustion of fossil fuels in urban and industrial areas [Currie et al. (1983)], [Klouda et al. (1986)], [Berger et al. (1986)], [Kuc (1986)], [Awsiuk et al. (1986)], [Currie et al. (1986)]. Pollution produced by fossil fuel combustion can be distinguished from biological sources using radiocarbon. [Berger et al. (1983)] collected smog particles at four sites in California on clean glass or quartz fibre paper and analysed them in a small-volume CO_2 proportional counter for the ^{14}C contents. The results yielded relative contributions of fossil or modern carbon from 0 to 74 % or 26 to 100 %, respectively.

[Freundlich (1979)] investigated the environmental pollution of the atmosphere by automobile exhaust gases using natural radiocarbon in carbon dioxide as a tracer (although in this case the "marked" substance is free of ^{14}C, whereas the atmosphere contains it). Carbon dioxide from automobile exhaust can be clearly distinguished from atmospheric carbon dioxide. In localities where the turbulent conditions are good, the local fossil CO_2 is very low because all fossil CO_2 is readily swept away by the high rate of exchange. On the other hand, in places with low ventilation or exchange, the area of contamination is restricted in size.

Radiocarbon was also used to monitor the emissions from nuclear power plants [Segl et al. (1983)], [McCartney et al. (1986)], [Obelić et al. (1986)], [Povinec et al. (1986)], [Walker et al. (1986)] and from other nuclear installations. The aim was to obtain data on gas dispersion patterns and on the effects of releases of ^{14}C into the atmosphere.

[Otlet et al. (1983)] used tree-ring measurements (dendrochronology) to establish the chronology of emissions of ^{14}C close to the nuclear reprocessing plant at Sellafield, Cumbria, UK. These measurements indicate the importance of low-level ^{14}C measurements on natural materials because they provide data that would be difficult to obtain by physical instrumentation.

After collecting samples of milk, grain, and potatoes from local farms and after considering various criteria of representativeness, availability, wind patterns, and natural occurrence, hawthorn fruits (*Crataegus Monogyna*) were selected for sampling. Direct air sample measurements were also made to determine the relationship of air concentration to plant uptake. [Walker et al. (1986)] summarized the results of the measurements of the ^{14}C activity of hawthorn berries and examined the dispersion pattern contours constructed from them. Meteorology and topography were found to be important in determining the shape of the observed pattern.

Radiocarbon and tritium, together with stable isotopes, were found useful in solving environmental and engineering problems.

One of the most topical environmental problems is the disposal of solid and liquid wastes and the handling of sludge after the treatment of wastes from settlements, industry and agriculture. Any of the wastes may become a source of pollutants, which can be transported by groundwater and depreciate valuable groundwater resources; in addition, such activities can change groundwater flow systems and depreciate agricultural lands and housing. The value of isotope techniques in solving such problems is in identifying modern groundwater of shallow origin with potential contamination and distinguishing it from earlier or fossil groundwater of deeper aquifers.

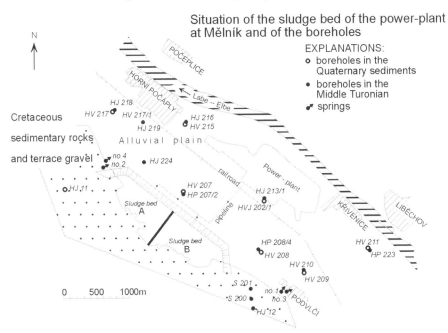

Figure 14: Map of the sludge bed of the power plant at Mělník and of the boreholes and springs, after [Šilar et al. (1993)].

During the hydraulic dumping of ashes in a sludge bed of the power plant at Mělník North of Prague, new springs emerged to the SE and NW of the sludge bed which dampened houses and properties. The sludge bed, as well as the springs, lie on the SW margin of the alluvial plain of the Labe River (Elbe) at the foot of a slope consisting of Cretaceous sedimentary rocks (Fig. 14, in the dotted area at lower left). Analysing the concentrations of radiocarbon and tritium permitted determination of the origin of the water and the ratio of the modern and original groundwater in the ground around the sludge bed [Šilar et al. (1993)].

A lower concentration of radiocarbon and a high concentration of tritium prevailed in samples collected in boreholes in the bedrock and in springs. This seeming contradiction indicates the presence of both old and modern groundwater components in the investigated samples and may be used for calculating the proportion of both. This is based on the mass balance of ^{3}H calculated from its concentration in both components and their resulting mixture. The prerequisite is the ability to determine the input concentration of ^{3}H in the modern groundwater. This concept corresponds to the two-component groundwater models, e.g., the model of base flow [Geyh (1980c)], and to the infiltration (injection) model [Fröhlich et al. (1977a)]. Attempts were made [Geyh (1980c)], [Balek (1983)], [Mazor et al. (1986)] and [Dietrich et al. (1986)] to calculate the proportions of the original components. Mixing of two water components having different concentrations of ^{14}C and ^{3}H was considered for the sludge bed of the power plant at Mělník:

$$C_{Ts} = a_m \cdot C_{Tm} \quad (19) \qquad \text{and} \qquad a_m = \frac{C_{Ts}}{C_{Tm}} \qquad (20)$$

where C_{Ts} is the concentration of ^{3}H in the investigated sample, a_m is the part of the modern component of water in the unit volume of the investigated sample, and C_{Tm} is the concentration of ^{3}H in the modern component of groundwater at the time of its infiltration. The concentration of ^{3}H in the fossil component of groundwater is zero.

It was advantageous set the modern tritium concentration equal to the value measured in a new spring that was active only during the hydraulic transport of ashes to the sludge bed. As much as 67 % of modern water was determined in the new springs in the affected village from the mass balance of ^{3}H in springs and boreholes.

In addition to the seepage from the sludge bed, the relationships between the groundwater in the Quaternary sediments and in the underlying Middle Turonian confined aquifer were studied by means of ^{18}O, ^{13}C, ^{3}H, and ^{14}C. Groundwater, exhibiting high concentrations of ^{3}H and ^{14}C, occurs in the upper aquifer in Quaternary sediments. Low ^{14}C concentrations and very low ^{3}H concentrations below the detection limit prevail in the lower Middle

Turonian confined aquifer, indicating a very long groundwater residence time with the exception of two boreholes (No. 14 and 16), in which mixed water was encountered. High $\delta^{13}C$ values (between -5.1‰ and +0.2‰) indicate carbon with an origin different from that in the shallow groundwater of the alluvial plain (Fig. 15).

Isotope methods were found useful when solving aspects of groundwater contamination by waste disposal and tailings of mines and when designing remediation activities [Jacob et al. (1998)], [Milintawisamai et al. (1998)], [Verhagen et al. (1998)].

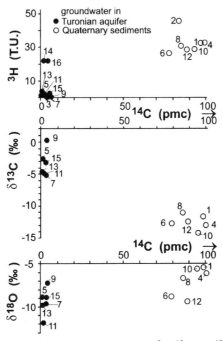

Figure 15: Correlation between ^{14}C and other isotopes (^{3}H, ^{13}C and ^{18}O) in boreholes (No. 1-15) in the vicinity of the power plant at Mělník in the alluvial plain and in the underlying bedrock, after [Šilar et al. (1993)].

4.6. Radiocarbon and Tritium in Technology

Measurements of ^{14}C and ^{3}H concentrations, either alone or in combination, when compared with atmospheric concentrations, are very effective means for discovering forgeries. ^{14}C dating can be used for dating writing materials such as paper, parchment or vellum in order to authenticate archival documents. The subject has been analysed in detail by [Burleigh et al. (1983)].

Among other applications, the dating of very small samples and the authentication of hitherto imprecisely dated archaeological finds and museum objects have become possible. The dating of paper entails some specific problems linked with the technology of its manufacturing. After about 1840, the use of rags for paper manufacture was superseded by the introduction of wood pulp. Wood pulp averages the age of the trees from which it is manufactured and again probably incorporates material representing up to 100 years or more of growth. Thus, it is evident that any apparent age obtained for paper will antedate its manufacture and use. From this it is clear that the only valid ^{14}C date for a paper document will be one that is significantly older than its purported historical date. In contrast, parchment and vellum were made from the skins of young animals (sheep or goats and calves, respectively), generally used soon after manufacture, and therefore can be expected to have much smaller, usually negligible, apparent age. However, parchment and vellum were often cleaned to remove the original ink and re-used. In view of these problems, it is essential that a detailed technical examination of any document be carried out before it is subjected to ^{14}C dating. In summary, the only unequivocal application of radiocarbon dating to documents appears to be for invalidation purposes, where a document purported to be much earlier was actually made from material incorporating artificial ^{14}C derived from testing of nuclear weapon. [Burleigh et al. (1983)] present a review of simple tests to determine other substances that may have been added during manufacture and which may contain carbon, in order to isolate the original cellulose or protein from the dated samples of paper and textiles.

The ^{14}C and ^{3}H concentrations can also be used for proving the origin and the authenticity of other organic materials, for instance wooden and textile artefacts (see 4.2) and food products.In addition, they have also been used in reconstructing the concentrations in the environment in the past.

A simple method for determining the ^{14}C concentration in ethanol was developed using an ultra low-level scintillation counter [Schönhofer (1989)]. Synthetic ethanol and other organic compounds synthesized from oil and coal-based materials can be distinguished from those of natural origin on the basis of the very low or zero ^{14}C activity. In this way, a large shipment of allegedly natural ethanol was discovered to be synthetic. Similarly, this method can be used in organic synthesis. For example, a sample of guayazulene, declared as synthetically produced, was proven to be of natural plant origin.

[Roether (1967)] measured ^{3}H concentrations in recent wines to find corresponding ^{3}H concentrations in the environment. He showed that the tritium content of a wine sample is not determined exclusively by water taken up by the roots, but is also influenced to a large extent by direct exchange with atmospheric moisture. The soil-water fraction normally amounts to not more than 40 %. Thus, wine is a sample partly of atmospheric moisture at ground level, and partly of soil moisture, integrated over a period of around three weeks before vintage. The first distinguishable influence of bomb tritium shows up in 1953 wine. From wine data, Roether calculated the average values of ^3H in rain for the pre-thermonuclear period at 6±1.5 T.U.

[Eichinger et al. (1980)] used wine of different regions in Germany and in South Africa, and the wood of tree rings to reconstruct the input of ^{14}C and ^3H for the previous years. To measure the ^{14}C activity, they extracted the ethanol of the wine by fractionated distillation. The ^3H age (not indicated in Fig. 16) was measured in the aqueous part of the sample.

The results have shown that the ^{14}C concentration in German wines in the years 1963/64 increased to 1.9 times the natural level. No significant difference among the individual wine regions was observed. In the wine from South Africa, the ^{14}C content reached a maximum one year later, as the major part of the nuclear tests took place in the Northern hemisphere. The results of ^{14}C and ^3H measurements in wine and wood samples are representative for the period of growth during the particular year.

[Wolf (1993)] investigated a sample of wine allegedly of 1787 vintage. The ^{14}C concentration of the ethanol fraction was measured using an accelerator mass spectrometer, and partly in a liquid-scintillation spectrometer. The mean ^{14}C activity was 131.6±1.6 pmc using 2σ deviation. The ^3H of the

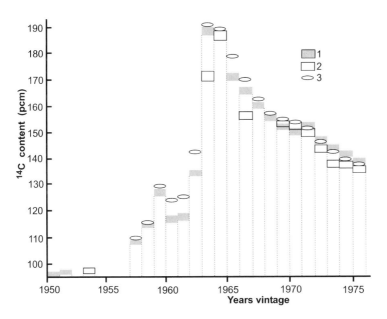

Figure 16: ^{14}C content of German wines of different vintages (1 – Franken, 2 – Rheinpfalz, Mosel and Baden) and 3 – of atmospheric CO_2 after [Rauert (1980)].

water fraction of wine was processed to propane and measured in a gas proportional detector. The mean ^{3}H activity was 73.2±9.3 T.U. (8.64±1.01 Bq/l) using 2σ deviation. It was concluded that the wine, if produced from a single vintage, must have originated in 1962, or, if mixed from the 205 year old wine, must contain more than 40 % of wine from the period 1963 to 1975.

5.   OTHER RADIONUCLIDES DATING METHODS
Jan Šilar

5.1.   Silicon ^{32}Si Dating

Silicon ^{32}Si is of cosmic-ray origin as a result of nuclear interactions of cosmic-ray particles with atmospheric ^{40}Ar nuclei. [Lal et al. (1970)] demonstrated that it could be used in groundwater studies, where a comparison of ^{32}Si and ^{14}C ages would be extremely valuable in the delineation of groundwater characteristics.

^{32}Si is washed out from the atmosphere by precipitation, in which it is dissolved as monomeric silicic acid. In European precipitation, the natural specific activity of ^{32}Si lies in the range of about 0.2 to 0.6 dpm (about 3 to 10 mBq) per cubic metre of water. This concentration increased in the 2nd half of the 1960s, indicating a contribution of bomb-produced ^{32}Si [IAEA (1983f)]. The half-life of ^{32}Si was estimated to lie in a very wide range [Lal et al. (1970)]. Following later determinations, the value of 105 years was adopted [Fröhlich et al. (1987)], see also Chap. 2, Table 2.

Due to the very low specific ^{32}Si activity, the ^{32}Si for groundwater dating must be extracted from groundwater samples with a volume of 5 to 20 m^3 [Geyh et al. (1980)]. This can be done either by scavenging with Fe(OH)$_3$ [Lal et al. (1970)] or by trapping on a strongly basic ion exchanger [Fröhlich et al. (1977b)]. Instead of the ^{32}Si activity, the activity of its daughter product ^{32}P is measured after storing the extracted sample for two to three months to reach equilibrium with the ^{32}Si activity and after preparing a radiochemically pure sample of ^{32}P in the form of Mg$_2$P$_2$O$_7$ [IAEA (1983f)].

[Fröhlich et al. (1987)] investigated the ^{32}Si concentration in soil water in the unsaturated zone, in plant material, and in groundwater in different types of aquifers (in sand, sandstone, and karst limestone). They concluded:
- ^{32}Si seems to be preferably fixed by clay, iron hydroxides and humic matter.
- ^{32}Si initial concentrations in uncovered sandy aquifers and limestone aquifers can be expected to reach values above 10 mBq·m^{-3}.
- Local variation of the ^{32}Si initial concentration can be understood as a consequence of variations of the thickness and composition of the unsaturated zone.

– In any case, non-radioactive losses must be taken into account in the interpretation of ^{32}Si groundwater studies.

It was hoped that study of ^{32}Si in groundwater could be used because of its half-life, which is between those of ^{3}H and ^{14}C. Its application, however, is limited partly by extraction and measurement techniques, partly by geochemical and biochemical processes in the soil. Therefore, it has been accepted with scepticism. [Fröhlich et al. (1987)] summarized that ^{32}Si may be used as a hydrological tracer in special aquifers. A further improvement of the ^{32}Si method would require a better specification of factors influencing the geochemical behaviour of ^{32}Si.

5.2. Argon ^{39}Ar Dating

^{39}Ar is of cosmic ray origin, produced mainly by the ^{40}Ar (n,2n)^{39}Ar process. Its half-life is 269 years and its atmospheric activity is 0.107±0.004 dpm (1.78±0.07 mBq) per litre of argon [IAEA (1983f)]. Anthropogenic contributions can be neglected and the major part of the ^{39}Ar activity is concentrated in the atmosphere, so that the input into other reservoirs can be assumed to be uniform.

After infiltration, groundwater containing ^{39}Ar of atmospheric origin becomes a closed system and can be dated according to the decrease in the ^{39}Ar activity of the dissolved argon.

The possible use of ^{39}Ar in isotope hydrology was investigated by [Oeschger et al. (1974)]. The ^{39}Ar half-life of 269 years allows dating of samples in the range of 50 to 1000 years, thus filling the gap in the range between ^{3}H and ^{14}C. As a noble gas isotope, ^{39}Ar is not involved in chemical processes. The following technical procedures are necessary for ^{39}Ar age determinations:

– Extraction of the gases contained in water to obtain the argon sample required for the radioactivity measurement.

– Separation of the argon from the collected gases, with special attention paid to the separation of krypton because of the relatively high ^{85}Kr activity from the nuclear industry.

– Measurement of the ^{39}Ar activity of the argon component. The details of the technical procedures were developed, tested and described by [Oeschger et al. (1974)].

To determine the ^{39}Ar activity in water samples, it is necessary to test the potential contamination of the extracted gas samples during collecting in the field and processing in the laboratory. This may be checked by measuring the ^{85}Kr activity of the same samples.

First, the ^{39}Ar dating method was successfully checked by measuring argon extracted from polar ice samples of known age.

The volume of water, which has to be processed, depends on the volume of the proportional gas counter. Originally, an argon sample of 3 litres had to be extracted from about 20 tons of water [Oeschger et al. (1974)]. For smaller high-pressure proportional counters, which only need about 0.2 litres of argon, degassing of only 1-1.5 m^3 of water yields enough gas for ^{39}Ar determination. The gases extracted from water in the field are compressed in cylinders and transported to the laboratory.

The ^{39}Ar dating method was applied in different aquifers mainly to compare the results of ^{39}Ar and ^{14}C dating. Usually, the ^{14}C groundwater age values are much higher. This can be explained when assuming that the water sample is a mixture of different components [Loosli et al. (1979)]. ^{39}Ar activities above the atmospheric level (100 %) were measured in thermal waters from granitic rock formations in Switzerland. This excess of ^{39}Ar is produced underground by the ^{39}K $(n,p)^{39}$Ar process. Apparently diffusion of the ^{39}Ar out of the grains into the water in these types of rocks is rapid enough to produce the measured ^{39}Ar excess. For other aquifers, it could be demonstrated that production of ^{39}Ar underground can be neglected [IAEA (1983f)].

5.3.   Krypton ^{85}Kr Dating

The natural level of ^{85}Kr in the atmosphere before the nuclear age resulted from the production of ^{85}Kr by the (n,γ) reaction of cosmic neutrons with the stable ^{84}Kr in the atmosphere and from the ^{85}Kr released from the earth owing to spontaneous fission of uranium and thorium. The expanding nuclear industry contributed considerably to the increase in the ^{85}Kr concentration [Rózański et al. (1979)] (see Chap. 2, 1.2.2, 3.4.5 and Chap. 3, 2.2). The production rate of ^{85}Kr in various rock types was described by [Florkowski (1992)]. However, it is negligible in comparison with the production rate in the atmosphere.

The half-life of ^{85}Kr (10.76 years) is similar to that of tritium. However, the ^{85}Kr concentration in the environment increases continually, unlike ^{3}H, which has been decreasing since 1963.

The atmosphere is the main reservoir of krypton, only 2.8 % of the total mass of krypton is contained in the hydrosphere. Precipitation and surface water are in equilibrium with the atmosphere. After infiltration, the activity of ^{85}Kr decreases due to its decay, which enables the dating of groundwater by measuring the ^{85}Kr concentration. The mean concentration of ^{85}Kr in the northern hemisphere is about 780 dpm/mmol (13 Bq/ mmol) Kr. The krypton concentration in the atmosphere is $1.1 \cdot 10^{-4}$ vol. %, and its solubility in water at 15°C is $6.9 \cdot 10^{-6}$ vol. %. This results in a concentration of 0.07 cm^3 krypton in 1 m^3 water, corresponding to an actual activity of 2.6 dpm (43 mBq). [Rózański et al. (1979)].

[Rózański et al. (1979)] summarized the conditions for the ^{85}Kr dating method in water:

- The only well-mixed reservoir of krypton is the atmosphere.
- The concentration of ^{85}Kr in the atmosphere is steadily increasing and is known with sufficient accuracy.
- Since krypton is a noble gas, it does not interact with the matrix of the aquifer.

Beyond that, [Rózański et al. (1979)] developed a procedure for collecting and processing water samples. In the field, 120 to 360 litres of water are collected in about 120-litre nitrogen-filled plastic containers. The water is pumped through a furnace where it is heated to 90°C and sprayed in an extraction container in which gases are separated from water and collected in a plastic bulb. The degassing of 360 litres of water yields about 7 litres of gases. From the plastic bulb, the gases are pumped by a vacuum oil rotary pump through traps cooled with liquid nitrogen to separate water and CO_2 and then through a charcoal column cooled with dry ice. In successive steps, water and CO_2 are separated in liquid nitrogen traps. Krypton and argon are selectively separated on dry-ice cooled charcoal, desorbed by heating to 350°C and passed at 500°C through a furnace containing metallic barium, which removes oxygen and nitrogen from the gases. After removing radon gas in metallic U-tubes before and after the furnace, and after completion of the reaction, about 0.7 cm^3 of argon and krypton mixture remains in the furnace, and is transferred to a miniature proportional counter. A mass spectrometer is used to determine the amount of krypton in the mixture. The sample gas is compared with standards of known krypton concentration in an argon-krypton mixture. The activity measured in the low-level counting system is related to the volume present in the sample.

Both authors tested the method by comparing ^{85}Kr, ^{3}H, and ^{14}C concentrations in 14 samples of surface and groundwater and concluded: The results prove that the krypton method works satisfactorily and can be used for dating young water bodies. It may provide more unequivocal values than the tritium method, where the input function is complicated. The application of the method is very limited due to the very complicated sampling and analytical work.

In a comparative study performed in a Tertiary fluvial aquifer in the Styrian Basin in Austria and in the East Midlands Triassic Sandstone in the United Kingdom, several available groundwater dating methods were compared, including that using ^{85}Kr [Andrews et al. (1984)]. ^{85}Kr was used as an indicator of atmospheric contamination to exclude discrepancies between ^{39}Ar and ^{14}C data. The study concluded that isotope techniques, which can yield direct information concerning groundwater residence times, consist in measurements of ^{3}H, ^{85}Kr, ^{14}C, and ^{39}Ar. The quantitative importance and significance of "age" estimates required for any specific groundwater investigation should be considered very carefully in selecting some or all of these isotope techniques, and interpretations should reflect all the uncertainties involved.

6.   K - Ar AND Ar - Ar DATING METHODS
Emil Jelínek

6.1.   Chemical Properties, Radioactive Decay and Isotopic Abundance

Potassium (atomic number Z=19) is a lithophile element, which belongs to the group of alkali metals. It is the eighth most abundant element in the Earth's crust and it is present as a major constituent in numerous minerals. Potassium has three natural isotopes: ^{39}K with abundance of 93.2581 %, ^{40}K of 0.01167 % and ^{41}K of 6.7302 %. Potassium 40 is radioactive and decays by branched disintegration (Fig. 17) to the stable isotopes ^{40}Ca (β^- decay) and ^{40}Ar (β^+ decay, electron capture). The decay constants describing the branched ^{40}K-^{40}Ar decay are designated as follows by [Steiger et al. (1977)]:

$$\lambda_{\beta-} \quad = 4.962 \cdot 10^{-10} \, y^{-1.}$$

$$\lambda_{EC, \, \beta+} = 0.581 \cdot 10^{-10} \, y^{-1} \text{ (also includes the } \lambda_{\beta+} \text{ decay constant)}$$

$$\lambda_{total} \quad = 5.543 \cdot 10^{-10} \, y^{-1}$$

Argon (Z=18) belongs to the group of rare gases. It is the third most abundant component of the Earth's atmosphere (0.934 vol. %). After He it is the most abundant noble gas in rocks and minerals. Argon has three natural stable isotopes: ^{36}Ar with abundance of 0.337 %, ^{38}Ar of 0.063 %, and ^{40}Ar of 99.6 %. The ratio of the abundance of N_{40Ar}/N_{36Ar}=295.5 is used as a constant to correct the Ar isotopic composition in minerals for the Ar contribution from the atmosphere [Nier (1950)], [Steiger et al. (1977)]. In addition, the short-lived isotopes ^{37}Ar and ^{39}Ar (half-lives 35 d and 269 y) are present in nature, mostly as products of cosmic irradiation.

While the K-Ca method of dating has received only little attention due to difficulties in distinguishing radiogenic and common ^{40}Ca in rocks and minerals, the K-Ar decay, both conventional K-Ar and Ar-Ar techniques, are suitable for dating the thermal (magmatic and metamorphic) and sedimentary histories of minerals and rocks.

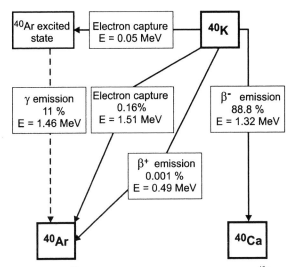

Figure 17: Decay scheme of ^{40}K showing branched disintegration to ^{40}Ar and ^{40}Ca, based on [Geyh et al. (1990)].

6.2. Concept of the Method

Growth of radiogenic products of ^{40}K decay in a closed system is described as

$$N_{40_{Ar^*}} + N_{40_{Ca^*}} = N_{40_K} \cdot \left(e^{\lambda_{total} \cdot t} - 1\right) \qquad (21)$$

where " * " denotes the radiogenic portion of number of Ar and Ca atoms at the present according to decay time t (see Eq. 2), ^{40}K is the number of the atoms at the present, and λ_{total} (see Fig. 18) is defined as:

$$\lambda_{total} = \lambda_{total} + \lambda_{EC,\beta^+} = \lambda_{total} + \lambda_{EC} + \lambda_{\beta^+}$$

The portion of N_{40Ar^*} which results from ^{40}K decay can be expressed as $[\lambda_{EC}+\lambda_{\beta^+}]/[\lambda_{total}]$. Accordingly, for the Ar branch of the decay scheme, Eq. 21 reduces to:

$$N_{40_{Ar^*}} = \frac{\lambda_{EC} + \lambda_{\beta^+}}{\lambda_{total}} N_{40_K} \cdot \left(e^{\lambda_{total} \cdot t} - 1\right) \qquad (22)$$

from which

$$t = \frac{1}{\lambda_{total}} ln\left(\frac{\lambda_{total}}{\lambda_{EC} + \lambda_{\beta^+}} \frac{N_{40_{Ar^*}}}{N_{40_K}} + 1\right) \qquad (23)$$

and, inserting the values of decay constants into Eq. 23,

$$t = 1.804 \cdot 10^9 \cdot \ln\left(9.54\frac{N_{40_{Ar}*}}{N_{40_K}} + 1\right) \qquad (24)$$

As the potassium isotopes do not fractionate during geological processes, the ^{40}K concentration (in wt. %) can be expressed as a portion of the total K content in the sample as

$$[^{40}K] = 0.0001167 \cdot [K] \qquad (25)$$

Similar to other radiogenic isotope dating techniques, the K-Ar method can yield reliable age values only assuming that:
- no initial excess of $N_{40_{Ar}*}$ was present in the sample at the time of mineral (rock) crystallisation or closure of the K-Ar system and
- the mineral or rock has behaved as a closed system with respect to both K and Ar.

6.3.   K - Ar Analytical Technique

Content of ^{40}K in a rock or mineral sample is usually determined by chemical analysis of the total K content. From the whole spectrum of methods (e.g., x-ray fluorescence, neutron activation, isotope dilution), flame photometry is the most suitable due to its simplicity, low cost of analysis and sufficient accuracy (ca ± 1 %). Ar is released from powdered rocks and minerals in a vacuum extraction line with an inductively heated furnace (usually Mo crucible). After spiking with ^{38}Ar, argon is separated from the gaseous mixture using hot copper oxide, cold traps, or molecular sieves, in a charcoal and titanium furnace. The isotopic composition of the unspiked portion (see Chapt. 5, 2.1 and 2.4) and concentration of the spiked portion are determined using mass spectrometric techniques. A crucial step in the Ar analysis is complete sample outgassing which should remove most of the atmospheric Ar. The total number of ^{40}Ar atoms in a sample is summed up as:

$$N_{40_{Ar_{total}}} = N_{40_{Ar_{atm}}} + N_{40_{Ar_i}} + N_{40_{Ar}*} \qquad (26)$$

where $N_{40_{Ar_{atm}}}$ is argon from the atmosphere, $N_{40_{Ar_i}}$ is initial excess argon, and $N_{40_{Ar}*}$ is argon from decay of ^{40}K.
Therefore, the radiogenic $^{40}Ar^*$ in Eq. 21 is given by:

$$N_{40_{Ar}*} = N_{40_{Ar_{total}}} - N_{40_{Ar_{atm}}} - N_{40_{Ar_i}} \qquad (27)$$

Often the extraction of all atmospheric Ar from the system is very difficult. Assuming that no initial excess Ar was present in the sample, the presence of ^{36}Ar indicates an atmospheric contribution. The present atmospheric N_{40Ar}/N_{36Ar} ratio is 295.5. In this case, ^{40}Ar* can be corrected for atmospheric argon Ar$_{atm}$ as follows:

$$N_{40_{Ar}*} = N_{40_{Ar_{total}}} - 295.5 \cdot N_{36_{Ar_{measured}}} \qquad (28)$$

where $N_{36Ar_{measured}}$ is the value measured in the sample.

The isotopic composition of the initial excess argon ($N_{40Ar_{init}}$) can be determined from the N_{40Ar}/N_{36Ar} vs. N_{40K}/N_{36Ar} isochron diagram. Eq. 27 can be normalised by non-radiogenic ^{36}Ar to yield

$$\left(\frac{N_{40_{Ar_{total}}}}{N_{36_{Ar_{total}}}} \right) = \left(\frac{N_{40_{Ar_{atm}}}}{N_{36_{Ar_{atm}}}} \right) + \left(\frac{N_{40_{Ar_i}}}{N_{36_{Ar_i}}} \right) +$$
$$+ \left(\frac{N_{40_K}}{N_{36_{Ar_{total}}}} \right) \cdot \frac{\lambda_{EC,\beta^+}}{\lambda_{total}} \left(e^{\lambda_{total} \cdot t} - 1 \right) \qquad (29)$$

6.4. Ar - Ar Analytical Technique

The classical K-Ar method of dating is based on the assumption of sample homogeneity, i.e., the same K/Ar ratio and Ar isotopic composition in all the analysed sample aliquots. The Ar-Ar technique is a modification of the classical K-Ar method. The stable ^{39}K in the sample is converted to ^{39}Ar by irradiation with fast neutrons (E>1 MeV) in a nuclear reactor:

$$^{39}\text{K (n,p) }^{39}\text{Ar}$$

The produced ^{39}Ar reflects the potassium content and can be analysed in the same aliquot as ^{40}Ar*. This procedure reduces the possible effects of lack of sample homogeneity. With a half life of 269 years, ^{39}Ar decays back to ^{39}K by β$^-$ emission and can be considered as a stable isotope for mass spectrometric determination. The production of ^{39}Ar in the sample by the irradiation of ^{39}K is described as:

$$N_{39_{Ar}} = N_{39_K} \cdot t_{ir} \cdot \int_{min_E}^{max_E} \phi_E \cdot \sigma_E \cdot dE \qquad (30)$$

where $N_{..}$ is the number of atoms of ^{39}Ar or ^{39}K, t_{ir} is the irradiation time, ϕ_e is the flux density of neutrons with the energy E, and σ_E is the effective capture cross section of ^{39}K. By dividing Eq. 22 by Eq. 30, we obtain:

$$\frac{N_{40Ar^*}}{N_{39Ar}} = \left[\frac{\lambda_{EC,\beta^+}}{\lambda_{total}} \cdot \frac{N_{40K}}{N_{39K} \cdot t_{ir} \cdot \displaystyle\int_{min_E}^{max_E} \phi_E \cdot \sigma_E \cdot dE}\right] \cdot \left(e^{\lambda_{total} \cdot t} - 1\right) \qquad (31)$$

Referred to as the irradiation parameter J^1, the expression in square brackets is similar for each sample or standard. According to Eq. 31, J can be defined using standards (monitors) of known age that are irradiated together with the samples as

$$J = \left(e^{\lambda_{total} \cdot t_{monitor}} - 1\right) \cdot \frac{N_{39Ar_{monitor}}}{N_{40Ar^*_{monitor}}} \qquad (32)$$

where $N_{39Ar_{monitor}}$ and $N_{40Ar^*_{monitor}}$ are the number of atoms in the monitored standard.

By combining with Eq. 31:

$$t = \frac{ln\left(1 + J \cdot \dfrac{N_{40Ar^*}}{N_{39Ar}}\right)}{\lambda_{total}} \qquad (33)$$

Accurate values of J for each sample are usually obtained by spatial extrapolation from several standards in the sample holder.

The irradiation of a sample by fast neutrons induces competitive Ar-reactions as described in Figure 18: ^{40}K (n,p) ^{40}Ar, ^{40}Ca (n,n'),α ^{36}Ar, ^{42}Ca (n,γ),α ^{39}Ar. [Mitchell (1968)], and [Turner (1971)], [Tetley et al. (1980)] have suggested several correction procedures but generally samples with low Ca/K ratios are more suitable for Ar-Ar dating.

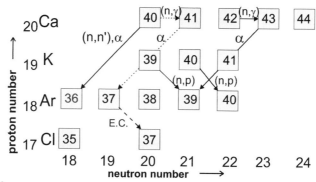

Figure 18: ^{39}Ar production and interfering reactions during irradiation of a sample by fast neutrons.

6.5. Step-Heating and Laser Probe Techniques

The main advantage of the Ar-Ar technique is that the age of the sample can be calculated solely on the basis on its N_{40Ar*}/N_{39Ar} ratio. Argon can be progressively liberated from different domains if the sample is incrementally heated in a furnace. Accordingly to different domains, a single sample can provide a series of N_{40Ar*}/N_{39Ar} ratios, which correspond to different age values. In the case of a disturbed K-Ar system, domains like the outer rims of mineral grains are more susceptible to Ar-loss; such systems should be outgassed during the first heating steps. The values of their N_{40Ar*}/N_{39Ar} ratios should correspond to lower age compared to argon liberated from the mineral cores. This technique is of great advantage for samples with a complex geological history, which also included periods of Ar-loss, e.g., samples from polyphase metamorphic terrains. Usually, the Ar diffusion rates for different temperatures are known with sufficient accuracy in the sample mineral. Therefore, by incremental heating, the resulting Ar spectrum (Fig 19a) can then be interpreted in terms of its thermal history. In this spectrum, periods of Ar loss are clearly distinguishable from a plateau whose age corresponds to the non-disturbed part of the sample. Alternatively, the Ar data obtained by step-heating for different temperatures from a single sample can be presented in the form of an N_{40Ar}/N_{36Ar} vs. N_{39Ar}/N_{36Ar} isochron (Fig. 19b), whose slope corresponds to the plateau age of the Ar spectrum diagram.

Complementary to the classical step-heating technique there are various applications of laser ablation in Ar-Ar dating. First, [Mergue (1973)] applied a Laser probe to the Ar-Ar technique. This application used a ruby laser to produce spots ca 100 μm in diameter. The resulting age values corresponded

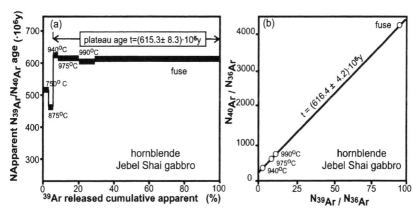

Figure 19: (a) Ar step-heating spectrum with a plateau age for hornblende from Jebel Shai gabbro (Saudi Arabia), (b) corresponding N_{40Ar}/N_{36Ar} vs. N_{39Ar}/N_{36Ar} isochron, based on [Geyh et al. (1990)]

to the total fusion analysis. With the progress in mass spectrometry and laser techniques, Ar released from smaller laser pits can now be analysed. The laser power can be increased during the analysis to produce Ar age spectra similar to the step-heating technique [Layer et al. (1987)], [Lee et al. (1991)], [Wright et al. (1991)]. The advantage of the laser probe Ar-Ar technique compared to the conventional method is that much smaller samples (down to 1 μg) can be analysed, the age can be measured for different domains of a sample grain and minerals can be dated in situ in a rock section, enabling age values to be related to the structures.

6.6. Applicability and Limitations of K - Ar and Ar - Ar Techniques

The K-Ar method represents a powerful tool for dating young volcanic rocks. Much attention has been paid to dating volcanic rocks of the ocean floor. The absolute age of the geomagnetic reversals have been calibrated using K-Ar and Ar-Ar techniques [Harland et al. (1982)]. Due to the gaseous nature of argon, only fresh magmatic rocks or their minerals that have cooled quickly below their respective blocking temperatures can yield magmatic crystallisation ages. In slowly cooled magmatic rocks, Ar can diffuse within mineral lattices, exchange between minerals or can be trapped by magmatic fluids. The resulting K-Ar and Ar-Ar age can only be interpreted as the cooling age, representing cooling through a particular blocking temperature T_B. Such temperatures were defined by [Dodson (1973)]

$$T_B = \frac{E}{R \cdot ln\left(\frac{A \cdot \tau \cdot D_0}{a^2} \right)} \qquad (34)$$

where A refers to mineral shape (55, 27 and 9 for sphere, cylinder and sheet, respectively), a is the average diffusion path, R is the gas constant, E is the activation energy for Ar diffusion, D_0 is the thermal diffusivity for a particular mineral and τ is a time constant, which reflects the cooling rate through the blocking temperature.

Blocking temperatures can be calculated from Ar step-heating data [Buchan et al. (1977)], [Berger et al. (1981)]. This enables direct calculation of blocking temperatures for dated minerals and reconstruction of the thermal history of the rock.

K-Ar and Ar-Ar methods also have great potential in dating the cooling of metamorphic rocks. There the concept of blocking temperatures is similar to that described above and the resulting age may be interpreted in terms of regional cooling associated with uplift. In addition, sedimentary rocks (shales) and K-bearing diagenetic minerals (e.g., glaucony) can also be dated using the K-Ar and (Ar-Ar) techniques.

The major limitation of both K-Ar and Ar-Ar methods is dating of rocks with polyphase history, such as those from major orogenic belts. The resulting age almost exclusively corresponds to the late stages of their evolution. In order to date the early phases, other methods of dating (e.g., Rb-Sr, Sm-Nd or U-Pb, see 7.-9.) must be employed.

6.7.   Representative Examples

The K-Ar method of radiogenic dating has been used extensively for dating young volcanic rocks. [Pankhurst et al. (1983)] have carried out K-Ar whole-rock dating of the tholeiitic and andesitic rocks of the South Shetland Islands (Fig. 20). The data have shown a shift to more alkaline magma compositions in recent eruptions. The progressive movement of volcanic centres towards the NE can be correlated with the plate movement in the area. Accordingly, K-Ar dating may form a basis for understanding the plate tectonic evolution.

Laser probe Ar-Ar dating of single crystals of sanidine, plagioclase and quartz in volcanic ash from the island of Kos (Greece) has been decoded as the history of a volcano. Four distinct volcanic events from 161 to $1728 \cdot 10^3$ y have been identified. The step-heated cooling of schist fragments from the ash provides evidence for a $250 \cdot 10^6$ y old crust under the volcano [Smith et al. (2000)].

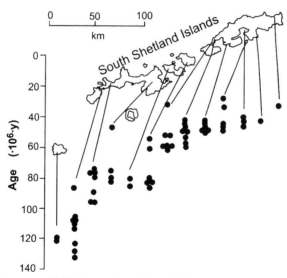

Figure 20: Outline map with K-Ar age (dots) from tholeiitic and andesitic rocks of the South Shetland Island, Lesser Antarctica, after [Pankhurst et al. (1983)].

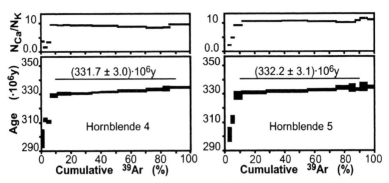

Figure 21: Ar age spectra for hornblende from Staré Sedlo orthogneisses, Bohemian Massif, after [Košler et al. (1995)].

In the deformed granitoid rocks (orthogneisses) of the Bohemian Massif the Ar-Ar (hornblende) and Rb-Sr (whole rock-biotite) dating techniques have yielded consistent age in the range of $336\text{-}331 \cdot 10^6$ y. This suggests rapid cooling through Ar-Ar and Rb-Sr blocking temperatures for hornblende (ca 500° C) [Košler et al. (1995)] (see Fig. 21) and biotite (ca 300° C), respectively. This has been explained on the basis of rapid exhumation of the rock that was temporally and spatially associated with intrusion of younger granitoid rocks of the Central Bohemian Batolith.

The high spatial resolution Ar-Ar dating of phlogopite grains in xenoliths in kimberlites from Malaiata (Solomon Islands) and the Elovy Islands (Kola Peninsula, Russia) indicate transport times of hours to days depending upon the kimberlite magma temperature. The phlogopite grains preserve the Ar-Ar age recorded at high temperature in the mantle [Kelley et al. (2000)].

7.   Rb - Sr DATING METHOD

Emil Jelínek

7.1.   Chemical Properties, Radioactive Decay and Isotopic Abundance

Rubidium ($Z=37$) belongs to the group of alkali metals (IA). Two isotopes of Rb (^{85}Rb and ^{87}Rb) occur naturally with isotopic abundances of 72.1654 % and 27.8346 %, respectively. Because of its large ionic radius, ($1.48 \cdot 10^{-10}$ m) rubidium commonly substitutes for potassium in rock-forming minerals (e.g., feldspars, micas, glauconite etc.).

The isotope ^{87}Rb (see Chapt. 2, Table 1) is radioactive and decays to stable ^{87}Sr by β^- emission. The half-life was estimated at $4.8813 \cdot 10^{10}$ y, corresponding to a decay constant of $\lambda = 1.42 \cdot 10^{-11}$ y^{-1} [Steiger et al. (1977)].

Strontium (Z=38, ionic radius $1.13 \cdot 10^{-10}$ m), on the other hand, is a member of group IIA. It has geochemical features similar to calcium and replaces Ca in many minerals (e.g., plagioclase, apatite, carbonates) (see Chapt.2, 3.4.8). Strontium has four natural isotopes: ^{84}Sr with an abundance of 0.56 %, ^{86}Sr of 9.86 %, ^{87}Sr of 7.0 % and ^{88}Sr of 82.58 %. The ratios of strontium isotopes varied over geological time, as a result of the formation of radiogenic ^{87}Sr by the radioactive decay of ^{87}Rb. The exact isotope composition of Sr depends on the [Rb]/[Sr] ratio of a geological material and on its age. The Rb content as well as [Rb]/[Sr] ratio varies in different Rb-bearing minerals, e.g., the Rb content in micas is much higher than in K-feldspar. Therefore, the abundances of radiogenic ^{87}Sr are also variable. Because the ratios of isotopes are more accurately measured than the absolute abundance, a stable isotope of ^{86}Sr is used to normalise the abundances of ^{87}Sr and ^{87}Rb.

7.2. Concept of the Method

The Rb-Sr method is used for dating minerals or whole rocks, whose age is usually greater than 10^7 y. It is suitable for dating of igneous rocks; however, the ages of sediments or metamorphic events can also be estimated. The method is useful for dating of rocks containing potassium-rich minerals (biotite, muscovite, K-feldspar, glauconite, smectite, illite, adularia, etc.). Therefore, granitoids and their metamorphic equivalents, glauconite and clay minerals bearing sediments, and ore deposits with syngenetic micas, are the main objects of study.

Two fundamental assumptions are important for successful dating:
- Dating minerals and rocks must be cogenetic
- All rocks from a homogenous source (e.g., magma chamber) have an identical initial N_{87Sr}/N_{86Sr_i} ratio. Because of the different [Rb]/[Sr] ratios of individual potassium-bearing minerals at the source material, their present day N_{87Sr}/N_{86Sr} ratios are different.
- The mineral must be a closed system throughout geological history with respect to Rb and Sr (i.e., ^{87}Sr and ^{87}Rb content).

The basic expression for the growth of radiogenic ^{87}Sr in Rb-bearing minerals over time is

$$N_{87Sr} = N_{87Sr_i} + N_{87Rb} \cdot \left(e^{\lambda_{87Rb} \cdot t} - 1 \right) \qquad (35)$$

where N_{87Sr} and N_{87Rb} are the total number of nuclei of these isotopes in the mineral or rocks at the present time, N_{87Sr_i} represents the atoms present in the sample at the time of formation, λ_{87Rb} is the decay constant of ^{87}Rb, and t is the age of the sample.

Because ^{86}Sr is a stable isotope and its abundance has not changed over time (i.e., $N_{86Sr} \approx N_{86Sr_i}$), it is possible to write Eq. 35 in the form

$$\frac{N_{87Sr}}{N_{86Sr}} = \frac{N_{87Sr_i}}{N_{86Sr_i}} + \frac{N_{87Rb}}{N_{86Sr}}\left(e^{\lambda_{87Rb} \cdot t} - 1\right) \qquad (36)$$

For age determination of a sample, we need to measure the present abundance of ^{87}Rb, ^{87}Sr and ^{86}Sr. For calculation of the value of the age (Eq. 36), it is necessary to estimate the initial N_{87Sr}/N_{86Sr_i} ratio, i.e., the ratio at time t_i. This problem can be solved by measuring this ratio in some coexisting undisturbed mineral that does not contain rubidium (e.g., apatite). However, the graphical "method of an isochron diagram" for determination of the initial N_{87Sr}/N_{86Sr_i} ratio is generally used today. The isochron is based on plotting N_{87Rb}/N_{86Sr} versus N_{87Sr}/N_{86Sr} for the various cogenetic minerals or whole rock samples (closed system) with different N_{87Rb}/N_{86Sr} ratios, respectively [Allsopp (1961)], [Nicolaysen (1961)], [York (1969)]. Apatites, plagioclases and K-feldspars are common minerals with low *[Rb]/[Sr]* ratios, while micas (muscovite and biotite) have high *[Rb]/[Sr]* ratios. The data points of individual samples with different *[Rb]/[Sr]* ratios (see Fig. 22) can be connected by a straight line (isochron). The slope of the isochron is $(e^{\lambda t}-1)$ and ordinate-intercept is the N_{87Sr}/N_{86Sr_i} ratio. The last Sr-isotope homogenisation occurred at time t_i (the initial strontium composition). For the isochron age determination, the slope of the line – which increases as a function of time – must be determined precisely. The system must be free of any gain or loss of Rb and Sr due to younger processes, like metamorphism, chemical exchange with circulating water or hydrothermal alteration.

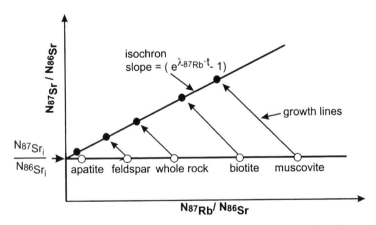

Figure 22: An isochron diagram of the Rb-Sr evolution of cogenetic suite of samples. The whole rock sample consisted of minerals apatite, feldspar, biotite and muscovite

[Provost (1990)] criticised the quality of numerical methods of calculation of isochron diagrams used during the past 25 years. He stressed that two main aspects have influenced the alignment quality: Tthe distance from the data points to the best-fit lines and the regularity of data point distribution along the line. Therefore, he introduced a new mathematical procedure BID ("the **b**est **i**sochron **d**iagram").

Given present-day analytical precision and the range of *[Rb]/[Sr]* ratios for unaltered common rocks and potassium-bearing minerals, the uncertainty in age determination is ± 2 % and analytical uncertainty for N_{87Sr}/N_{86Sr} ratios is ± 0.1 %.

7.3.   Rb - Sr Analytical Techniques

To obtain the age of a sample, it is necessary to determine the present number of atoms N_{87Rb}, N_{87Sr} and N_{86Sr} or the N_{87Rb}/N_{86Sr} and N_{87Sr}/N_{86Sr} ratios, as well as the Rb and Sr concentrations. Fresh rock samples are essential for age estimation. [Geyh et al. (1990)] recommended a weight of 20-50 kg for the whole rock sample. X-ray fluorescence (XRF), isotope dilution analysis or inductively coupled plasma analysis (ICP) are commonly used to estimate the Rb and Sr concentrations.

The use of super-pure acids and teflon or platinum crucibles is necessary for sample treatment. Due to the mass interference of ^{87}Rb and ^{87}Sr (same atomic mass), chemical separation of Rb and Sr before the analysis is required, e.g. chromatography with ion exchange resins.

7.4.   Whole-Rock and Mineral Dating of Magmatic and Metamorphic Events

The dating of igneous rocks is based on the assumption that all minerals or rocks formed from a common magma had the same initial N_{87Sri}/N_{86Sri} ratio and all samples belonging to a comagmatic suite will plot on the same isochron. The crystallisation of the magmatic liquid produces solid phases (minerals) and the remaining liquids give rise to another, chemically different, comagmatic suite of igneous rocks. The cooling time of such a magmatic system was relatively short. Therefore, the rocks have nearly the same geological age. For these reasons, the whole-rock sample instead of individual minerals (e.g., intrusion of magma) can be used for the age determination. This is done by using isochron diagrams. The individual whole-rock samples must be collected to cover a wide range of *[Rb]/[Sr]* ratios.

If the individual samples have different *[Rb]/[Sr]* ratios, the isochron provides the age of their crystallisation from a magma by its slope and their initial strontium isotope ratio by its intercept on the y-axis. The minerals of igneous rocks that have not been altered (e.g. by a temperature effect or chemical exchange with circulating water, fluid phase, etc.) are plotted on the same isochron as the whole-rock samples from which they separated. However, the Rb-Sr system in the whole-rock samples is more stable than in the minerals.

Study of magma mixing, modelling of magma formation – partial melting, fractional crystallisation, assimilation, contamination etc. – can be approached through Sr isotopes on the condition that the *[Rb]/[Sr]* and N_{87Sr}/N_{86Sr} ratios of the end members or of the source material are known [Faure (1986)], [Langmuir et al. (1978)].

Mostly due to a temperature and pressure effects, metamorphism results in recrystallisation of the primary mineral assemblage and formation of new minerals, stable under new pressure-temperature (p-T) conditions. If some fluid phases (e.g., water) are present, the primary chemical composition of the minerals and rock may also be changed. Consequently, metamorphism can affect the Rb-Sr system and redistribution of these elements among the new mineral phases. Usually, partial or complete homogenisation of Sr isotopes is achieved within the amphibolite and higher facies conditions of the metamorphic process (Fig. 23).

In this case, all the primary $^{87}Sr/^{86}Sr$ ratios are changed and the "radiogenic clocks" are restarted. The age information stored in the primary minerals disappears and only the age of the last metamorphic event can be estimated. The new Sr initial ratio is higher than the original one. Therefore, the metaigneousminerals cannot be used for dating their magmatic origin. However, for a closed system during metamorphism, the cogenetic whole-rock samples can still retain the primary *[Rb]/[Sr]* ratios and the slope of their isochron will still give the primary magmatic age and the initial strontium isotope ratio. In this case, the mineral isochron represents the age of the metamorphic event and the whole-rock isochron dates the time of crystallisation of its igneous precursor (Fig. 24).

In case of partial homogenisation of minerals, dating of the metamorphic event from them is not possible. However, the whole-rock isochron still can provide information about the age of the initial crystallisation of the lithological unit. In the case of chemical alteration of rocks, during which Rb and Sr were either added or lost due to metasomatic processes, another dating method can be more successful [Faure (1986)].

Regional metamorphism due to a temperature effect can cause opening of Rb-Sr mineral systems, which are again closed to element mobility. Different minerals are stable in the sense of Rb-Sr mobility, depending on the metamorphic grade (blocking temperature). E.g., biotites have different blocking temperature than white micas [Dickin (1995)]. The cooling history of metamorphic events can be studied by dating the blocking temperatures of different mineral systems.

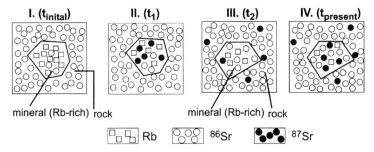

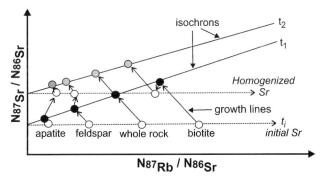

Figure 23: A relationship of common and radiogenic Sr in metamorphic rocks.

I. At time t_{inital}, the igneous rock consisted of two phases: mineral with high content of Rb, and host rock with high content of ^{86}Sr.

II. At time t_1, the mineral is enriched in radiogenic ^{87}Sr* and the N_{87Sr}/N_{86Sr} ratios were changed according to the primary Rb content in both phases.

III. At time t_2, the metamorphic event homogenised Sr composition.

IV. At time $t_{present}$, the contemporary distribution of Sr. The Rb-rich mineral is enriched by newly formed ^{87}Sr. The new Sr initial ratio is higher than the original.

Figure 24: Two-stage evolution of N_{87Rb}/N_{86Sr} and N_{87Sr}/N_{86Sr} ratios: the initial values *i* at time t_i were changed and Sr-homogenised t_1 after the metamorphic event. The result is a new initial N_{87Sr}/N_{86Sr} value and new t_2 isochron representing the time of the metamorphic event.

7.5. Sr Isotopic Composition of Sediments and Ocean Water

The Rb-Sr dating of sedimentary rocks is more difficult than for igneous rocks, even thoughmany sedimentary rocks contain minerals with high enough Rb contents as well as *[Rb]/[Sr]* ratios, e.g. glauconite, K-feldspars, micas, some clay minerals, sylvite and other evaporites. It is necessary to decide which of them actually originated in the sedimentary environment (authigenic minerals). Detritic (allogenic) minerals like a mica, K-feldspar, etc., have isotopic signatures which are not necessarily associated with the time of deposition. Their dating may provide information about the proto-lithic character, or may be the result of diagenesis, structural deformation or metamorphic recrystallisation.

Authigenic minerals, glauconite, illite, etc., are deposited directly from water and hence display good initial Sr isotope homogeneity [Dickin (1995)] and they are useful for Rb-Sr dating of the time of sedimentation.

Biogenic and chemogenic seawater sediments (carbonates, evaporates, etc.) have been used to interpret the variation of strontium isotopic composi-tions in paleo-oceans through time. For example [Burke et al. (1982)] and [Kaufman et al. (1993)] presented a time-line for the Sr composition in sea-water from the Late Proterozoic. These studies are based on the fact that the N_{87Sr}/N_{86Sr} ratio of terrestrial water depends on the ratio of ^{87}Sr and ^{86}Sr in the rocks through which the water passes. Isotopic homogenisation of Sr during transportation and attaining of closed basins is followed by co-precipitation and is fairly resistant to diagenetic alteration.

[Geyh et al. (1990)] gave a homogeneous value of 0.7090 for the N_{87Sr}/N_{86Sr} ratio in present-day sea water. This value is influenced by three sources with different isotopic compositions: marine sediments (intermediate N_{87Sr}/N_{86Sr}), continental crustal rocks (high ratio ~0.720) and young volcanics (mean ratio ~0.704). Fig. 25 shows the variation in the Sr isotope ratio in Phanerozoic car-bonates, reflecting different proportions of these reservoirs in the oceans dur-ing the Earth's history.

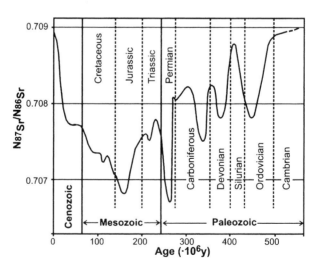

Figure 25: Variation of the N_{87Sr}/N_{86Sr} ratios in marine carbonates during the Phanerozoic, af-ter [Faure (1986)].

7.6.   Evolution of Sr Isotopes in Time and the Sr Model Ages

The initial N_{87Sr_i}/N_{86Sr_i} ratios found for rocks with different tectonomagmatic positions, for example continental, island arc and oceanic, have been used to interpret the Earth's history and to define source regions of rocks (e.g., igneous). The evolution of Sr isotope ratios have been based on the assumptions that the Earth was formed from solar nebula before about 4.5±0.1 billion years (10^9 y) and that the initial N_{87Sr_i}/N_{86Sr_i} ratio was close to 0.699, which corresponds to the Sr ratio value of chondrite meteorites. Rubidium is geochemicaly fractionated between the Earth's mantle and crust with a higher *[Rb]/[Sr]* ratio. Thus, rocks derived from the mantle will have a different initial Sr isotope ratio than rocks formed from crustal material. The amount of radiogenic $^{87}Sr*$ as well as Sr isotope ratios in the mantle and crustal rocks increased through time due to decay of ^{87}Rb.

The upper mantle initial N_{87Sr_i}/N_{86Sr_i} ratio at the time of formation of the Earth was close to the chondritic value 0.699. The present-day value for the upper mantle rocks is assumed to have evolved to 0.704 [Papanastassiou et al. (1969)], [Cox et al. (1979)].

The initial N_{87Sr_i}/N_{86Sr_i} ratios of the oldest (~$3.7 \cdot 10^9$ y) continental crustal rocks of granitic composition are usually very low – about 0.701 [Moorbath et al. (1975)]. This supports the idea that these old gneisses were derived from the mantle rocks. The evolution of the Sr isotope was different in the

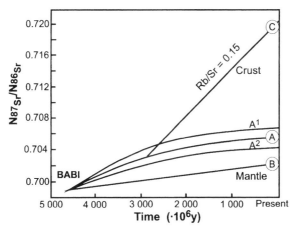

Figure 26: Evolution of the Sr - isotopic composition in individual terrestrial reservoirs (Earth's mantle and crust). BABI - basaltic achondrite best initial (N_{87Sr}/N_{86Sr}=0.698990). The curves represent hypothetical evolution of the Sr ratio, A, A^1, and A^2 in the mantle under the continents, B in the mantle region depleted in Rb, C development of Sr in the crust, that was withdrawn from the mantle about $2.9 \cdot 10^9$ y ago, after [Faure (1986)].

mantle and in the crustal rocks (Fig. 26). The initial ratios in the crustal material have increased more quickly than in the upper mantle due to the higher average *[Rb]/[Sr]* ratio.

The contemporary upper mantle rocks exhibits low variability of Rb/Sr ratios. This heterogeneity appears to be the result of depletion of some parts of the mantle in Rb relatively to Sr early in its history. Because the initial N_{87Sr}/N_{86Sr_i} ratio of magma formed in the mantle or in the crust depends on the original ratio in the source material, time of melting and other evolutionary processes, the rock initial ratio can be used for identification of the source. An example of such applications is establishing of the provenance of granitic magma. The initial N_{87Sr}/N_{86Sr_i} ratio can distinguish granite derived by partial melting of the upper mantle (low initial N_{87Sr}/N_{86Sr_i} ratio) from granite derived by crustal anatexis (relatively high initial ratio). At least three important assumptions must be taken into consideration: rocks (granite) derived from the mantle or crust are the result of single-stage processes from an isotopically homogeneous mantle and post-crystallisation processes do not affect this isotopic ratio [Henderson (1982)].

The Sr model-age of a rock can be determined by calculation of the time when the rock had an Sr isotopic composition equal to its assumed source or an uniform reservoir (UR). The calculation of the model-age can only be successful when the [Rb]/[Sr] ratio was efficiently decoupled during the rock-forming processes (e.g., partial melting). This assumption is correct for the most of the crustal rocks. In this case, the usual procedure is the calculation of the Sr model age ($T_{Sr_{UR}}$) relative to a uniform reservoir (e.g., the upper mantle):

$$T_{Sr_{UR}} = \lambda_{87Rb}^{-1} \cdot ln\left(1 + \frac{\dfrac{N_{87}Sr_{sample}}{N_{86}Sr_{sample}} - \dfrac{N_{87}Sr_{UR}}{N_{86}Sr_{UR}}}{\dfrac{N_{87}Rb_{sample}}{N_{86}Sr_{sample}} - \dfrac{N_{87}Rb_{UR}}{N_{86}Sr_{UR}}}\right) \tag{37}$$

where $N_{87Sr...}$, $N_{86Sr...}$ and $N_{87Rb...}$ are the numbers of atoms in a sample or in UR at the present time.

Sr model ages are significant only if the following assumptions are fulfilled:
– the isotopic evolution of the source is known,
– there was only a short time period between the emplacement of the sample in the crust and acquisition of a crust-like *[Rb]/[Sr]* concentration ratio,
– acquisition of a crust-like *[Rb]/[Sr]* concentration ratio,
– the *[Rb]/[Sr]* ratio of the sample did not change since the time of Sr separation from the mantle and
– all the material in the sample must have come from the mantle during a single event.

UR parameters used for calculation are $N_{87Sr_{UR}}/N_{86Sr_{UR}}=0.7045$ and $N_{87Rb_{UR}}/N_{86Sr_{UR}}=0.0827$ *for* present-day mantle [De Paolo et al. (1976ab)].

The ε parameter for the Sr system is widely used to describe the variation of the initial N_{87Sr}/N_{86Sr_i} ratios compared to a primitive chondritic mantle or a model bulk Earth UR:

$$\varepsilon_{Sr}(T) = \left(\frac{\dfrac{N_{87Sr_{sample}}(T)}{N_{86Sr_{sample}}(T)}}{\dfrac{N_{87Sr_{UR}}(T)}{N_{86Sr_{UR}}(T)}} - 1 \right) \cdot 10^4 \qquad (38)$$

where *T* is the age.

The interpretation of initial N_{87Sr}/N_{86Sr_i} ratios (ε_{Sr}) in combination with ε_{Nd} (see 8.6, Eq. 42) has been very useful for the geochemical characterisation of the mantle heterogeneity, metasomatic processes, differentiation, interaction between mantle and crust, history of rock units derived from mantle, etc.

7.7.   Representative Examples

The Rb-Sr method has been used mostly for dating of granitoids and their metamorphic equivalents. However, Sr dating of granitoids can be difficult in relation to discrimination among several possible sources, alteration etc. for these rocks.

Within the Bohemian Massif, [Siebel (1993)] dated the Leuchtenberg granite, which represents early Variscan magmatic activity, with conflicting results. The isochron parameters of garnet-muscovite granite gave an intrusion age of $(317\pm2)\cdot10^6$ y and crystallisation time of $10\cdot10^6$ y (Fig. 27), but this result disagrees with the K-Ar mineral pattern $(325\cdot10^6$ y). The Rb-Sr system of fin-grained garnet-muscovite granite was affected by postmagmatic partial Sr-rehomogenisation, which rotated the isochron towards a younger age and higher initial Sr isotope ratio. On the other hand the biotite granite Rb-Sr age of $(326\pm2)\cdot10^6$ y is in a good agreement with K-Ar mica dates and is interpreted as the intrusion age (Fig. 28).

[Verschure et al. (1993)] show that large crystals may be isotopically inhomogeneous enough to enable Rb-Sr isochron dating. They used the "internal isochron method" to date crystallisation of a single biotite crystal from an undeformed granite pegmatite in metagabbro near Buras (S. Norway). These authors estimated an age of $(987\pm62)\cdot10^6$ y for the closure of the Rb-Sr isotopic system of biotite megacryst at about 400° C (Fig. 29). The Rb-Sr isotope system of the biotite remained open during the period $1060\text{-}980\cdot10^6$ y ago.

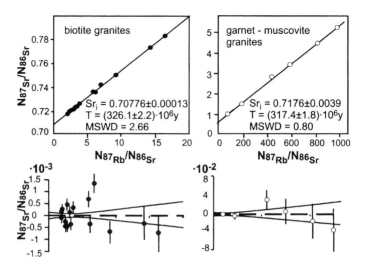

Figure 27: The ratios of N_{87Sr}, N_{87Rb}, and N_{86Sr} into whole rock for Leuchtenberg (Germany) and deviations of data points from the isochrone (below) at error levels of $1\cdot\sigma$, after [Siebel (1993)]: (Sr_i is the initial Sr isotope ratio, T is the age of the rock and MSWD is the mean squared weighted deviation).

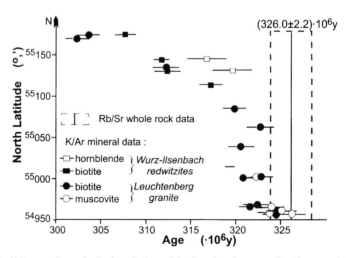

Figure 28: K/Ar geochronological evolution of the Leuchtenberg granite (Germany). Dashed lines represent the Rb-Sr whole rock data for the biotite granites, after [Siebel (1993)].

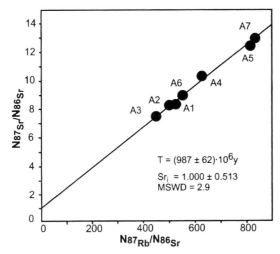

Figure 29: Rb/Sr isochron diagram of the Buras (S. Norway) biotite megacryst (A1-A7) initial ratio of Sr isotope $Sr_i=1.000\pm0.513$, age $T=(987\pm62)\cdot10^6$ y and mean squared weighted deviation $MSWD=2.9$, after [Verschure and Mauer (1993)].

A technique of texturally controlled microsampling from thick rock sections and direct Rb-Sr dating of minerals (e.g., white mica) unequivocally related to deformation has been developed by [Muller et al. (2000)]. The Rb-Sr age and the field observation in the Eastern Alps are fully consistent and demonstrate the reliability of the method for sub-millimetres scale geochronology.

Sr isotope geochemistry has been used to investigate the paleohydrogeological conditions in fractured crystalline rocks [Bottomley et al. (1992)]. The authors have studied N_{87Sr}/N_{86Sr} ratios in fracture calcite in Precambrian gneisses (Chalk River pluton, Canada). The gneiss samples Rb-Sr isotope plot is scattered around $1.3\cdot10^9$ y with an initial N_{87Sr}/N_{86Sr} concentration ratio of 0.704. The N_{87Sr}/N_{86Sr} ratios of 0.7090 in fractured calcite are significantly less radiogenic than the present day whole-rock values (Fig. 30), but more radiogenic than the initial Sr ratio of the host gneisses. The N_{87Sr}/N_{86Sr} ratio of hydrothermal calcite (low Rb content) records the Sr isotopic composition of the wall rock at the time of fracturing mineral formation. Because the gneisses are relatively rich in Rb, it is possible to estimate an age for the calcite and a minimum age for the fracturing. Development trends of Chalk River pluton are depicted in Fig 31. The average N_{87Sr}/N_{86Sr} ratio indicates that the calcite formed at least $400\cdot10^6$ y ago. If the minimum gneiss N_{87Rb}/N_{86Sr} ratio is considered, the calcite is more likely 700-800$\cdot10^6$ y old. Calcite probably formed several hundred millions of years after the Grenville orogeny. The formation age of the hydrothermal calcite in the Chalk River pluton approximates the age of initial rifting in the graben estimated at 700-600$\cdot10^6$ y ago during opening of a proto-Atlantic Ocean.

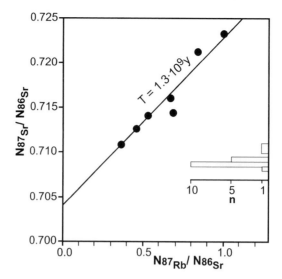

Figure 30: The N_{87Sr}/N_{86Sr} concentration ratios of Chalk River (Canada) gneisses with $1.3 \cdot 10^9$ y reference line and "initial" ratio of 0.704. The distribution (*n=number* of samples) of N_{87Sr}/N_{86Sr} in fracture calcite is shown by the histogram, after [Bottomley et al. (1992)].

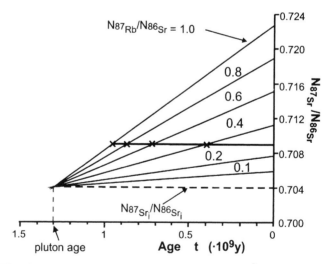

Figure 31: The development trends in Sr ratios of the $1.3 \cdot 10^9$ y old Chalk River pluton (dashed line is the "initial" Sr isotope ratio of the pluton) as a function of possible N_{87Sr}/N_{86Sr} ratios of gneisses. The solid horizontal line is the average present-day ratio for fracture calcite, after [Bottomley et al. (1992)].

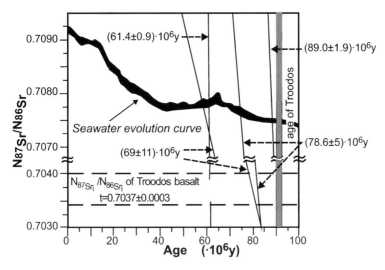

Figure 32: Sr evolution lines from selected celadonites (Troodos ophiolite complex, Cyprus) in comparison with the seawater evolution curve, after [Booij et al., (1995)].

The Rb-Sr dating of the hydrothermal oceanic crust alteration is manifested by study of celadonites from Troodos ophiolite, Cyprus [Booij et al. (1995)]. The Rb/Sr model crystallisation age for celadonites from the drillcore samples show only a small variation between 83 and $89 \cdot 10^6$ y, whereas use of the model age of the outcrop samples yields more variable values between 58 and $84 \cdot 10^6$ y. Drillcore samples suggest a duration of low-temperature alteration of ca. $10 \cdot 10^6$ y, whereas outcrop samples yield longer a duration of ca. $30 \cdot 10^6$ y (Fig. 32).

8.   Sm - Nd METHOD
Jan Košler

8.1.   Chemical Properties, Radioactive Decay and Isotopic Abundance

Neodymium ($Z=60$) and samarium ($Z=62$) belong to the lithophile light rare earth elements (REE, group IIIB of the periodic table). Neodymium has seven natural isotopes: masses 142 with isotopic abundance of 27.13 %, 143 (12.18 %), 144 (23.80 %), 145 (8.30 %), 146 (17.19 %), 148 (5.76 %) and mass 150 (5.64 %). They represent both the stable radiogenic products and radioactive nuclides with long half-life. Samarium also has seven naturally occurring isotopes with masses of 144 (3.1 %), 147 (15.0 %), 148 (11.3 %), 149 (13.8 %), 150 (7.4 %), 152 (26.7 %) and 154 (22.7 %). While the alpha decay of ^{147}Sm to ^{143}Nd (half-life $1.06 \cdot 10^{11}$ y, see Chapt. 2, Table 1) has several geochronological applications, alpha decay of ^{148}Sm to ^{144}Nd (half life $7 \cdot 10^{15}$ y) is too slow to be geologically useful. The alpha decay of ^{144}Nd to ^{140}Ce is also slow (see Chapt. 2, Table 1). In fact, the decrease in the isotopic abundance of ^{144}Nd due to its radioactive decay during the geological history of the Earth is negligible compared to the commonly achieved precision of a mass-spectrometric measurement. Accordingly, in geological applications isotope ^{144}Nd is considered to be stable and, as such, and because of its relatively high abundance, it is used to normalise the measured Nd isotopic ratios. The decay scheme of the Sm-Nd isotopic system and related rare earth elements is illustrated in Fig. 33.

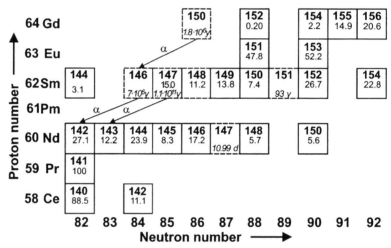

Figure 33: Decay scheme of rare earth elements, atomic abundances are in percent, half life in italics, after [De Paolo (1988)].

8.2.   Concept of the Method

The Sm-Nd method of dating is particularly suitable for determining the crystallisation ages of igneous rocks (whole-rock isochron), the average source ages of sedimentary rocks (Nd model ages) and igneous and metamorphic mineral crystallisation ages (mineral-mineral and mineral-whole-rock isochrons). The method is especially useful for dating mafic and ultramafic rocks which are generally difficult to date using other techniques. The underlying assumption is that, up to time T_x in the history of the rock, the neodymium isotopic compositions of all cogenetic samples TR, A, B and C were identical. Following T_x, however, due to closure of the Sm-Nd isotopic system during igneous crystallisation, the Nd isotopic composition in individual samples was modified as a result of radioactive decay of ^{147}Sm. The present N_{143Nd}/N_{144Nd} ratios of a number of atoms in these samples are different from the same ratios prior to T_x. The Sm and Nd isotopic ratios for minerals A, B and C from rock TR in Fig. 34 plot on a straight line (isochron, see 7.2). Similar to Eq. 35, the straight line is given by:

$$\left(\frac{N_{143Nd}}{N_{144Nd}} \right) = \left(\frac{N_{143NdT_x}}{N_{144NdT_x}} \right) + \left(\frac{N_{147Sm}}{N_{144Nd}} \right)\left(e^{\lambda_{147Sm} \cdot T_x} - 1 \right) \qquad (39)$$

where the decay constant $\lambda_{147Sm} = 6.539 \cdot 10^{-12} \ y^{-1}$, the isotopic ratios of the number of atoms $N_{...}$ are the values at the present or at time T_x B.P.

For the age determination, the slope of the line must be determined precisely. Similar to the Rb-Sr method (see 7.2), it is of key importance that the individual samples, be it rocks or minerals, which define the line, must have sufficient spread in their N_{147Sm}/N_{144Nd} ratios. In common igneous rocks, the low Sm/Nd phases are feldspars, apatite and monazite, high Sm/Nd minerals are pyroxene, garnet, titanite, zircon and amphibole [Faure (1986)]. Given the range of Sm/Nd ratios for minerals in a common igneous rock (ca 0.1-0.5) and the contemporary analytical precision of $2 \cdot \sigma \sim 0.003$ %, the uncertainty in the age determination is usually less than 5 % [De Paolo (1988)].

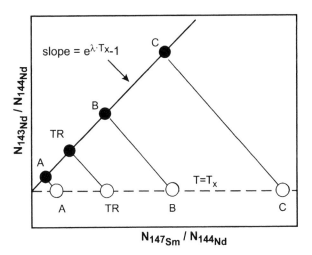

Figure 34: Sm - Nd isotopic evolution of a cogenetic suite of samples; TR is whole-rock sample consisting of minerals A, B and C, T_x is time of igneous crystallisation.

8.3.   Sm - Nd Analytical Techniques

The solid source mass spectrometry (see Chapt. 5, 2) is the most common technique for determining both the Nd isotopic ratios and Sm and Nd element concentrations using isotope dilution analysis. However, due to mass interference between Nd and other REE (see Fig. 33), as well as other elements (e.g., Ba), chemical separation is required prior to mass spectrometric determinations. Up to 500 mg of sample are dissolved, often in a $HF-HNO_3$ mixture, aliquoted and spiked with Sm and Nd (see Chapt. 5, 1.7.2 and 2.1), usually ^{149}Sm and ^{145}Nd or ^{150}Nd. The bulk REE fraction is separated using an ion-exchange technique; several methods are available to extract Sm and Nd from the bulk REE fraction [Hooker et al. (1975)], [Richard et al. (1976)], [De Paolo et al (1976ab)]. Sm and Nd are run from a double (triple) Re or Re-Ta filament assembly as metal or oxide ions [Thirlwall (1982),(1991)]. Alternatively, the Sm and Nd concentrations can be determined from a sample aliquot using ICP-MS or laser ablation (LA ICP-MS) technique [Prince et al. (2000)]. Where possible, the data are corrected for mass interference and fractionation using compatible sets of isotopic ratios [Faure (1986)]. The *[Sm]/[Nd]* concentration ratio is recalculated to $N_{147_{Sm}}/N_{144_{Nd}}$ ratio as follows:

$$\frac{N_{147_{Sm}}}{N_{144_{Nd}}} = \frac{[Sm]}{[Nd]}\left(0.5315 + 0.14252\frac{N_{143_{Nd}}}{N_{144_{Nd}}}\right) \qquad (40)$$

8.4.   Nd in Rock-Forming Minerals

Light rare earth elements have relatively large 3+ ionic radii (Nd^{3+} in 8 coordination=$1.109 \cdot 10^{-10}$ m) similar to Ca^{2+} and Th^{4+} and often substitute for each other in minerals. Substitutions for other than trivalent ions are often coupled, and with charge compensation or vacancies present in the crystal lattice [Clark (1984)], [Burt (1989)]. During igneous processes, Nd, which is always present together with other rare earth elements, is dispersed in the mineral phases as

- trace constituents whose concentrations may be inferred from the distribution coefficients (e.g., most major rock-forming phases),
- minor constituents (> 0.1 wt.% REE, e.g., in common accessory phases -
  zircon, titanite) and
- major constituents of REE-bearing minerals (e.g., allanite, xenotime,
  monazite, bastnaezite).

Accordingly, while, in primary, e.g., mantle-derived magmas, the Nd and REE distribution is mostly controlled by the composition of the source, in more differentiated magmas of crustal or mixed crustal-mantle origin, the behaviour of the mineral phases plays a key role in REE distribution. [Gromet and Silver (1983)] first demonstrated that, in granitoid rocks, the REE distribution is controlled by the accessory phases of the above groups (II) and (III).

8.5. Whole-Rock and Mineral Dating of Magmatic and Metamorphic Events

In magmatic rocks, the slope of an isochron constructed from the whole-rock data is usually interpreted as corresponding to the time of closure of the Sm-Nd isotopic system on the scale of individual samples. This usually corresponds to conditions when temperature and/or fluid activity became so low that the decoupling of Sm and Nd and/or isotopic equilibration of Nd became ineffective. Although it is generally accepted that closure temperatures for the Sm-Nd system are high, it has also been shown that, in pillow lava under conditions of ocean-floor metamorphism, the mobility of REE can be effective on the scale of individual pillows [Jelínek et al. (1984)]. On the other hand, the age derived from isochrons constructed from whole-rock-mineral or mineral-mineral data are interpreted as corresponding to closure of the Sm-Nd system on the scale of a mineral grain, i.e., corresponding to the time when diffusion of Sm and Nd became ineffective and the exchange of Sm and Nd between mineral and whole-rock or two minerals has ceased. Although little is known about the closure temperature of Sm and Nd in rock-forming minerals, it is generally accepted to lie between 500 and 700° C [Cliff et al. (1983)].

The interpretation of whole-rock-mineral and mineral-mineral isochrons in metamorphic rocks is similar to that in igneous rocks. However, the concept of metamorphic closure temperatures is different from the igneous concept in that the pressure-temperature-time (p-T-t) evolution of metamorphic rocks is often not just simple cooling but some kind of loop in p-T space [Cliff et al. (1985)]. Accordingly, also the residence time of a rock under certain p - T and fluid activity conditions must be taken into account and special attention must be paid to sample homogeneity both for whole rocks and especially for minerals, as the metamorphic overprint on relic igneous zoning is a common feature in many mineral phases.

8.6. Evolution of Nd Isotopes in Time

The amount of radiogenic ^{143}Nd, and also N_{143Nd}/N_{144Nd} ratio, of the Earth as a whole has increased through time due to the decay of ^{147}Sm (see Eq. 38). The increase in the Nd isotopic ratio through time is exponentially related to the age, the [Sm]/[Nd] ratio of a particular reservoir, be it mineral, rock unit or part of the Earth (geosphere), and the initial, primordial $(N_{143Nd}/N_{144Nd})_i$ ratio. The isotopic evolution of the Earth is often described using the "Chondritic Uniform Reservoir" concept (CHUR) [De Paolo et al. (1976ab)]. [Jacobsen et al. (1980)] have estimated N_{143Nd}/N_{144Nd} and N_{147Sm}/N_{144Nd} ratios of 0.512638 and of 0.1967, respectively, at the present time. The model assumes that any rock within the Earth was derived from this reservoir, and permits the calculation of $N_{143NdCHUR}/N_{144NdCHUR}$ in CHUR at any decay time t using

$$\frac{N_{143}Nd_{CHUR}}{N_{144}Nd_{CHUR}} = \left(\frac{N_{143}Nd_{CHUR_i}}{N_{144}Nd_{CHUR_i}}\right) + \frac{N_{147}Sm_{CHUR}}{N_{144}Nd_{CHUR}}\left(e^{\lambda_{137Sm} \cdot t} - 1\right) \qquad (41)$$

where $N_{143NdCHUR}/N_{147SmCHUR}$ is the ratio of numbers of atoms at the present time and index i refers to the initial values at the decay time t in CHUR.

The Nd isotopic evolution of the Earth is schematically shown in Fig. 35, based on the CHUR model. Due to the more incompatible behaviour of Nd relative to Sm and decoupling of Nd and Sm during partial melting, the melts have lower [Sm]/[Nd] ratios compared to CHUR, resulting in lower contemporary N_{143Nd}/N_{144Nd} values. And vice versa, the residual solids have higher Sm/Nd and higher contemporary N_{143Nd}/N_{144Nd} ratios compared to CHUR. This principle is reflected in the contrasting compositions of the Earth's crust and the upper mantle: the crust, which was derived from the upper mantle through partial melting processes, has grossly lower N_{Sm}/N_{Nd} and N_{143Nd}/N_{144Nd} ratios compared to both the CHUR and the residual rocks of the upper mantle depleted in large-ion lithophile elements (i.e., "depleted mantle", DM). The current estimations of the N_{143Nd}/N_{144Nd} and N_{147Sm}/N_{144Nd} ratios of the depleted mantle are 0.513114 and 0.222, respectively [Michard et al. (1985)]. For the purpose of better comparison, the Nd isotopic composition is often referred to in the form of epsilon values (analogy to 7.6, Eq. 38), defined as the relative deviation of $N_{143Ndsample}/N_{144Ndsample}$ in the samples in parts per 10^4 from the $N_{143NdCHUR}/N_{144NdCHUR}$ composition of CHUR at a given decay time T:

$$\varepsilon_{Nd}(T) = \left(\frac{\dfrac{N_{143}Nd_{sample}}{N_{144}Nd_{sample}}}{\dfrac{N_{143}Nd_{CHUR}}{N_{144}Nd_{CHUR}}} - 1\right) \cdot 10^4 \qquad (42)$$

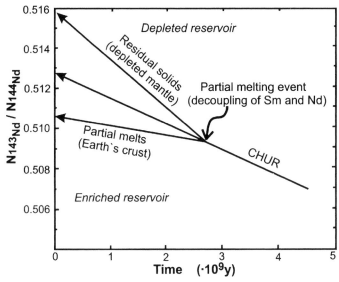

Figure 35: Evolution of the Nd isotopic composition in individual reservoirs after decoupling
of Sm and Nd due to partial melting of the upper mantle, after [Faure (1986)].

8.7.   Age Calculation Based on Nd Model

Based on the model of Nd isotopic evolution in time, the model Nd age is
determined by calculation of the time when the rock had an Nd isotopic
composition equal to that of its assumed source. This calculation can only be
successful when the *[Sm]/[Nd]* ratio of the source is different from that of
the sample, i.e., only if the rock-forming processes were efficient in decoupling Sm and Nd (e.g., partial melting). For most of the crustal rocks, the
usual procedure is to calculate the age relative to the chondritic uniform
reservoir t_{NdCHUR} and depleted mantle t_{NdDM}:

The calculation of the age t_{NdCHUR} or t_{NdDM} of a sample relative to CHUR
or DM ($t_{NdCHUR,DM}$) is as follows:

$$t_{Nd_{CHUR,DM}} = \frac{1}{\lambda_{147_{Sm}}} \cdot ln\left(1 + \frac{\dfrac{N_{143}Nd_{sample}}{N_{144}Nd_{sample}} - \dfrac{N_{143}Nd_{CHUR,DM}}{N_{144}Nd_{CHUR,DM}}}{\dfrac{N_{147}Sm_{sample}}{N_{144}Nd_{sample}} - \dfrac{N_{147}Sm_{CHUR,DM}}{N_{144}Nd_{CHUR,DM}}}\right) \qquad (43)$$

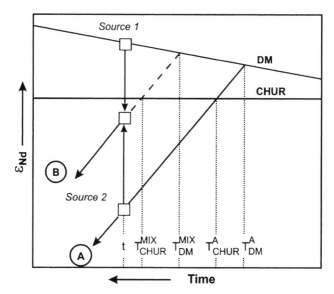

Figure 36: Schematic principles of Nd isotopic evolution based on a one-stage model. Age calculated for rock A corresponds to the time of separation from the mantle, model age for rock B - a mixture (MIX) of two sources – does not represent separation from the mantle reservoir, nor does it correspond to the crystallisation age t; (DM = depleted mantle, CHUR = chondritic uniform reservoir), after [Arndt et al. (1987)].

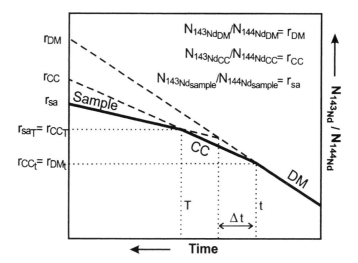

Figure 37: Nd evolution diagram showing how the age is calculated from the two-stage Nd model: Δt corresponds to the time difference between a single-stage and two-stage model age, after [Liew et al. (1988)]. (CC-composition of average crust, DM-depleted mantle).

The age values from the Nd model can have geological significance (e.g., as crust-formation age) only if the following assumptions are fulfilled [Arndt et al. (1987)] (Fig. 36):
- The isotopic evolution of the source is known.
- There was only a short time period between the emplacement of the sample in the crust and acquisition of a crust-like N_{Sm}/N_{Nd} ratio
- The *[Sm]/[Nd]* ratio of the sample did not change since the time of Nd separation from the mantle.
- Each material in the sample must have come from the mantle during a single event.

A more complex approach to the Nd model age calculation is to assume sample derivation from a source that has itself been derived from another reservoir; for example derivation of a granite from the continental crust which has been derived from the mantle (two-stage model; Fig. 37) [Liew et al. (1988)], [De Paolo et al. (1991)]:

$$ t_{Nd_{DM}} = \lambda_{147_{Sm}}^{-1} \cdot ln\left(1 + D \middle/ \left(\frac{N_{147\,Sm_{CC}}}{N_{144\,Nd_{CC}}} - \frac{N_{147\,Sm_{DM}}}{N_{144\,Nd_{DM}}}\right)\right) \qquad (44) $$

where D is

$$ D = \frac{N_{143\,Nd_{sample}}}{N_{144\,Nd_{sample}}} - \left(e^{T \cdot \lambda_{147\,Sm}} - 1\right)\frac{N_{147\,Sm_{sample}}}{N_{144\,Nd_{sample}}} - $$
$$ - \frac{N_{147\,Sm_{CC}}}{N_{144\,Nd_{CC}}} - \frac{N_{143\,Nd_{DM}}}{N_{144\,Nd_{DM}}} $$

In this case, $N_{147Sm_{CC}}/N_{144Nd_{CC}}$ stands for the contemporary composition of an average crustal reservoir, for instance a value of 0.219 [Liew et al. (1988)], t is the crystallisation age of the sample, and the other symbols are as above.

8.8. Nd Isotopes: Key to the Petrogenesis of Igneous Rocks

Modelling of the magma formation (e.g., by partial melting), fractionation, mixing and wall-rock assimilation processes can be based on Nd isotopes on the condition that the Sm and Nd contents and Nd isotopic compositions of the source material and the end-member(s) are known. However, this can only be successful if Sm and Nd are effectively decoupled during the process and separate reservoirs are formed. The entire mechanism of magma formation or modification of the magma composition can then be modelled using the distribution coefficients and equations describing the trace element behaviour in magma petrogenesis [Cox et al. (1979)], [De Paolo (1988)]. In addition, an assumption of equilibrium has often to be made, and independent geochronological data about the timing of the process are needed to time-correct the Nd isotopic ratios. As the process of wall-rock assimilation often causes magma cooling, creates a thermal gradient and induces magma fractionation, a more complex set of equations describing concurrent assimilation and fractional crystallisation has been developed [De Paolo (1981)], [Aitcheson et al. (1994)].

The isotopic compositions of Nd and Sr are often combined in order to discriminate between rocks of different origin (see Fig. 38). In the plot of N_{143Nd}/N_{144Nd} vs. N_{87Sr}/N_{86Sr} or ε_{Nd} vs. ε_{Sr} [Faure (1986)], [De Paolo (1988)] the rocks whose source had both *[Sm]/[Nd]* and *[Rb]/[Sr]* ratios greater than the respective values in CHUR or Uniform Reservoir UR – in the case of Sr - plot in quadrant I. In fact, only the data for a few rocks plot in this quadrant, as coupled enrichment of both Sm and Rb acts against their geochemical properties. Quadrant II contains rocks from sources with high N_{Sm}/N_{Nd} and low N_{Rb}/N_{Sr} ratios relative to CHUR or UR. These are typically residual solids after partial melting of the upper mantle and their derivates. The values of modern volcanic rocks (MORBs and some OIBs – see Fig 38) also plot in this quadrant and often exhibit negative ε_{Nd}-ε_{Sr} correlation (mantle array). This correlation is explained on the basis that sources with higher N_{Sm}/N_{Nd} have lower N_{Rb}/N_{Sr} and vice versa. Quadrant III of the diagram in Fig. 38 contains data for rocks from sources with both low N_{Sm}/N_{Nd} and N_{Rb}/N_{Sr} ratios. Although such characteristics may seem inconsistent with the geochemical properties of these elements, some mantle derivates, which have been contaminated by the Rb-depleted lower crust may evolve in such an isotopic pattern. Finally, rocks depleted in Sm and enriched in Rb, which were isolated from their source for a sufficient time period, evolve isotopic characteristics typical of quadrant IV. Such a pattern is common for most rocks within the continental crust, which, as the final product of mantle differentiation, is depleted in Sm, enriched in Rb and forms a reservoir separate from the mantle. Accordingly, compared to the Phanerozoic crustal rocks, comparable rocks within the older parts of the continental crust have more evolved Nd isotopic composition, i.e., lower N_{143Nd}/N_{144Nd} ratio and more negative ε_{Nd} value (Fig. 38).

Neodymium and strontium isotopes are particularly useful for studying petrogenetic processes above subduction zones in both the oceanic and continental magmatic arcs. The evidence based on Nd and Sr isotopes suggests involvement of both the subducted oceanic crust, including the sediments, and the overhanging mantle wedge, as well as the role of crustal contamination in island arc magmas [Saunders et al. (1991)], [Hawkesworth et al. (1991)], [Wilson (1989)]. The Nd isotopic composition of a particular rock suite in the magmatic arc depends on the composition of the subducted crust, mantle wedge, crust of the magmatic arc, the proportions in which they were mixed and the time elapsed since the formation of the magma. Although the Nd isotopic composition of magmas reflects the proportion of contributions from different sources, it does not reveal the mechanisms of their involvement, such as mass transfer via partial melt or fluids. In addition, magmas produced in continental magmatic arcs (active continental margins) have some isotopic signatures corresponding to those of the continental lithospheric mantle and the crustal contamination is often more pronounced.

Within the continental crust, the Nd isotopic composition is a powerful tool for identifying the sources of granitic rocks and it helps in establishing the contribution of crustally recycled and undifferentiated material to the granite [McCulloch et al. (1982)], [Chappell et al. (1992)]. On the other hand, provided the composition of the source and the age of the granite are known, it is often possible to infer which processes, like magma fractionation, assimilation, and magma mixing, operated during the formation of the granitic rock. However, as the REE distribution in granitic rocks is often controlled by REE-rich accessory minerals, at least part of their Nd isotopic signature may result from an inherited Sm - Nd component from refractory phases [Paterson et al. (1992)]. This is particularly true for crustally-derived anatectic granitic magmas.

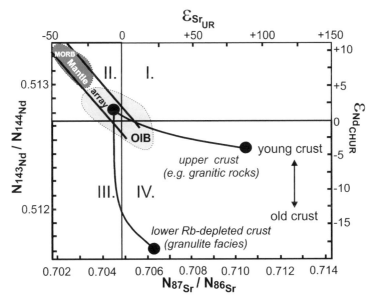

Figure 38: Nd-Sr correlation diagram showing four quadrants corresponding to reservoirs with different Sm/Nd and Rb/Sr ratios; also shown are two mixing lines between a hypothetical ocean island basalt (OIB) end-member and two different crustal rocks, after [Faure (1986)]. MORB=mid ocean ridge basalts, CHUR=chondritic uniform reservoir, UR=uniform reservoir.

8.9.   Representative Examples

The Sm-Nd method has mostly been used to date mafic and ultramafic rocks in cases where other techniques failed due to low Rb content, small proportion or lack of zircon, etc. Within the Moldanubian zone of the Bohemian Massif, the Sm-Nd data for garnet and clinopyroxene pairs from eclogites yielded highly variable age values of 377±7, 342±9, 336±16 and $(323±7) \cdot 10^6$ y. [Beard et al. (1995)] suggesting that the eclogites were derived from the inhomogeneous mantle and tectonically emplaced during the Hercynian orogeny. In the Saxothuringian zone of the Bohemian Massif, the

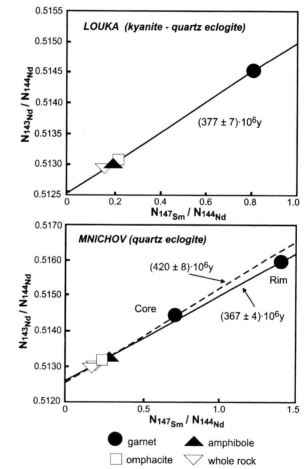

Figure 39: Sm-Nd isochrons for minerals and whole-rocks from the Saxothuringian zone (Mariánské Lázně complex) of the Bohemian Massif, after [Beard et al. (1995)]

timing of high-pressure eclogite facies metamorphism was studied on the basis of garnet-omphacite pairs [Beard et al. (1995)]. They have found that, in places with dominant peak metamorphic assemblage, the garnet-omphacite ages correspond to the garnet-whole-rock age and reflect the timing of eclogite facies metamorphism of $(377\pm7)\cdot10^6$ y. Elsewhere, the rims of the garnet have preserved the peak metamorphic ages but the garnet core-omphacite age corresponds to an amphibolite facies event at ca $420\cdot10^6$ y, i.e., prior to the eclogite facies metamorphism (Fig. 39).

Coupled Nd-Sr isotopic has been used by many authors to explain the petrogenesis of granitoid rocks. An extensive study has been carried out on granites in the British Caledonides [Halliday (1984)] as well as on the Hercynian granites of Europe (see references in [Janoušek et al. (1995)], [Janoušek et al. (2000)]). In both cases, the Nd and Sr isotopes have enabled discrimination between several sources for granitoid rocks in both the Caledonian and Hercynian orogenic belts.

9. U - Th - Pb DATING METHODS
Jan Košler

9.1. Chemical Properties, Radioactive Decay and Isotopic Abundance

Uranium, atomic number $Z=92$, is a litophile element which belongs to the actinides (group III of the periodic table). It has three natural isotopes with mass numbers of 234 (natural abundance of 0.005 %), 235 (0.770 %) and 238 (99.275 %). None of the three U isotopes is stable. ^{235}U and ^{238}U are the initial nuclides of two different decay series for which ^{207}Pb and ^{206}Pb are the respective end products (see Chapt. 2, Fig. 1). The half-lives of ^{238}U and ^{235}U are $0.704\cdot10^9$ and $4.468\cdot10^9$ y, respectively. ^{234}U is an intermediate member of the ^{238}U decay series with a half-life of $2.5\cdot10^5$ y.

Thorium ($Z=90$) is also a lithophile element which belongs to the same chemical group as U. Th has six natural isotopes: ^{232}Th is a long-lived radio-active isotope (see Chapt. 2, 2.1). It is the initial nuclide of a decay series with half-life $14.01\cdot10^9$ y and a stable end product ^{208}Pb. Th isotopes 227, 228, 230, 231, and 234 are short-lived nuclides which are members of the ^{235}U, ^{238}U and ^{232}Th decay series.

Lead is a chalcophile element which belongs to group IVB of the periodic table. It has four naturally occurring isotopes with masses 204, 206, 207 and 208 with respective average abundances of 1.4 %, 24.1 %, 22.1 % and

52.4 %. Of the four Pb isotopes, only ^{204}Pb is not radiogenic. Accordingly, the amount of ^{204}Pb is considered to be indicative of initial (common) lead component in the sample. It is also used for normalising the amount of radiogenic Pb in rocks and minerals.

9.2.   Concept of the Method

U-Th-Pb dating methods are suitable for establishing the crystallisation ages in both the magmatic and metamorphic rocks. This can be achieved either by using the standard isochron method (see 7.2 and 8.2) or by a combination of two different U decay schemes (the concordia method). In addition, based on the models of Pb isotopic evolution, the age can be calculated from the natural (common) lead composition in rocks and minerals. The U-Th-Pb dating methods are mostly used for fractionated rocks due to their high U and Th contents, although U-Pb dating of mafic rocks has also been successfully performed. The underlying assumption is the same as that for the other radiogenic isotope dating techniques, i.e., that, at some time t B.P. in the history of the rock, the initial lead isotopic compositions (^{207}Pb/^{204}Pb, ^{206}Pb/^{204}Pb and ^{208}Pb/^{204}Pb) of all cogenetic samples were identical. However, due to geochemical fractionation of elements, the *[U]* and *[Pb]* concentrations of the individual samples were modified and the present isotopic ratios of the number of atoms of ^{207}Pb, ^{206}Pb, and ^{208}Pb are different in different samples. Similar to the methods described in 8.2 and 7.2, the isotopic ratios to N_{204Pb} of several cogenetic samples lie on a straight line (isochron) in the three plots: N_{207Pb}/N_{204Pb} vs. N_{235U}/N_{204Pb}, N_{206Pb}/N_{204Pb} vs. N_{238U}/N_{204Pb} or N_{208Pb}/N_{204Pb} vs. N_{232Th}/N_{204Pb}. The slope of this line is a function of age t. The ordinate-intercept gives the initial Pb isotopic ratio at time t_i B.P. Accordingly, for each sample, the present-day Pb isotopic compositions are:

$$\frac{N_{207\,Pb}}{N_{204\,Pb}} = \frac{N_{207\,Pb_i}}{N_{204\,Pb_i}} + \frac{N_{235U}}{N_{204\,Pb}}\left(e^{\lambda_{235U}\cdot t} - 1\right) \qquad (45)$$

$$\frac{N_{206\,Pb}}{N_{204\,Pb}} = \frac{N_{206\,Pb_i}}{N_{204\,Pb_i}} + \frac{N_{238U}}{N_{204\,Pb}}\left(e^{\lambda_{238U}\cdot t} - 1\right) \qquad (46)$$

$$\frac{N_{208\,Pb}}{N_{204\,Pb}} = \frac{N_{208\,Pb_i}}{N_{204\,Pb_i}} + \frac{N_{232Th}}{N_{204\,Pb}}\left(e^{\lambda_{232Th}\cdot t} - 1\right) \qquad (47)$$

where $\lambda_{..}$ are the decay constants for ^{235}U, ^{238}U and ^{232}Th, respectively, $N_{...Pb, ...U}$ are number of atoms at the present time in a sample and $N_{...Pb_i}$ are the initial numbers of atoms at time t_i B.P.

For the U and Th decay constants, the following values have been recommended by the International Union of Geological Sciences (IUGS) [Jaffey et al. (1971)], [Le Roux et al. (1963)]:

$$\lambda_{235U} = 9.845 \cdot 10^{-10} \ y^{-1}$$
$$\lambda_{238U} = 1.55125 \cdot 10^{-10} \ y^{-1}$$
$$\lambda_{232Th} = 4.9475 \cdot 10^{-11} \ y^{-1}$$

By combining Eqs. 45 and 46, an equation independent of the U concentration in the sample is obtained:

$$\frac{N_{207Pb}}{N_{204Pb}} = \frac{N_{207Pb_i}}{N_{204Pb_i}} + \frac{N_{235U}}{N_{238U}} \cdot \frac{e^{\lambda_{235U} \cdot t} - 1}{e^{\lambda_{238U} \cdot t} - 1} \tag{48}$$

where $N_{235U}/N_{238U} = 1/137.88$. Assuming that the sample contains only radiogenic Pb*, Eq. 48 simplifies to:

$$\frac{N_{207Pb^*}}{N_{204Pb}} = \frac{1}{137.88} \cdot \frac{e^{\lambda_{235U} \cdot t} - 1}{e^{\lambda_{238U} \cdot t} - 1} \tag{49}$$

The radiometric age derived from Eq. 48 or 49 is called the "lead-lead age". For dating U-rich minerals with no or negligible initial Pb component, two different chronometers, ^{235}U-^{207}Pb and ^{238}U-^{206}Pb, are often combined in a concordia diagram [Wetherill (1956)] (Fig. 40). Alternatively, a ^{232}Th-^{208}Pb chronometer can be combined with either of the two U-Pb chronometers. The assumption is that, in a closed system, ^{235}U and ^{238}U decay independently to their end products, ^{207}Pb and ^{206}Pb and, accordingly, the two respective age values should be identical (concordant). In the diagram in Fig. 40, the composition of a closed system must plot on the curve for which

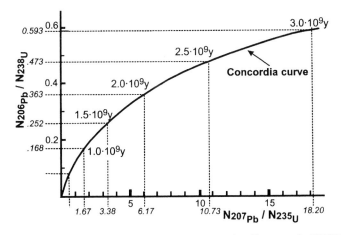

Figure 40: Concordia curve for the U-Pb isotopic system, after [Jäger et al. (1979)].

$t(N_{206Pb*}/N_{238U})=t(N_{207Pb*}/N_{235U})$. The mathematical expression of the concordia curve can be derived from combination of two equations:

$$\frac{N_{206Pb^*}}{N_{238U}} = e^{\lambda_{238U} \cdot t} - 1 \quad (50) \quad \text{and} \quad \frac{N_{207Pb^*}}{N_{235U}} = e^{\lambda_{235U} \cdot t} - 1 \quad (51)$$

as

$$\frac{N_{206Pb^*}}{N_{238U}} = \frac{N_{207Pb^*}}{N_{235U}} \cdot \left(\frac{e^{\lambda_{238U} \cdot t} - 1}{e^{\lambda_{235U} \cdot t} - 1} \right) \quad (52)$$

The combination of two different U-Pb systems provides a tool for testing whether the system has remained closed, i.e., whether there has been a loss or gain of U, Th or Pb during the history of a rock or a mineral. As, in most mineral phases, the daughter Pb isotopes are present in the U (Th)-radiation-damaged domains of the crystal structure, Pb-loss is a common feature. In the diagram in Fig. 41, the composition of samples that have lost some radiogenic lead (discordant samples) plots below the concordia. Various degrees of lead loss at the same time are reflected by a linear array of their respective data points. For discordant samples (below the concordia) and the U-Pb system, it always holds that $t(N_{207Pb}/N_{206Pb})>t(N_{207Pb}/N_{235U})>t(N_{206Pb}/N_{238U})$. The interpretation of discordant age patterns is discussed in more detail later in this section. Most applications of the U(Th)-Pb methods of dating are useful for U(Th)-rich mineral accessory phases such as zircon, baddeleyite, monazite, allanite, titanite, rutile, apatite, or xenotime, although major rock-forming silicates such as staurolite or garnet, as well as tantalo-niobates (columbite) [Romer et al. (1992)], have been also successfully dated by these methods. The uncertainty in age determination from the concordia diagram is usually less than 2 %.

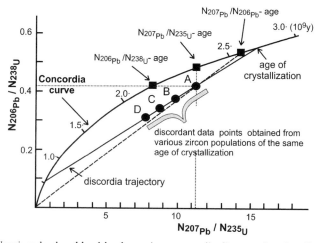

Figure 41: Uranium-lead and lead-lead ages in a concordia diagram showing discordia of zircon populations with different Pb-loss, after [Jäger et al. (1979)].

9.3. U(Th) - Pb Analytical Techniques

As the most critical step in U(Th)-Pb dating is the isotopic lead determination, the amount of sample needed for the analysis is subject to the U(Th) content and the age of the U(Th)-Pb system. Table 7 gives the ranges for U and Th concentrations in the commonly dated minerals.

Table 7: U and Th concentration in commonly dated accessory minerals

Mineral	U (µg/g)	Th (µg/g)
Zircon	10 – 4000	5 – 1000
Monazite	2000 – 10000	10000 – 100000
Titanite	2000 – 10000	70 – 600
Xenotime	5000 – 50000	500 – 10000

Minerals are separated from the rocks using a combination of sieving, heavy liquid and magnetic separation in order to obtain a concordant or near-concordant mineral population. Therefore, only the least magnetic fraction of minerals is further processed [Krogh (1982a)]. Size fractions or morphological fractions or a combination of both are used for U-Pb analysis. In either case, only the best high-quality grains are hand-picked in ethanol under a binocular microscope. In order to increase the concordance of data points, the hand-picked grains are air-abraded following the technique described in [Krogh (1982b)]. The samples are dissolved in acids prior to the analysis, for example zircon samples in pressure vessels take up to four days in HF-HNO$_3$ to pass into solution [Krogh (1973)], [Parrish (1987)]. Dissolved samples are aliquot and spiked for U, Th and Pb (see Chapt. 5, 2.3). The most commonly used spikes are enriched in ^{235}U, ^{230}Th, ^{208}Pb, ^{205}Pb and ^{202}Pb. ^{205}Pb and ^{202}Pb do not require aliquoting and can be added prior to the dissolution. Standard anion-exchange column chromatography is used to separate U, Th and Pb from the samples. The U, Th and Pb isotopic ratios and element concentrations (isotope dilution technique) are most commonly determined using solid-source mass spectrometry (see Chapt.5, 2). Samples are run from a single Re filament using a silica gel technique (U, Th, Pb) [Cameron et al. (1969)] as oxide and metallic ions or from a double Re filament assembly (U, Th) as metallic ions. Alternatively, U can be loaded onto a single W filament with silica gel and Ta$_2$O$_5$ and run as an oxide. U and Th ions are usually collected in the peak-jumping or static acquisition modes. As the most critical step is the analysis of the Pb isotopic composition, the Pb ions are collected in modern multicollector mass spectrometers in a static mode and the smallest ^{204}Pb peak is often collected on a more sensitive detector (Daly, secondary electron multiplier – SEM; see Chapt. 5, 2). This technique has been used successfully to determine the isotopic composition of sub-nanogram lead samples from small portions of mineral separates or even single zircon grains [Roddick et al. (1987)], [Vance et al.(1998)]. Recently, advances in solid-source mass spectrometry have permitted direct measurement of the N_{210Pb}/N_{206Pb} ratios in zircons [Wendt et al. (1993)]. The ^{238}U daughter isotope ^{210}Pb is generally in radioactive equilibrium with ^{238}U, although its very short half-life (22.3 years) results in very low N_{210Pb}/N_{206Pb} ratios (ca 10^{-8}) and consequently low precision of N_{238U}/N_{206Pb} and N_{210Pb}/N_{206Pb} in the age determination. The main advantage of this technique is that it does not require spiking for U and Pb. However, at present, only U-rich samples can be dated with satisfactory precision.

A large group of instrumental techniques focuses on single-grain dating of U(Th)-bearing minerals. The evaporation method [Gentry et al. (1982)], [Kober (1986)] uses a solid-source mass spectrometer to measure ^{207}Pb/^{206}Pb ages. The method has mostly been applied to zircon. Zircon grains are mounted onto a Re filament and heated. Increased temperature induces structural changes in zircon (zircon→baddeleyite) and enhances lead diffusion in the crystal lattice. Heat-released Pb is focused into the mass spectrometer and detected. Although this technique is relatively simple, it provides no information on discordance of samples. A sensitive high mass-resolution ion microprobe (SHRIMP) is used for isotopic analyses of parts of individual mineral grains [Compston et al. (1983)]. A beam of about 20 μm in diameter of high-energy primary ions causes erosion of the mineral surface and sputtering of secondary atoms. As some of the ejected particles are ionised, they can be focused into a mass spectrometer. High mass resolution is required for detection of these secondary ions to discriminate between often coinciding masses of previously chemically untreated samples. The U, Th and Pb concentrations are determined against external standards. Recently, similar spot isotopic analyses of U and Pb have been performed on ICP-MS linked to a laser ablation system [Fryer et al. (1993)], [Hirata et al. (1995)], [Horn et al. (2000)], [Košler et al. (2001)], [Košler et al. (2002)]. A UV laser probe can produce craters down to 5 μm in diameter and the ablated material is then ionised in Ar plasma and introduced into a mass spectrometer. Finally, the U- and Th-rich minerals, for instance monazite, can also be dated using an electron microprobe [Montel et al. (1994)]. However, this technique relies on the assumption that all the Pb in the sample is radiogenic and is can be employed only for materials older than ca $500 \cdot 10^6$ y.

9.4.   U-Pb Dating - Concordia Diagrams, Models of Lead Loss and Intercept Age

The concordia diagram combines two decay systems and is used for graphical presentation of U(Th)-Pb isotopic data. It should be noted that due to the different geochemical behaviour of U and Th, the combination of the U and Th decay systems is much less practical and is rarely used in comparison with the $^{207}Pb/^{235}U$-$^{206}Pb/^{238}U$ system. The evolution of the U and Pb isotopic composition for a hypothetical sample is illustrated in 9.2, Fig. 41. Assuming no initial Pb component in the sample, at the time of mineral crystallisation and closure of the U-Pb system shortly after crystallisation, the sample contains no radiogenic lead and its composition plots at 0.0 in the $^{207}Pb/^{235}U$-$^{206}Pb/^{238}U$ coordinates. As time proceeds, and in the closed system, the decay of ^{235}U and ^{238}U produces radiogenic ^{207}Pb and ^{206}Pb and the sample composition moves upwards along the concordia. The sample displacement along the concordia is related to its age. However, real minerals in real rocks often experience a complex history, which may include several episodes of increased temperature, pressure and fluid activity. This may result in loss of radiogenic lead, which, in the concordia diagram, is reflected by a shift in the sample composition from the concordia towards the origin (0,0). Such compositions are known as discordant (e.g. A, B, C and D in Fig. 41) and the amount of discordance is proportional to the lead loss. In most cases the lead loss is associated with mineral recrystallization, although loss of Pb from the crystal structure by simple volume diffusion has also been documented. Loss of lead is more pronounced in the radiation-damaged mineral grains or grain domains. In magmatic rocks, for example, the smaller and U-rich zircon grains often show more discordant composition compared to the large zircon crystals. For the first time, a discordant composition was attributed to the loss of Pb by [Holmes (1954)] and a linear array of variably discordant data points (discordia) has been described by [Ahrens (1955)]. [Wetherill (1956)] has developed a model of episodic lead loss in which lead is released from U-bearing minerals as a result of a secondary event (e.g., metamorphism) and minerals with variable degrees of lead loss plot along a discordia (see Fig. 42).

The upper intersection of discordia and concordia defines the age of mineral crystallisation, while their lower intersection corresponds to the age of a secondary event. In the model proposed by [Tilton (1960)], Pb is released from U-bearing minerals by continuous diffusion, which is characterised by curved discordia, although it approaches a straight line for older age values (Fig. 43). The lower intersection in this case has no geological meaning. In the dilatancy model defined by [Goldich et al. (1972)] the loss of lead takes place through water-filled microchannels in the metamict crystal lattice.

In this model, lead dissolved in water escapes from the mineral grains as a result of pressure release (e.g., due to crustal uplift) and, accordingly, the lower intersection of the discordia with the concordia dates the timing of the uplift.

Real rocks, however, often experience more than one metamorphic event and, in such cases, interpretation of the discordant age may be very difficult. Alternatively, linear arrays of discordant data points may be interpreted as lines of mixing between two end members, e.g., in sedimentary provenance studies (see Fig. 44) or as reflecting the presence of an inherited component, e.g., older cores in magmatic zircon [Paterson et al. (1992)], or metamorphic overgrowth on igneous zircon grains in the samples.

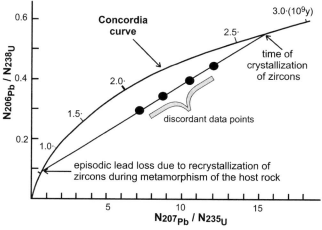

Figure 42: Episodic lead loss in zircon fraction plotted in a concordia diagram, after [Jäger et al. (1979)].

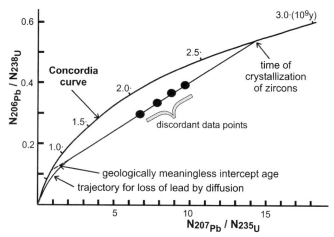

Figure 43: Continuous lead loss in zircon fraction plotted in a concordia diagram.

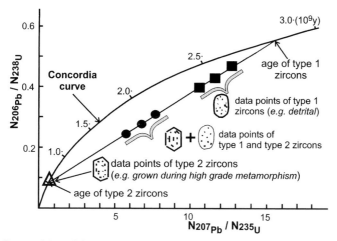

Figure 44: Composition of detrital zircon in a mixed population, after [Jäger et al. (1979)].

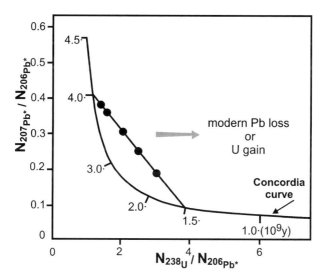

Figure 45: Concordia plot, after[Tera - Wasserburg (1972)].

More details and historical background of various models of lead loss are given in [Jäger et al. (1979)] and [Dickin (1995)]. Discordant data points should always be interpreted with caution and independent evidence, for example cathodoluminiscence or back-scattered electron studies coupled with electron or ion microprobe analyses, is essential to fully reveal the origin and history of dated minerals.

Alternative presentation of U-Pb data in the concordia diagram in which N_{207Pb}/N_{206Pb} is plotted against N_{238U}/N_{206Pb} (Fig. 45) has been proposed by [Tera et al. (1972)]. Unlike in the conventional concordia diagram with highly correlated errors (typically $\rho=0.8$-0.9), error correlation is avoided in the Tera-Wasserburg projection. Due to the different curvature of concordia, this method is usually preferred for dating young rocks. A three-dimensional concordia diagram, in which both the concordia and discordia are represented by planes, has been proposed by [Wendt (1984)]. He added a N_{204Pb}/N_{206Pb} axis to the Tera-Wasserburg projection and this method is especially useful for visualising the common lead component in the samples.

9.5.   Single-Zircon Evaporation Dating, ^{207}Pb/^{206}Pb Apparent Age

Zircon is probably the most common mineral used for U-Pb dating. However, the occurrence of older Pb components in many zircons, lead loss and the presence of more than one zircon population in one rock are common features. Accordingly, only the analysis of small and well-defined zircon samples can produce age data that are geologically meaningful. This can be achieved by analysing individual spots within zircons using either SHRIMP or laser-probe ICP-MS techniques. The single-zircon evaporation technique [Gentry et al. (1982)], [Kober (1986), (1987)] is an alternative and cheaper approach to this problem, although only the ^{207}Pb/^{206}Pb (see 9.6) apparent age can be determined. In an older version of the technique, zircon grains were mounted onto a single-Re filament and heated. High temperatures of ca 1500 $^{\circ}$C induce structural changes from zircon to baddeleyite [Ansdell et al. (1993)], [Chapman et al. (1994)]. This reaction progresses as a sharp reaction front, which migrates from the margins towards the centre of the grain and enhances lead diffusion in the crystal lattice. The isotopic ratios of the heat-released Pb are measured in a mass spectrometer, preferentially using a Daly or SEM detector (see Chapt. 5, 2). A newer version of the evaporation technique utilises a double-Re filament assembly [Kober (1987)]. Zircon grains are mounted onto a Re side filament and heated in steps. HfO$_2$ and SiO$_2$ together with emitted radiogenic Pb* are deposited on the second filament, from which the Pb isotopic ratios can be measured in a mass spectrometer separately for each heating step. SiO$_2$ and HfO$_2$ bedding deposited on the filament forms a thermally stable compound which retains Pb* up to 1400-1500 $^{\circ}$C and serves as an efficient Pb ion emitter, similar to the Si-gel technique [Cameron et al. (1969)]. Accordingly, only U-bearing minerals that contain Si can be successfully dated by the evaporation technique. In the evaporation method of dating, zircon grains are heated from margins towards the core. If the heating is performed in steps, the Pb isotopic composition of individual growth zones can be recorded and the corresponding age can be calculated. This can be used as a tool for identification of older inherited cores in zircons. Although this technique is relatively simple, it provides no information on lead loss and discordance of samples.

9.6. Common Lead Method of Dating, Age Data from Pb Model

The common lead method of dating is the reverse to the previously de-scribed dating methods, which used radiogenic lead in U- and Th-bearing minerals to calculate the crystallisation or metamorphic age. In the common lead method, the Pb-rich and U- and Th-poor mineral phases are used to measure the isotopic composition of lead. Here, the underlying assumption is that, due to negligible content of the parent isotopes of U and Th in the sam-ple, its Pb isotopic composition has not changed significantly since the separation from its source (e.g., the Earth's reservoir, such as the upper mantle). Unlike in the radiogenic lead dating methods where the resulting age usually corresponds to the time elapsed since the crystallisation or metamorphic event, the common lead method yields information on the time period the sample has spent in the particular reservoir. Accordingly, the calculated age values are dependant upon certain models of lead isotopic evolution and should only be considered as "model ages" which may not necessarily have any geological meaning.

Models of single-stage lead evolution, i.e., a single event of lead separation from its parent isotopes [Holmes (1946)] and [Houtermans (1946)], assume that the U, Th and Pb distribution in the Earth was originally homogeneous and that the Pb isotopic composition was also uniform. With the formation of separate reservoirs, regions of different U/Pb and Th/Pb ratios formed and, with the crystallisation of any common Pb mineral (galena, feldspar), the isotopic composition of Pb separated from its parent isotopes remained constant over time. Separation of lead from its reservoir *t* years ago can be described using a modified Eq. 45 as:

$$\frac{N_{206Pb}}{N_{204Pb}} = \frac{N_{206Pb_i}}{N_{204Pb_i}} + \frac{N_{238U}}{N_{204Pb}} \cdot \left(e^{T \cdot \lambda_{238U}} - 1 \right) - \frac{N_{238U}}{N_{204Pb}} \cdot \left(e^{t \cdot \lambda_{238U}} - 1 \right) \quad (53)$$

or

$$\frac{N_{206Pb}}{N_{204Pb}} = \frac{N_{206Pb_i}}{N_{204Pb_i}} + \frac{N_{238U}}{N_{204Pb}} \cdot \left(e^{T \cdot \lambda_{238U}} - e^{t \cdot \lambda_{238U}} \right) \quad (54)$$

where t refers to the time elapsed since the lead separated from its reservoir, T is the age of Earth, N_{206Pb}/N_{204Pbi} is the primordial lead isotopic composition and N_{238U}/N_{204Pb} is the ratio which defines a particular reservoir.

Similar equations can be derived for N_{207Pb}/N_{204Pb} and N_{208Pb}/N_{204Pb} ratios. By introducing:

$$N_{206Pb}/N_{204Pbi} = a_0 \qquad\qquad N_{238U}/N_{204Pb} = \mu$$

$$N_{207Pb}/N_{204Pbi} = b_0 \qquad\qquad N_{235U}/N_{204Pb} = \omega = 137.88^{-1}\cdot\mu \quad (\text{see } 9.2)$$

$$N_{208Pb}/N_{204Pbi} = c_0 \qquad\qquad N_{232Th}/N_{204Pb} = \kappa$$

the following equations are obtained:

$$\frac{N_{206\,Pb}}{N_{204\,Pb}} = a_0 + \mu\cdot\left(e^{T\cdot\lambda_{238U}} - e^{t\cdot\lambda_{238U}} \right) \tag{55}$$

$$\frac{N_{207\,Pb}}{N_{204\,Pb}} = b_0 + \omega\cdot\left(e^{T\cdot\lambda_{235U}} - e^{t\cdot\lambda_{235U}} \right) \tag{56}$$

$$\frac{N_{208\,Pb}}{N_{204\,Pb}} = c_0 + \kappa\cdot\left(e^{T\cdot\lambda_{232Th}} - e^{t\cdot\lambda_{232Th}} \right) \tag{57}$$

from which parameter μ can be eliminated by division of Eq. 55 by Eq. 56

$$\frac{\dfrac{N_{207\,Pb}}{N_{204\,Pb}} - b_0}{\dfrac{N_{206\,Pb}}{N_{204\,Pb}} - a_0} = \frac{1}{137.88}\cdot\frac{e^{T\cdot\lambda_{235U}} - e^{t\cdot\lambda_{235U}}}{e^{T\cdot\lambda_{238U}} - e^{t\cdot\lambda_{238U}}} \tag{58}$$

This equation describes a straight line (isochron) in the N_{206Pb}/N_{204Pb} vs. N_{207Pb}/N_{204Pb} coordinates, from which the age can be determined without knowing the μ parameter. For $t=0$ (present time) this isochron is called "present-day geochron" and theoretically the present-day compositions of all the Earth's reservoirs should lie on this line (Fig. 46). The primordial isotopic composition of lead is derived from meteorites. The most frequently used values are derived from troilite from the iron meteorite found in Canyon Diablo in Arizona and are as follows: $a_0=9.3066$, $b_0=10.293$, $c_0=29.475$ [Chen et al. (1983)].

Lead with isotopic composition corresponding to a growth curve for a particular reservoir characterised by a certain μ value is called "ordinary lead". However, most lead compositions, especially from Phanerozoic samples, do not lie on single-stage growth curves and do not plot along a geochron; in fact most Phanerozoic samples yield negative model age values. The composition of such samples cannot be explained by single-stage models of lead evolution and their Pb isotopic composition is referred to as "anomalous". Some anomalous lead compositions can be explained by two-stage models of lead evolution. Such models assume a change in parameters μ, ω and κ during the lead isotopic evolution as a result of geochemical differentiation of the Earth. A common assumption is that the primordial reservoir, which compositionally corresponds to the mantle, differentiated into two separate reservoirs: depleted mantle and crust, from which Pb can separate and form minerals. This situation is illustrated in Fig. 47. The most commonly used model is the two-stage Pb evolution model proposed by [Stacey et al. (1975)], the parameters of which are given in Table 8.

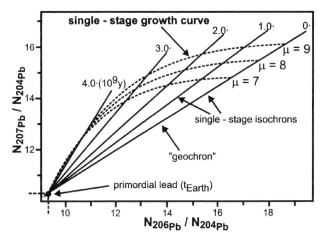

Figure 46: Lead growth curves in a single stage Pb evolution model of Earth's reservoirs with different composition $N_{238U}/N_{204Pb}=\mu$.

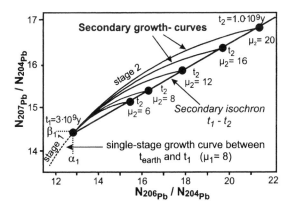

Figure 47: Lead growth curves in a two stage Pb evolution model of Earth's reservoirs with different composition $N_{238U}/N_{204Pb}=\mu$. First and second evolution steps are separated by age $t_1=3 \cdot 10^9$ y, after [Stacey et al. (1975)].

The Pb growth curves can be calculated using Eq. 54 to 56 for each stage. Several poly-stage models of lead isotopic evolution have been developed, although their assumptions usually fit only very special situations. For a more comprehensive description of these models, the reader should refer to [Faure (1986)] and [Dickin (1995)].

Table 8: Lead isotopic composition for two stage Pb evolution model, after [Stacey and Kramers (1975)].

	Start ($\cdot 10^9$ y)	$\dfrac{N\ _{206\ Pb}}{N\ _{204\ Pb}}$	$\dfrac{N\ _{207\ Pb}}{N\ _{204\ Pb}}$	$\dfrac{N\ _{208\ Pb}}{N\ _{204\ Pb}}$	μ	ω
stage 1	4.57	9.307	10.294	29.476	7.192	32.208
stage 1	3.70	11.152	12.998	31.230	9.735	36.837
recent	0	18.700	15.628	38.630	9.735	38.837

9.7.   Representative Examples

The U-Pb method of dating is mostly used for dating acidic and intermediate rocks, although basaltic rocks have also been successfully dated [Heaman and Tarney (1989)]. To illustrate the age differences following from various U-Pb analytical techniques, we chose the example of the Ben Vuirich granite in the Scottish Highlands. This granite intruded between two deformational events into the Dalradian metasediments. The previously obtained U-Pb zircon lower intercept age [Pankhurst et al. (1976)] from the zircon size population suggested its emplacement at ca. $514 \cdot 10^6$ y. However, new high-precision U-Pb dating of abraded zircons (morphological populations) [Rogers et al. (1989)] yields an upper intercept age of $(590 \pm 2) \cdot 10^6$ y, strongly contrasting with the previous dating and suggesting that the sedimentation of the Dalradian occurred in the Precambrian. [Rogers et al. (1989)] have also identified an older (ca $1450 \cdot 10^6$ y) lead component in zircons and suggested that the discrepancy between the two datings might have been due to both the inheritance and surface-correlated lead loss in zircons dated by [Pankhurst et al. (1976)]. The new dating has further been confirmed by SHRIMP dating of zircons at $(597 \pm 11) \cdot 10^6$ y [Pidgeon et al. (1992)]. This case illustrates the importance of modern, highly accurate and precise dating for establishing the timing of magmatic events.

10. Re - Os DATING METHOD
Emil Jelínek, Jan Košler

10.1. Chemical Properties, Radioactive Decay and Isotopic Abundance

Rhenium atomic number $Z=75$ and osmium ($Z=76$) belong to the transitional metals (group VIIB of the periodic table). Rhenium is usually not a major constituent of minerals but it is commonly dispersed in sulphides. Osmium belongs to the platinum group elements (PGE), it is strongly siderophile and, together with Ir and other PGEs, forms natural alloys (e.g., osmiridium). Rhenium has two natural isotopes with masses 185 and 187 whose natural abundances are 37.398 and 62.602 %, respectively.

Osmium has seven stable natural isotopes: 184 (0.02 %), 186 (1.58 %), 187 (1.6 %), 188 (13.3 %), 189 (16.1 %), 190 (26.4 %) and 192 (41.0 %). ^{187}Re decays (see Chapt. 2, Table 1) by β^- emission to stable ^{187}Os with a half-life of $4.16 \cdot 10^{10}$ y which corresponds to a decay constant of $1.666 \cdot 10^{-11}$ y^{-1} [Smoliar et al. (1996)]. This decay scheme is utilised in geochronological and cosmochronological determinations.

10.2. Concept of the Method

The siderophile and chalcophile nature of both Re and Os, together with the low decay constant of ^{187}Re, makes this method suitable for dating metallic phases in meteorites as well as terrestrial sulphide deposits. The concept of the Re-Os method is similar to other parent-daughter techniques of dating (see 7.-9.). Assuming that no initial Os was present in the sample at the time of closure of the Re-Os system, the ^{187}Os radiogenic growth of the concentration is described as:

$$N_{187_{Os}} = N_{187_{Re}} \cdot \left(e^{\lambda_{187_{Re}} \cdot t} - 1 \right) \tag{59}$$

where λ_{187Re} is the ^{187}Re decay constant, t is the decay time, and N_{187Os} and N_{187Re} are the numbers of atoms in the sample at the present time. Eq. 59 solved for age t yields

$$t = \frac{ln\left(\dfrac{N_{187_{Os}}}{N_{187_{Re}}} + 1 \right)}{\lambda_{187_{Re}}} \tag{60}$$

Similar to the Rb-Sr or Sm-Nd methods (see 7.2, 8.2) of dating, an age corresponding to closure of the Re-Os isotopic system can be calculated for a suite of cogenetic samples (e.g., sulphide minerals that had identical N_{187Os}/N_{186Os} isotopic compositions at some time t in their history but, due to the different *[Re]/[Os]* concentration ratios of the individual samples, their present-day N_{187Os}/N_{186Os} ratios are different and plot on an isochron. The equation of this isochron has the form

$$\frac{N_{187Os}}{N_{186Os}} = \frac{N_{187Os_i}}{N_{186Os_i}} + \frac{N_{187Re}}{N_{186Os}} \cdot \left(e^{t \cdot \lambda_{187Re}} - 1 \right) \qquad (61)$$

where the ordinate-intercept corresponds to the initial value of N_{187Os}/N_{186Os_i} and $e^{\lambda_{187Re}t}-1$ corresponds to the slope of the regression line.

Eq. 61 can be valid for age determination only if the proportion of ^{186}Os in nature is constant. However, ^{186}Os is also produced by α-decay of ^{190}Pt with a decay constant of ca $1 \cdot 10^{-12}$ y^{-1} [MacFarlane et al. (1961)], [Walker et al. (1991)]. For the purpose of meteoritic dating, the effect of radiogenic ^{186}Os is negligible, as the proportion of ^{190}Pt is only 0.0122 %. But it may play an important role in dating of PGE ores with high Pt contents. [Walker et al. (1991)] used an equation for the Pt-Os isochron in the form

$$\frac{N_{186Os}}{N_{188Os}} = \frac{N_{186Os_i}}{N_{188Os_i}} + \frac{N_{190Pt}}{N_{188Os}} \cdot \left(e^{t \cdot \lambda_{190Pt}} - 1 \right) \qquad (62)$$

where the symbols are similar to Eq. 60 and λ_{190Pt} is the decay constant of ^{190}Pt.

The isotopic composition of Os can also be presented in γ values which, similar to the ε values (see 7.6, 8.6) used for Sr and Nd, reflect the deviation from the Os isotopic composition of a chondritic reservoir. The proposed present-day values for this reservoir are $N_{187Os}/N_{186Os}=1.06$ and $N_{187Re}/N_{186Os}=3.3$ [Walker et al. (1991)]. However, as the present-day composition of the whole Earth does not perfectly correspond to that of the chondritic reservoir, γ Os values have rather limited applicability.

10.3. Re - Os Analytical Techniques

The analytical difficulties in analysing Os have limited the widespread application of this method. The difficulties arise from the high Os ionization potential (approx. 9 electron-volts). The Os concentration and isotopic ratios can be measured using ion-sputtering mass spectrometry [Luck et al. (1980)], inductively coupled plasma mass spectrometry [Russ et al. (1987)], [Houk et al. (1980)] or accelerator mass spectrometry [Fehn et al. (1986)]. The Re concentration can be determined by neutron activation techniques [Herr et al. (1967)]. Negative-ion thermal ionization mass spectrometry (N-TIMS) yields a precision that is an order of magnitude better than that of classical TIMS

[Volkening et al. (1991)]. Using this technique, the Os isotopic composition is measured as OsO_3^- ions. The formation of this atomic species is further enhanced by sample loading with Ba-nitrate and by bleeding oxygen into the source chamber of the mass spectrometer [Creaser et al. (1991)], [Walcyk et al. (1991)]. This method can also be applied to other PGEs including Re, which forms ReO_4^- ions. Os and Re concentrations can also be determined by isotope dilution using ^{190}Os and ^{185}Re spikes, respectively. Fractionation effects in the course of mass spectrometric determination are corrected for by normalising the measured Os isotopic ratios to $N_{192Os}/N_{188Os}=3.0827$ [Luck et al. (1983)]. Because mass 187 occurs in both elements, Re and Os must be quantitatively separated prior to analysis by N-TIMS isotope dilution. The method includes distillation of osmium tetraoxide, organic solvent extraction of Re and ion-exchange purification [Luck et al. (1980)].

A new approach, which uses neutron activation of a sample prior to mass spectrometric analysis of Os, has been published by [Yin et al. (1993)]. Similar to the Ar-Ar activation technique, they converted ^{185}Re and ^{187}Re in a sample into ^{186}Os and ^{188}Os, respectively. Calibration of Os production after neutron irradiation allowed the Re/Os ratio to be determined as part of the Os isotopic analysis. However, because interfering Os isotopes are also produced by neutron irradiation of Os, this technique can only be used for samples that contain only a small amount of initial non-radiogenic osmium (i.e., samples with high Re/Os ratios, e.g., molybdenite).

10.4.   Applicability and Limitations of the Re - Os Method

Slow decay of ^{187}Re and the siderophile/chalcophile nature of both elements make this method an excellent tool for dating iron meteorites and sulphide phases in stone meteorites. In terrestrial materials, the Re-Os method is suitable for dating PGE-rich minerals of sulphide mineralization, especially molybdenite for its high Re/Os ratio. Strong decoupling of Re and Os during the crust-formation processes, such as partial melting of upper mantle rocks, results in crustal enrichment in radiogenic Os. This process makes this element a powerful petrogenetic tracer, similar to Nd, Sr and Pb isotopes. The Os isotopic composition can also be used as a tracer for the origin of PGE enrichment associated with impact activity.

The major drawback of the Re-Os method lies in the analytical requirements on Re and especially Os analysis, which are still beyond the capabilities of most geochronological laboratories. This is directly linked to the limited availability, compared to other elements, of Os analytical data from various types of geological samples. The wider applicability of the method is also hampered by limited knowledge of Re and Os behaviour during the magmatic, metamorphic and sedimentary processes.

10.5.   Representative Examples

[Luck et al. (1980)] performed the Re-Os isochron age determination of five iron meteorites and one chondrite (Fig. 48). As a result of this study, they suggested that the meteorites, despite having different composition and probably also different origin, were formed during a narrow time span at around $4.55 \cdot 10^9$ y. The initial N_{187Os}/N_{186Os_i} ratio of 0.805 calculated for this suite of meteorites is interpreted as an Os initial isotopic ratio for our Solar system. Further implications from this study suggest that the age of our galaxy is between 13.3 and $22.4 \cdot 10^9$ y.

An excellent Re-Os isochron fit for Re-rich whole rock samples from the Stillwater Complex in Montana was obtained by [Lambert et al. (1989)]. For sulphide cumulates, bronzite pegmatite and the chromite-rich layer, they obtained an age of $2.66 \cdot 10^9$ y, which was in good agreement with previously obtained Sm-Nd and U-Pb age data for this intrusion.

[Ellam et al. (1992)] used Os and Nd isotopic data to study the mixing processes that led to the formation of picritic Karoo flood basalts in southern Africa (Fig. 49). They demonstrated that the use of Os isotopes is a powerful tool in elucidating the evolution of basaltic magma, especially when studying the potential sources of magmatic rocks.

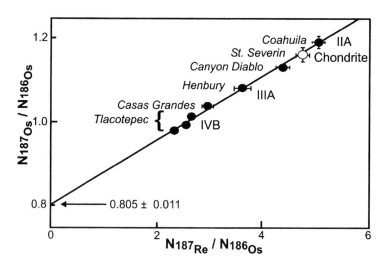

Figure 48: Re-Os isochron diagram for five iron meteorites and one chondrite, after [Luck et al. (1980)].

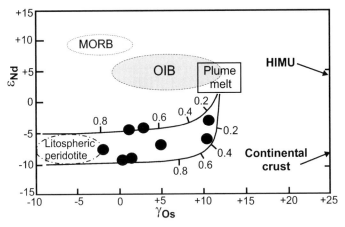

Figure 49: Nd-Os diagram showing the composition of Karoo basalts and their potential
sources and two calculated mixing lines between the plume melt and lithospheric
components, after [Ellam et al. (1992)] OIB = ocean islands basalt, MORB = mid-
ocean ridge basalt, CHUR = chondritic uniform reservoir, HIMU = subducted
oceanic crust. The numbers along the mixing curves correspond to the proportion
of peridotite component in the magma.

11.   Lu - Hf DATING METHOD
Emil Jelínek

11.1.   Chemical Properties, Radioactive Decay and Isotopic Abundance

Lutetium, with an atomic number of $Z=71$, is the heaviest of the lithophile
rare earth elements (group IIIB of the periodic table). It has two natural
isotopes ^{175}Lu and ^{176}Lu (see Chapt. 2, Table 1) with abundances of 97.4 and
2.6 %, respectively. ^{176}Lu decays by branched decay to ^{176}Hf (β^-, 97 %, excited
state) and ^{176}Yb (electron capture, 3 %), but only the former is used for geochronological determinations. ^{176}Hf produced by β^- decay attains a stable state
through further γ emission. The decay constant for β^- decay is $1.94 \cdot 10^{-11}$ y^{-1},
corresponding to a half-life of $3.57 \cdot 10^{10}$ y [Patchett et al. (1980a)].

Hafnium ($Z=72$) belongs to the transition metals (group IVB of the periodic table). Unlike Lu, which is incompatible during most magmatic processes, Hf strongly resembles the geochemical behaviour of Zr and, together
with Nb, these elements are often classified as being "high-field-strength",
i.e., bound in mineral phases with a substantial proportion of covalent bonds.
Hf has five stable natural isotopes with masses of 176 (5.2 %), 177 (18.6 %),
178 (27.1 %), 179 (13.74 %) and 180 (35.2 %). ^{174}Hf (relative abundance
0.16 %) is unstable and decays by alpha emission to ^{170}Yb.

11.2.   Concept of the Method

Similar to the Rb-Sr or Sm-Nd method of dating (see 7.2, 8.2), decay of ^{176}Lu to ^{176}Hf is described as

$$N_{176_{Hf}} = N_{176_{Lu}} \cdot \left(e^{\lambda_{176_{Lu}} \cdot t} - 1 \right) \qquad (63)$$

where $\lambda_{176_{Lu}}$ is the ^{176}Lu decay constant, t is the decay time (the age), and $N_{176_{Hf}}$ and $N_{176_{Lu}}$ are the numbers of atoms in the sample at the present. Eq. 63 solved for age yields:

$$t = \lambda_{176_{Lu}}^{-1} \cdot ln \left(\frac{N_{176_{Hf}}}{N_{176_{Lu}}} + 1 \right) \qquad (64)$$

As the minerals usually contain primary Hf, its isotopic composition must be known for age calculations. Similar to Rb-Sr or Sm-Nd dating methods, the initial Hf isotopic composition can be derived from the isochron equation:

$$\frac{N_{176_{Hf}}}{N_{177_{Hf}}} = \frac{N_{176_{Hf_i}}}{N_{177_{Hf_i}}} + \frac{N_{176_{Lu}}}{N_{177_{Hf}}} \cdot \left(e^{t \cdot \lambda_{176_{Lu}}} - 1 \right) \qquad (65)$$

where $N_{176_{Hf}}/N_{177_{Hf_i}}$ gives the initial value and corresponds to the ordinate-intercept, and $(e^{\lambda_{176_{Lu}} t} - 1)$ corresponds to the isochron slope.

This equation can be valid only for a suite of cogenetic samples which have the same age and initial $N_{176_{Hf}}/N_{177_{Hf_i}}$ isotopic ratio. The Lu-Hf method is especially useful for dating REE-rich and Hf-poor minerals (i.e., gadolinite, xenotime) but it has also been used for dating whole-rock samples as well as meteorites [Patchett and Tatsumoto (1980a), (1981)].

11.3.   Lu - Hf Analytical Techniques

For Lu-Hf dating, it is necessary to determine the Lu and Hf concentrations and $N_{176_{Hf}}/N_{177_{Hf}}$ isotopic ratios. The method has several difficulties associated both with the sample preparation and with mass spectrometry:

- As the rare earth and hafnium in most rocks are contained in accessory mineral phases which are difficult to dissolve (e.g., zircon, titanite, monazite, xenotime), sample homogeneity and complete sample dissolution are crucial for the analysis.
- Because mass 176 is contained in both Lu and Hf and also in Yb, Lu and Hf must be quantitatively separated prior to mass spectrometric determinations to avoid mass interference. This is achieved through ion-exchange separation techniques [Patchett et al. (1980b)], [Hooker et al. (1975)].

– The high ionization potential of Hf results in low ion production and high filament temperatures are therefore required. A triple filament assembly is often used for Hf analysis, although techniques which use Ir or Mo to promote Hf ionization and graphite to suppress ionization of REEs have also been described [Corfu et al. (1992)]. Lu is analysed using a triple filament assembly [Thirwall (1982)].

11.4. Evolution of Hf Isotopes in Time

During most magmatic processes (e.g., partial melting of upper mantle rocks), Hf enters the melt more rapidly than Lu. This leads to lower Lu/Hf ratios in crustal rocks compared to both the depleted and the undepleted parts of the Earth's mantle. Accordingly, similar to the Sr or Nd isotopes, since the formation of the Earth's crust, the crustal and mantle reservoirs have different Hf isotopic compositions. Zircon is an especially useful material for studying the evolution of the Hf isotopic composition in time, because it contains almost no lutetium, its hafnium content is up to 3 wt % and the age of zircon crystallisation can be determined independently from its U-Pb system. [Patchett et al. (1981)] found that rocks of different ages of presumably mantle origin differ in their N_{176Hf}/N_{177Hf} ratios. Their data array roughly corresponds to the chondritic evolution calculated by [Patchett et al. (1980a)], suggesting that the present-day N_{176Hf}/N_{177Hf} and N_{176Lu}/N_{177Hf} ratios of the chondritic reservoir are 0.28286 and 0.0334, respectively (see Fig. 50).

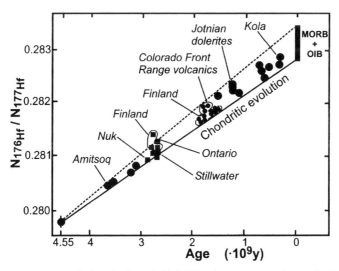

Figure 50: Hf isotopic evolution in time. Initial Hf ratios correspond to a depleted mantle source, after [Patchett et al. (1981)]. OIB = ocean island basalt, MORB = mid-ocean ridge basalt.

Similar to the Sr and Nd systems (see 7.6, 8.6), the Hf isotopic composition can also be described in terms of ε_{Hf} values. i.e., as deviations from the chondritic evolution in parts per 10^4 as

$$\varepsilon_{Hf}(t) = \left(\frac{\dfrac{N_{176\,Hf\,sample}}{N_{177\,Hf}}}{\dfrac{N_{176\,Hf_{CHUR}}}{N_{177\,Hf}}} - 1 \right) \cdot 10^4 \qquad (66)$$

where N_{176Hf}/N_{177Hf} refers to the Hf isotopic ratio of a sample and the chondritic reservoir (CHUR) at decay time t.

Analogous to the Nd isotopic system, Hf model age values can also be calculated as corresponding to the time when the isotopic composition of Hf in the sample was identical to that of its original reservoir (e.g., depleted mantle). However, the applicability of the age determination using the Hf model to real rock samples is limited, similar to the Nd model.

11.5.   Applicability and Limitations of the Lu - Hf Method

The applications of the Lu-Hf method of dating are similar to those of the Rb-Sr method but, because both Lu and Hf are preferentially bonded in more resistant mineral phases, this method is more suitable for dating rocks which have experienced post-crystallisation disturbances. Because the Lu/Hf ratio decreases with increasing magma differentiation, the method can also be employed for basic rocks, where the Rb-Sr method usually fails because of the high Lu contents. Extraterrestrial samples are usually difficult to date using the Lu-Hf technique because relatively large samples (grams) are required in order to extract a sufficient amount of radiogenic Hf.

In general, because the chemical separation of Lu and Hf and mass spectrometric determination of Hf are more complicated compared to the Rb-Sr technique, the Lu-Hf dating method is often used in cases where the Rb-Sr method is not applicable. However, the information on the initial Hf isotopic compositions (e.g., from zircons) constitutes a powerful tool in studying the magma petrogenesis as well as evolution of the Earth's reservoirs.

11.6.   Representative Examples

A suite of whole-rock samples and zircon separates from Amitsoq gneisses of western Greenland yielded an errorchron corresponding to an age of $(3.58\pm0.22)\cdot10^9$ y [Pettingill et al. (1981)], (Fig. 51). This age is in agreement with previous U-Pb and Rb-Sr ages derived from this rock. As the compositions of zircon separates from the Amitsoq gneisses form a linear array with whole-rock samples, the dating confirms the usefulness of Hf isotopic studies on zircon.

The behaviour of Hf isotopes in basaltic rocks of oceanic islands (OIB) and mid-ocean ridges (MORB) has been studied by [Patchett (1983)]. He has found in the studied OIB samples that the N_{176Hf}/N_{177Hf} ratios strongly resemble the N_{143Nd}/N_{144Nd} ratios and exhibit the opposite behaviour compared to Sr isotopes. Considerably greater spread of the Hf isotopic composition compared to Nd and Sr isotopes in MORB samples is attributed to a more effective Lu/Hf fractionation compared to that of Sm/Nd or Rb/Sr in the depleted parts of the upper mantle (see Fig. 52).

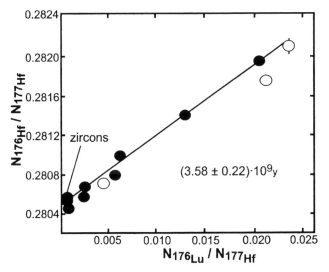

Figure 51: Whole-rock and zircon Lu-Hf isochron for Amitsoq gneisses, western Greenland, after [Pettingill et al. (1981)].

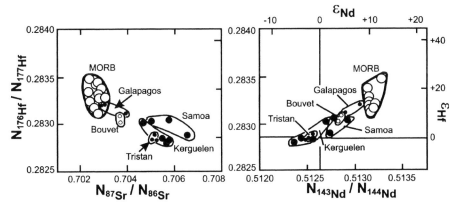

Figure 52: Hf-Sr-Nd isotopic composition of ocean island basalt and mid-ocean ridge (MORB) basalt, after [Patchett (1983)].

Chapter 5

RADIONUCLIDE ANALYSES

Richard Tykva
Academy of Sciences of the Czech Republic, Institute of Organic Chemistry and Biochemistry, Department of Radioisotopes. CZ – 16610 Prague 6, Czech Republic

Jan Košler
Charles University, Faculty of Science, Institute of Geochemistry, Mineralogy and Mineral Resources. CZ – 12843 Prague 2, Czech Republic

1.　ACTIVITY MEASUREMENTS
Richard Tykva

1.1.　Basic Terms

The most often used procedure to analyze environmental radioactivity is measurement of activity connected with spectrometry in the case of more present radionuclides. In these arrangements, the measurement is carried out by means of detectors in which interactions take place between the input particles emitted from the analyzed source and the detector mass within its effective volume. The detectors are divided according to the character of these interactions (see 1.5). The following processing of the detector voltage response are carried out by appropriate commercially available electronic circuits usually realizing amplification and amplitude analysis. The record is mostly evaluated by PC, applied often also for a measurement control. An important condition for radionuclide detection and/or spectrometry is the elimination of disturbances leading to creation of false counts. A comprehensive description of different arrangements for activity measurements of radionuclides are to be found in the specialized monographs [Knoll (1979)], [Tsoulfanides (1983)].

R. Tykva and D. Berg (eds.), Man-Made and Natural Radioactivity
in Environmental Pollution and Radiochronology, 273-335.
© 2004 *Kluwer Academic Publishers. Printed in the Netherlands.*

As it is described in Chap. 1, 4., the level of activity is given by the number of disintegrations of nuclei per second (Bq), i.e., by the decay rate. The determination of activity is based on measurement of the number of registered decays (counts) per time unit, i.e., on the counting rate (cps = counts per second, often cpm = counts per minute, in low levels also cph = counts per hour are used). The ratio of the counting rate to the disintegration rate represents the efficiency of measurement (counting or detection efficiency).

The condition for successful determination of a certain activity of a radionuclide is the attainment of a corresponding detection sensitivity. For its simple expression the figure of merit T can be used, for which

$$T = E^2 \cdot B^{-1} \qquad\qquad (1)$$

where E means the detection efficiency and B the background.

The value of the background (e.g., in cpm) can be obtained using just the same measuring arrangement without the measured radionuclide(-s) in the sample. In order to enable the measurement of low activity and in view of the quadratic dependence, the decisive factor is primarily the detection efficiency of the radiation emitted by the measured radionuclide. However, a lowering of the background is also desirable. Figure of merit (Eq. 1) is decisive to optimize the detection conditions (see, e.g., Fig. 4.).

Usually relative measurements of activity are carried out which represent a determination relative to the defined activity of a reference source of the same radionuclide (see 1.7.2). In this case, the counting efficiency is given by measurement of the counting rate of a standard source with a known decay rate. The absolute determination of activity is used for environmental radioactivity quite exceptionally. On the contrary, for samples with identical counting conditions only comparison of counting rates is often performed without any real activity measurements.

It is always necessary to choose the detection arrangement according to the conditions analyzed in the following Part 1.2 to obtain the best measuring parameters.

1.2.   Detection Efficiency and Background

After the determination of the counting rate of a reference source in the measuring arrangement used, we express the detection (counting) efficiency E in a given time interval in percents, using the expression

$$E = \frac{number\ of\ recorded\ counts}{number\ of\ decays\ in\ the\ source} \cdot 100\% \qquad\qquad (2)$$

To achieve a high efficiency of the detection it is indispensable to assure that the highest possible number of particles emitted from the source enters into the effective volume of the detector. This can be achieved:

- by the choice of the maximum possible measurement geometry, i.e., the maximum part from each emitted particle enters the effective detection volume,
- by exclusion or limitation of emitted particles in the sample itself (selfabsorption),
- by exclusion or limitation of the kinetic energy losses of the detected particles between the sample and the effective volume of the detector, e.g., during their passage through the air and/or the input substructures of the detector which do not contribute to the generation of the detector response.

In the interaction of the detected particle inside the effective detector volume it is indispensable, for the achievement of a high detection efficiency, to ensure:
- the maximum particles generated in interactions of the detected particle with the efficient medium of the detector should take part in the formation of the corresponding response,
- in the active part of the detector, the generation of the response should exceed the noise equivalent of the detector, i.e., the detector noise expressed in units of the magnitude in which the result of the primary interaction of the detected particle is expressed,
- connection of the detector to a suitable electronic device, in order to secure the transmission of the response generated in the detector for its further processing and recording.

From the above summary it is evident that the influencing of the detection efficiency is primarily possible by the detector. The relevant steps concerning the detector are analysed in Part 1.5. From the relationship (Eq. 1) it follows that low background is also decisive for the achievement of high detection sensitivity. But, unlike detection efficiency, background depends on other parameters.

A permanent source of the registered detector background is the cosmic radiation [Janossy (1950)]. From the point of view of the effect on detector background, it is desirable to take into consideration the particle and energy composition of cosmic radiation after passage through the earth's atmosphere with a predominance of μ-mesons, nucleons, electrons and photons [May et al. (1964)]. These elicit the detector background either by direct response after penetration into the effective detector volume or by secondary particles emitted by interactions of incident cosmic radiation in substances into which the incident cosmic radiation has entered, e.g., in the wall of the ionization detector, respectively. For common view, cosmic radiation incident on the detector can be divided roughly in two parts: a soft component, representing about 2/3, and a hard one composed predominantly of high energy μ-mesons.

A further background source is created by contamination of natural or man made radionuclides in the environment of the detector, determined primarily by the geochemical composition of the laboratory surroundings, especially by the presence of primordial radionuclides and their decay products (Chap. 2, 1). For this reason, it is, e.g., useful to check the ^{238}U, ^{232}Th and ^{40}K contents in the mountain rock samples sometimes, along several km adit of the underground laboratory [Kovalchuk et al. (1982)]. In addition to the emission of high-energy photons, penetrating through the ground or walls of the laboratory, the main danger is produced by contamination with radon (^{222}Rn), escaping from the soil or rock. Mathematical models have been constructed [Revzan et al. (1991)] for the description of the penetration of radon of terrestrial origin into buildings. In buildings on the Earth surface, the highest concentration of radon is, naturally, in the cellars which are, for reasons of screening cosmic radiation and bearing capacity of the floor, most frequently used for the measurements of lower activities. It is evident that in the elaboration of a laboratory project, designed for the measurement of low activities of radionuclides, it is desirable to chose a place where there is no danger of background increase in this way (for indoor and outdoor evaluations of ^{222}Rn see 1.5.1 and 1.6.2, Chap. 2, 3.4.6). Desorption of ^{222}Rn from metal surfaces was also apllied [Bign (1992)]

A further source of radionuclide contamination, contributing to an increase of the detector background, is the material of the building and the construction material of the laboratory equipment, mainly that used for the manufacture of the detector shielding and of the detector itself. The chemical composition itself does not guarantee radionuclide purity. A certain material may be variously contaminated even if produced by the same technology. Pollution by radioactive fallout can also play an important role.

In addition to external contamination in metallurgical or technological processing, a number of materials contain also naturally present radionuclides. Of the uranium decay series this is mainly ^{214}Bi as a daughter product of ^{226}Ra, which occurs, e.g., in zinc ores and predetermines thus the inadequacy of zinc or brass for use in an equipment for the determination of lower activity. Of the thorium decay series (see Chap. 2, Fig. 1b), ^{208}Tl is mainly important. A very distinct natural contaminant is radionuclide ^{40}K, the natural presence of which in 1 g of natural potassium is 30.9 Bq represented by about 25 electrons per second (spectrum beta with maximum energy 1.33 MeV) and approximately 210 photons of 1.4 MeV per minute. The natural presence of radionuclide ^{14}C (0.34 Bq·g^{-1}) affects, for example the area activity of plastic foils prepared from recent raw materials (averagely 50 mBq per 100 cm^{2}). In close proximity of the active volume of the detector for measuring low activities it is therefore necessary to use plastic materials made of crude oil or coal. The dispersion of the values of contamination for the same kind of material can be considerable. For a preliminary choice, it is suitable to select a construction material with a low natural radionuclide contamination, originating from the time before the first explosions of nuclear weapons in the atmosphere.

In all cases it is essential to check the radionuclide purity of every material before use because the specific activity of building [Wogman et al. (1982)] or construction [Kamikubota et al. (1989)] materials are very variable. A possible arrangement to check the radionuclide concentrations in

Table 1: Selected natural radionuclides in building materials, based on [Wirdzek et al. (1985)].

Material	Specific Activities (mean; min-max) (Bq·kg^{-1})			
	^{40}K	^{226}Ra	^{232}Th	RE[1]
Granites	680 383-1043	58 10-101	59 17-137	191 64-336
Sands	197 65-328	8 3-15	10 4-15	38 25-47
Gravels	152 64-271	10 4-15	12 3-17	38 13-51
Limestones	63 28-137	10 5-14	4 2-8	21 15-32
Marbles	132 103-148	12 11-14	4 3-5	28 24-30
Cements	185 104-271	20 12-28	12 7-16	51 30-61
Bricks	610 477-804	45 28-64	45 22-70	154 123-191
Clinkers	613 435-819	57 20-91	67 34-113	194 116-263
Aerated concretes	368 171-557	40 8-71	33 6-61	114 30-95
Coal fly ashes	443 194-726	86 42-129	72 41-107	215 139-273
Azbestos	191 118-230	9 5-13	14 7-19	43 31-56
Pearlites	1149 1125-1172	53 39-67	76 66-87	248 219-277
Ceramics	543 302-770	29 20-39	39 20-53	125 66-156

[1] $RE = C_{Ra226} + 1.26\, C_{Th232} + 0.086\, C_{K40}$ (radium equivalent)

Table 2: Comparison of primordial radionuclide levels in the background of a 132 cm^3 intrinsic germanium gamma spectrometer before and after rebuilding using selected construction materials, based on [Brodzinski et al. (1988)].

Primordial Radionuclide	Energy	Background		Improvement Factor
		Before	After	
	(keV)	(cph)	(cph)	
^{235}U	186	73	< 0.0048	> 15000
^{228}Ac (^{232}Th)	911	9.0	< 0.0034	> 2700
^{234m}Pa (^{238}U)	1001	3.4	< 0.0024	> 1400
^{40}K	1461	22	0.017	1300
^{208}Tl (^{228}Th)	2614	1.0	< 0.0011	> 910

such materials by the use of a special germanium spectrometer system has been described in detail [Arthur et al. (1987)]. For illustration of the above data, the specific activities of some frequently used building materials are summarized in Table 1 and the effect of selection of the starting material of the detector on background is shown in Table 2.

A serious natural contamination of ^{40}K occurs in a standard glass. Therefore, glass parts of equipment for environmental radionuclide analyses (e.g., vials, walls of photomultipliers, etc.) are made of borosilicate (^{40}K and ^{226}Ra free) or low-potassium glass, teflon, polyethylene, exceptionally from quartz or other materials [Polach et al. (1983)]. It has been also shown [Brodzinski et al. (1990)] that not only the contamination with primordial radionuclides plays an important role in a detector background but the cosmogenic radionuclides as well. Taking this fact into consideration [Brodzinski et al. (1990)] reduced germanium detector background using material selections and careful control of fabrication. The radionuclides ^{57}Co, ^{58}Co and ^{65}Zn, formed cosmogenically in the crystals, were reduced by minimizing the time between the final zone refinement, crystal growth, installation in the cryostat, and placement underground. An attempt was made to reduce the background from the decay of ^{68}Ge in the detectors by deep mining the ore, rushing it through the refinement, crystal growing, and detector fabrication processes, and storing the germanium underground at all times it was not "in process". Cosmogenically formed ^{54}Mn, ^{59}Fe and $^{56-60}$Co in a cryostat were minimized by electroforming the cryostat parts. The ubiquitous background from primordial ^{40}K in electronic components was virtually eliminated by selecting low-background components and by hiding the first-stage preamplifier behind 2.5 cm of 450 years old lead in one unit and special low-background lead in the other. In the same facility the underground detectors have copper cryostats completely electroformed from low-background copper (electroforming is a process analogous to zone refining in its ability to remove chemical impurities).

The influence of sources outside the detector on detector background can be limited by shielding. Firstly, the locality of the workplace itself contributes to considerable absorption of kinetic energy of the cosmic radiation particles. Cosmic ray background is increased with increasing altitude above sea-level. Therefore, the laboratories for the determination of low activity of environmental radionuclides are often located in cellars, where the shielding effect of the building is useful, and also the effect of the radiation dispersed on surrounding objects is lowered, e.g., trees in the vicinity of the building. Moreover, the necessary carrying capacity of the floor is also available for the shielding arrangement, the weight of which is usually several hundreds or even thousands of kilograms. When an especially intense suppression of the background is necessary, the workplace is located either in an especially constructed underground bunker [Winn et al. (1988)] or in abandoned limestone or salt mines [Zdesenko et al. (1985)]. In all cases, it is necessary to measure the activity of the radionuclides present in the neighbourhood (e.g., the building base) and at the measuring site (e.g., the building and equipment materials). The effects of shielding are shown in Table 3.

Table 3: Background of a germanium detector (35 cm^3) using different shielding: A - ground level without shielding, B - underground laboratory in a salt mine (430 m deep) without shielding, C - arrangement B and shielding box, based on [Zdesenko et al. (1985)].

Radio-nuclide	Gamma Energy (keV)	Background (cph)			Ratio		
		A	B	C ·10^{-3}	A/B	B/C ·10^2	A/C ·10^3
^{212}Pb	238.6	877.7±7.1	15.3±1.6	150±40	57.4±4.4	1.0	6
^{214}Pb	295.2	222.7±4.7	5.4±1.0	28±20	41.2±4.7	1.0	8
^{214}Pb	352	387.9±0.4	10.2±1.0	46±23	38.0±4.4	2.2	8
^{208}Ti	511	189.5±3.1	2.4±0.6	-[1]	79.0±5.2	-	-
^{208}Ti	583.1	284.6±2.9	3.8±0.6	26±20	74.9±5.0	1.5	10
^{214}Bi	609.3	345.3±3.0	8.4±0.7	32±15	41.1±4.3	2.6	10
^{228}Ac	911.2	191.4±2.1	2.1±0.5	20±10	91.1±4.2	1.0	10
^{214}Bi	1120.3	80.8±1.9	2.0±0.4	-	40.4±4.0	-	-
^{40}K	1460.8	1280.0±3.0	16.5±0.6	13±6	77.6±5.0	12.7	100
^{214}Bi	1764.3	69.4±0.9	1.9±0.2	4±4	36.5±4.5	4.6	17
^{214}Bi	2204.2	18.2±0.7	0.6±0.1	-	32.5±5.4	-	-
^{208}Tl	2614.5	166.1±1.0	2.1±0.2	8±4	79.1±5.0	2.6	20

[1] Indicates no detectable peak

In some measuring facilities existing areas are used, characterized by strong natural shielding, for example transport tunnels in the Alps. A systematic study [Aglietta et al. (1992)] of the background radiation has been performed for three different experimental sets under changing conditions in the underground experiment at Mt. Blanc laboratory. It was found that the variations in background is due to the presence of ^{222}Rn in the laboratory room. Another analysis in similar experimental conditions has been concentrated on alpha-ray induced background [Hubert et al. (1986)]. Background spectra of several spectrometers have been recorded in a deep underground laboratory located in the Frejus tunnel. The results show that an alpha ray induced background from the ^{210}Pb decay is observed. A possible explanation could be related to the adsorption of the Rn gas on the surfaces of the Ge crystal and/or other parts during the assembly of the used spectrometer.

The basis of the mechanical shielding consists in materials with a high density (e.g., Pb, Fe, Hg), a high proton number and a high radionuclide purity, guaranteeing lowering of the background caused by soft cosmic radiation and radionuclide contamination in the vicinity of the measuring arrangement. This basic material in a thick layer is supplemented on the inner side, for some types of detectors, by a thin layer of an additional shielding material (e.g., Cu) absorbing secondary particles set free in the basic shielding material by interaction with cosmic radiation. Another independent shielding layer in the proximity of the detector itself is also used for the shielding against thermal neutrons in order to prevent the activation of the detector material or the emission of a false impulse by the reflected hydrogen nucleus, forming a part of the effective detector volume. For the absorption of neutrons radionuclide-free material is used inside the outer shielding, displaying a high concentration of hydrogen for neutron scattering (polymers or paraffine) and containing boric acid to absorb neutrons.

Among the basic shielding materials, lead takes first place. For the determination of low activity and in view of the proportion of radionuclide ^{210}Pb with half-life of approximately 20 years, sufficiently old lead should be used, adjusted to the required shapes in a special melt. A radionuclide free lead is commercially available, e.g., from old wrecked ships. The low content of antimony in lead is also important to decrease the shielding effect on background. In laboratories the lead shielding absorbs mainly the soft component of cosmic radiation, while small loss of kinetic energy of its penetrating component does not prevent penetration into the detector. Comparative measurements have shown that, for this reason, it is not useful to increase the thickness of the lead shielding to over 10 - 15 cm. A comparison of the influence of a shield arrangement was carried out in detail [Lindstrom et al. (1990)]. In addition to lead, non-alloyed, preferably old iron of a high radiochemical purity, or mercury are also used as basic shielding materials. The high radiochemical purity of mercury can be achieved by distillation.

The application of inner linings must be carefully considered. A Japanese group [Shizuma et al. (1987)] measured background spectra with a Ge detector in a low-background shielding using various inner linings; Lucite, aluminium, iron, copper and lead. The experimental results show that the background counting rate in the energy region 0 - 500 keV and backscattered gamma-ray increase in a shielding with low Z linings. This can be qualitatively understood through the gamma-ray absorption and scattering processes in the shielding material.

For the suppression of the effect of the hard background component a further shielding detector is used or a system of shielding detectors, surrounding the measuring detector proper, and connected with it in anticoincidence [Schönhofer (1991)]. The anticoincidence arrangement permits the background particles causing the response in the detector for the proper measurement to elicit the response in the shielding detector as well. Such a pair of impulses is then not counted. Due to different specific ionization of the detected particles in the primary interactions, the effect of discrimination using different rise time of the voltage response of the detector is sometimes applied.

Some results obtained in a relatively simple arrangement in a surface laboratory [Grinberg et al. (1961)] are presented in Table 4. An interesting comparison of influencing the detector background by different changes in the measuring arrangement is shown in Fig. 1 [Brodzinski et al. (1988)].

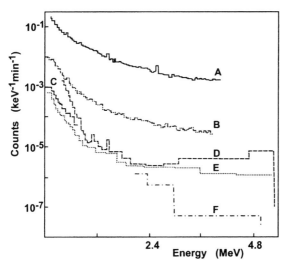

Figure 1: Background spectra of "radiopure" germanium spectrometer: A- in 10 cm thick lead shield, B- cryostat assembly rebuilt with radiopure material and anticoincidence shielding added, C- 7.3 cm thick copper inner shield added, D- indium electrical contact removed, E- copper shield replaced with 10 cm thick shield of 448 year old lead, F- solder electrical connection removed, based on [Brodzinski et al. (1988)].

Finally, it must be noted that optimizing of a low level detector for decreased background is an iterative procedure. With every step a new generation of background counts can show up [Verplancke (1992)].

Table 4: The influence of different simple shielding arrangements on a wide - channel background of Na(Tl) detector (4x3 in), after [Grinberg et al. (1961)].

Measurement Conditions	Background
	(cpm)
Ground level laboratory (A)	36
Underground laboratory (B)	20
B + 10 cm Pb	7
B + anticoincidence	2.2
B + 10 cm Pb + 5 cm Hg	0.5

1.3.   Counting and Spectrometry

Different demands are applied in experimental arrangements for environmental radioactivity according to the resolved radionuclides in the sample. The Fig. 2 shows an example of an arrangement of a device to measure radioactivity. In Part 1.5 the different detectors are described and discussed. If we should measure activity of only one radionuclide or the total activity of a mixture of radionuclides (counting), a suitable detector with simpler electronic circuits is sufficient. If activities of several present radionuclides should be determined separately or a radionuclide and/or background spectrum should be analyzed (spectrometry), not only another electronics is necessary (multichannel amplitude analyzer) but sometimes also another detector may be more convenient. So, e.g., for one gamma emitter a scintillator detector can be better than a semiconductor due to its higher detection efficiency. However, the lowering of background in a semiconductor spectrometry has to be considered due to very narrow peaks, Moreover, a scintillation detector does not enable a higher resolution of a complex gamma spectra and must be removed by a germanium detector with a different electronics.

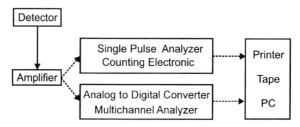

Figure 2: Arrangement of a counting and spectrometric device.

The spectra of alpha or gamma mixtures having monoenergetic components are analyzed using different detectors with the voltage response proportional to the kinetic energy of the particle and electronics including a multichannel amplitude analyzer. Such spectra consist of individual peaks. The resolving of beta radionuclides is more complicated because all beta spectra are continuous from zero energy and, therefore, different beta emitters cannot be separated into small regions of individual channels. There exist three following possibilities: to resolve the spectra by means of absorption of the low-energy emitter or by chemical separation before measurement, or by measurement in different wide-channel windows of an amplitude analyzer, respectively. In such a case, for each measured radionuclide – usually only two but not more than three – the lower discrimination level of the counting channel for the beta nuclide with a higher energy spectrum corresponds to the level excluding the impulses elicited by a radionuclide with a lower energy. For this radionuclide such a level represents simultaneously the higher discrimination of its counting channel. The two other discrimination levels are selected to get the highest figure of merit (Eq. 1) for both radionuclides measured (see also Fig. 5). Smoothing [Tykva et al. (1986)] and/or deconvolution [Jisl et al. (1986)] are often used to evaluate either the measured spectra or the planar distribution of radionuclides with higher accuracy. An example of such measurement analyzing the distribution of simultaneously present ^{36}Cl and ^{14}C in plant material by the use silicon detectors [Tykva et al. (1992)] is shown in Fig. 3.

It is evident that the spectrometric analysis, using for counting always only a part of the total sample activity, is more complicated than only the detection. To make the resolution of several radionuclides easier, various sample treatments are frequently applied (see 1.4).

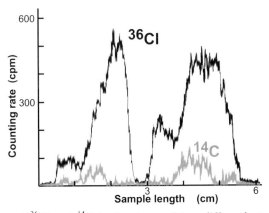

Figure 3: Resolving of ^{36}Cl and ^{14}C in the stems of two differently treated plants, after [Tykva et al.(1992)].

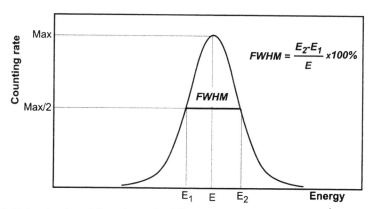

Figure 4: Determination of the noise as the full width at half maximum (*FWHM*).

In the detection and spectrometric systems, great attention has to be paid to noise [D'Amico et al. (1986)]. The signal-to-noise ratio is crucial, under the working conditions of the detector, both for the quality of the detection and for spectrometry as well. The magnitude of noise can be expressed in several different manners. Usually, it is determined by the use of monoenergetic radiation (for example conversion electrons, alpha particles, gamma rays, characteristic X-radiation) on the basis of the determination of the full width of the corresponding peak at its half maximum (*FWHM*). The noise expressed in this way is usually determined in eV, or also in % of energy corresponding to the peak (Fig. 4).

The noise of the detector itself, FHHM$_d$ is determined from the relation:

$$FWHM_c^2 = FWHM_d^2 + FWHM_e^2 \qquad (3)$$

where $FWHM_c$ is the measured value and $FWHM_e$ means the noise of the electronic devices recording the detector response. This value is usually measured by means of signals from the pulse generator, introduced onto the input of recording electronics. At the same time the detector is exchanged for a condenser of a capacity equivalent to the sum of the detector capacity at the working bias and parasitic capacity at the input member of the recording electronics. It should be stressed that the detector noise usually assumes a higher value than the noise of the electronics, so that it is decisive for the total noise.

The importance of the decrease of the detector noise for an improvement of the energy resolution in alpha and gamma spectrometry is evident. With decreasing noise the counting rate in each channel (or a certain energy), relatively increases, thus increasing the signal-to-noise ratio. When recording a continuous energy spectrum of beta particles the noise is also important for a low activity, especially in the case of a low-energy emitter. The value of the noise determines the lower discrimination level of the recording channel, i.e., the lowest level of the detector response which can still be recorded. Then in detectors with a linear relationship between the amplitude of the voltage response and the kinetic energy loss in the effective detector volume the energy interval of the recorded particles in the lower energy part increases and so the efficiency of the detector as well.

When measuring low activities of beta emitters, measurement from the discrimination level is used as a rule, guaranteeing the exclusion of noise from the record. However, in all instances the spectral composition of the background must be taken into consideration simultaneously, the counting efficiency alone is not decisive. The aim is to set the counting channel in such a way as to achieve a maximum value of the figure of merit (Eq. 1). For the sake of clarity, Fig. 5 shows the spectra of the background and of the counts evolved by ^{3}H in the liquid scintillation spectrometer. It demonstrates that the maximum detection efficiency doesn't correspond to the optimal detection sensitivity.

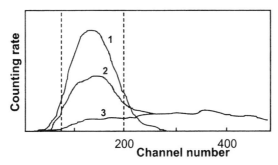

Figure 5: Choice of the discrimination levels for the highest figure of merit: 1 - ^{3}H reference source, 2 - ^{3}H water sample, 3 - background, based on [Schönhofer et al. (1987)].

1.4. Sample Treatment

An important part of radionuclide analyses using activity measurements is sample treatments. Without a suitable treatment procedure, no successful measurement can be realized. These procedures include two different groups. On one side, sample preparation to the measurement according to the detection arrangement used and, on the other, previous operations connected with the sample withdrawal, the transportation from locality to laboratory processing and the storage to the time of processing.

If the sample treatment was carried out incorrectly, between the sample withdrawal and the measurement, considerable distortion of the true value of measured activity may take place, both by its increase or decrease. The reason for increase consists in the contamination of the sample from the surrounding medium, e.g., radioactive fallout from air, radionuclide contaminated soil, natural radionuclides in the surrounding material or chemicals used, etc. Therefore, it is indispensable to protect the sample perfectly and, if possible, to exclude permanently the contact with the outside medium during all operations (e.g., by using separate working tools, by selecting suitable or previously adjusted protecting packings etc.). A decrease of the specific activity can be caused mainly by sorption of the radionuclide labelled molecule, or its considerable part (e.g., after dissociation in solution) onto the walls of the vessel in which the sample is stored, transported or measured. The importance of this effect is affected both by the nature of the compound and the medium in which it is kept, and by the material on which sorption takes place. Depending on the character of the sample, sorption from solution may be combined with sedimentation to the bottom, which also contributes to the change of the activity into the solution. The simplest methods leading to a decrease of sorption are usually increase of the acidity of the solution and the selection of a suitable material of the container. Teflon may be used with preference, but polyethylene may also do, while untreated glass surface is unsuitable for a direct contact.

A prolonged measurement requires a guarantee of stable measurement conditions. The geometry "sample to detector" is essential in arrangements of radioactive measurements. Therefore, it must also be guaranteed that no change takes place in the sample till the end of the measurement. As an example let us mention liquid scintillators, where under certain conditions we observed sorption of the dissolved sample on the wall of the measuring vial or its partial precipitation followed by sedimentation, respectively. Such undesirable phenomena can be prevented by an adequate preparation of the sample [Tykva (1980)].

The problems of the determination of low activity of environmental radionuclides is often considerably affected by the level of its specific activity. The requirements on measuring arrangements increase with decreasing specific activity. This depends primarily on the fact that the effective volume of the detector increases with the increasing amount of the measured sample, and thus its background as well. This leads to a decrease in the attainable detection sensitivity. Simultaneously, with increasing detector volume the requirements on the screening arrangement increase considerably from the point of view of expenses, volume and weight. However, the dimensions of the detector proper are limited, e.g., by the increase of the operation voltage guaranteeing the collection of the generated charge, increased noise, availability of sufficiently pure crystals etc. An increase of the amount of sample is also limited in number of cases, e.g., in the case of solid phase samples predominantly by selfabsorption in the sample, especially, at alpha-spectrometry.

For lower specific activities which occur currently by environmental radionuclides, the increasing of specific activities of the measured radionuclides before measurement is necessary using different radiochemical procedures such as precipitation, filtration, extraction, sorption, crystallization, evaporation in the case of non-volatile samples, distillation in the case of volatile samples, ion exchange during the passage through a bed of ion exchanger or chromatographic separation [Cheronis (1954)], [Bubner et al. (1966)], [Faires et al. (1973)], [Shono (1990)]. As an example of a gradual use of several working procedures for the transfer of the analyzed radionuclide from an environmental sample into the detector, the obtainment and transfer of krypton dissolved in water into an internal gas detector may serve [Held et al. (1992)]. A water sample of about 200 L is flushed with helium to extract the dissolved gases. Krypton is then extracted from the mixture by repeated adsorption on charcoal, fractionated by desorption and subsequent gas chromatography.

Among the individual laboratory procedures, the separation of the measured radionuclide from the rest of the sample is most commonly applied. Its use can be demonstrated on the separation of strontium from a large amount of calcium in the determination of low level ^{90}Sr in environmental materials, especially in sea water [Bojanowski et al. (1990)]. As an example of the separation by precipitation from gas phase the determination of the specific activity of $^{14}CO_2$ in air may be mentioned, used for the assessment of the exposure to ^{14}C produced in the nuclear fuel cycle in radiation protection [Tschurlovits et al. (1982)]. To the most frequently applied preparations of low level samples belongs concentration used, for example, in measurements of ^{137}Cs in lacrustine sediments and its pore water [Das (1992)], and also as preconcentration applied for low specific activities before separation, for example, during the determination of plutonium in water [Yu-Fu et al. (1991)]. Another important procedure is sorption using sometimes ion exchangers [Loosneskovic et al. (1999)]. For the measurement of the ^{60}Co concentration in sea water, a special device was used consisting of a pumping system combined with a cartridge packed with the adsorbent [Hashimoto et al. (1989)]. Extraction is also used frequently. Its application for uranium was analyzed by the spectrophotometric method [Almerey et al. (1999)].

For the preparation of samples for alpha spectrometry, which require very thin uniform samples owing to the very high specific ionization of the analyzed particles, electrodeposition from suitably prepared solution is used, which was described, e.g., for the determination of isotopic composition of enriched or depleted uranium samples. Such electrodeposition is frequently preceded by several preparatory steps, e.g., in the determination of plutonium and americium in environmental samples it is preconcentration of the nuclides by coprecipitation and ion exchange [Pillai et al. (1987)].

As an introductory step for the preparation of the filling of the internal gas detector in radiocarbon dating various types of oxidation are usually used [Šilar et al. (1977)]. Oxidation to the preparation of carbon dioxide is used also as the starting substance for the synthesis of benzene as a component of a liquid scintillator at very low levels of ^{14}C specific activity [Polach et al. (1983)]. Carbon dioxide may be used for filling of internal gas detectors using sources of negligible natural ^{14}C content to prepare a background filling. In view of its high affinity to electrons, it is important to decrease the presence of electronegative admixtures before the introduction into the detector, using special traps [Šilar et al. (1991)]. For the mentioned reasons hydrocarbons (e.g., methane) are frequently used as the detector filling. Moreover, they enable the determination of lower specific activity while using the same volume of the measured gas, with respect to the increased number of atoms of the corresponding element in the molecule (e.g., four atoms of hydrogen in the molecules of methane). As an example of such a synthesis of hydrocarbons we can examine the preparation of propane via hydrogenation of propadiene for measurement of low activity of tritium [Wolf et al. (1991)]. For determinations of low levels of tritium enrichment is sometimes applied to increase its specific activity in the electrolyte at the preferential dissociation of water molecules [Moser et al. (1980)].

Some special procedures for collection storage and measurement preparation of ^3H and ^{14}C environmental samples are described in Chap. 4, 2.11 and 4.4 in detail.

1.5.   Categories of Detectors

1.5.1.   Choice of Working Conditions

From Chapter 2 it follows that environmental radionuclides represent a different relatively broad interval of activity values and specific activities. Therefore, specific criteria should be applied for different problems. This principle applies in a distinct manner mainly for laboratories with high investment requirements. It is essential, for certain analyses to establish a corresponding workplace with appropriate equipment and working regimen.

An important role is played by the localization of the laboratories. They should be separated from all areas in which a distinctly higher radioactivity level or sources of disturbances might occur. All new parts of the equipment must be controlled with respect to their radionuclide contents before their introduction into the laboratories. In some cases it is useful to divide the room into several compartments, for example, to separate the measurement proper from the pre-treatment procedure of the samples. The constant temperature and humidity of the laboratory air contributes to the stabilization of the measuring conditions, especially during long-term measurements, indispensable for low-levels. From the point of view of radionuclide purity it is suitable to use the air under control in a closed cycle (especially Chap. 2, 3.4.6) ^{222}Rn can be removed.

The choice of a detector for certain measurements forms one of the basic conditions for successful work. Localization of the workplace, selection of radionuclide-free surroundings and materials, choice of suitable electronic circuits, stability of the equipment and suitably combined shieldings create the conditions of measurements. However, the attainable efficiency of detection, which affects the desirable increase in detection sensitivity significantly (Eq. 1), is determined predominantly by the detector itself, which also predetermines the processing and the arrangement of the sample. Therefore great attention should be devoted to the detector choice.

The selection of detectors for certain uses is given by four fundamental viewpoints:
 − nature of the sample,
 − emitted radiation,
 − detection/counting or spectrometry,
 − requirements on the precision of the measurements.

Table 5: Brief overview of detectors for routine analyses of environmental radionuclides.

Emitted Radiation	Detector Type	Application[1]
Alpha	Silicon Liquid scintillation Ion. chambers, prop. det. GM counters	S, limited C C, limited S C, limited S C
Beta with a lower energy spectrum	Gas − filled Liquid scintillation Silicon	C, S below app. 200 keV C, limited S S, limited C
Beta with a higher energy spectrum	Crystal scintillators Gas − filled Silicon	C, limited S C S to appr. 2 MeV
Gamma	Crystal scintillators Germanium	C, limited S S

[1] C - counting; S – spectrometry

Table 6:   A mean energy necessary to generate a unit charge response $\bar{E}$ and the total charge equivalent K_{eq} for different values of the multiplication factor M.

Detector Type		$\bar{E}$ (eV)	K_{Eq} (pC·MeV^{-1})		
			M=1	M=10^3	M=10^6
Semiconductor	Si (300 K)	3.62	0.044	-	-
	Ge (100 K)	2.96	0.054	-	-
Gas - filled	Air	32	0.005	5	-
	Argon	26	0.006	6.2	-
	Krypton	23	0.007	7	-
	Xenon	21	0.008	7.6	-
Scintillation	NaI(Tl)	110	-	-	1450
	CsI(Tl)	280	-	-	570
	LiI(Eu)	330	-	-	490
	anthracene	250	-	-	620
	stilbene	500	-	-	320
	plastic and liquid	400-800	-	-	200-400

The detectors for activity measurements are based on different principles [Knoll (1979)], [Tsoulfanides (1983)]. However, for the determination of activities of environmental radionuclides three groups of detectors are usually applied, namely ionization-, scintillation- and semiconductor-detectors. Their basic overview is given in Table 5.

In Table 6 the values of a mean energy $\bar{E}$ are given which is necessary to generate a unit charge in interaction of the detected particle during forming of an appropriate detector response. Therefore, this value determines the total number of particles generated during the whole energy loss of the detected particle in the effective detector volume. So, from the statistical point of view (see 1.7.1), the value of a mean energy is decisive for the intrinsic component of detector noise and, in this way, for the energy resolution which determines the spectrometric ability (see 1.3). From the Table 6 it is evident, that semiconductor detectors are extremely suitable for spectrometry due to relatively very small $\bar{E}$ (Table 6).

1.5.2.   Gas-Filled Detectors

Gas-filled detectors are based on the ionization of molecules of their filling after interactions with an incident radiation. They are among older types of detectors for low radioactivity and their present development is practically negligible. As it is shown in Table 5, their use can be considered for measurements of the environmental radioactivity for emitters of alpha particles, electrons and low energy photons. Due to the low density of the counting-volume they are unsuitable for pure gamma or high energy X-ray emitters, respectively, with regard to low detection efficiency (lower than 1 %). Ionization detectors are only utilizable to a limited extent for spectrometry, because the average energy required for the formation of the positive ion-electron pair by ionization in a common gas filling (approx. about 30 eV, Table 6) causes bad energy resolution of the detector and thus it decreases the counting rate belonging to the channel (or keV). For this reason, ionization detectors are used predominantly for the measurement of activity of a single radionuclide. The essence of gas-filled detectors is described in detail in relevant monographs [Rossi et al. (1949)], [Wilkinson (1950)], [Knoll (1979)].

In principle, all gas-filled detectors represent condensers in which the space between electrodes is filled with a suitable gas. When putting voltage onto the detector, an electric field is created between its electrodes. The magnitude and the distribution of the electric field has a decisive influence on the detector properties. On impact of a particle emitted by a radionuclide into the gas filling, ionization of gas molecules takes place: interaction of the incident particle causes a part of its kinetic energy to be lost (Table 6) to release the electron from the gas molecule so that a positive ion is formed. The charge of each of the generated parts of the molecule is $1.602 \cdot 10^{-19}$ C.

At the relatively small intensity of the electric field the current of the detector practically does not depend on its working voltage. Its value corresponds to the number of ionic pairs formed in the detector volume in one time unit. This operational regimen is characteristic of ionization chambers which are used only quite exceptionally for environmental radionuclide analyses. Their application is limited practically only to the form of large-area gridded chambers for very low alpha activities in spectrometric measurements. They are used especially for the environmental samples or mixtures of transuranium elements because of their large active detector area (parallel-plate type up to 0.05 m^2 and cylindrical type up to 2 m^2). So, e.g.,

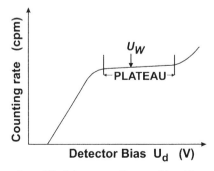

Figure 6: Counting curve of gas-filled detectors, U_w - working bias.

after deposition of aerosols on a circular tin plate dish of 20 cm diameter by electrostatic precipitation, energy resolution of 35 to 70 keV for the ^{210}Pb line of a layer thickness of between 30 and 100 µg·cm^{-2} has been obtained to reduce selfabsorption [Hötzl et al. (1984)]. Using this uncomplicated procedure for detection of artificially produced alpha emitters (^{241}Am, ^{239}Pu, ^{242}Cm) in environmental air, levels of units of µBq·m^{-3} can be estimated. Measurements of actinides in water samples from the primary circuit of a nuclear power plant have been also carried out [Rösner (1981)]. If lower specific activities have to be analyzed, recourse must be made to chemical separation or extraction techniques (see 1.4), with subsequent electrodeposition on small area planchets.

By increasing the voltage, the situation on the detector changes and the gas amplification takes place, caused by the fact that the molecules of the gas filling are ionized by the impacts of the ions from the primary ionization, which are sufficiently accelerated in the electric field of the detector. Therefore the gas amplification increases with increasing field intensity. This has the consequence that the current increases, which is higher than it would correspond to the formation of ionic pairs in consequence of primary ionization. If the charge response resulting in this way is proportional to the charge generated in the primary ionization, the detector is called proportional detector. The gas amplification M (see Table 6) ranges from 10^3 to 10^4. The voltage response, i.e., the amplitude of the output signal is thus proportional to the energy of the detected particle which was absorbed during its interaction with the gas filling. When all the kinetic energy of the detected particle is absorbed in the effective detector volume and the collection of the generated charge is perfect, a proportional detector may be used for spectrometric determinations. It is evident that the probability of secondary ionization depends on the intensity of the electric field. Therefore, only a very limited area around the central anode with high intensity exists in the cylindrical proportional detector, where the ionization by accelerated electrons generated in the primary ionization is considerable, while in the rest of the volume this phenomenon practically plays no role. Only the area of very close surroundings of the central electrode contributes significantly to the magnitude of the gas amplification. So it is independent of the place of the inlet of the detected particle and also of the localization of the primary ionization.

In contrast to proportional counters, in another operational field of the gas-filled detectors, called Geiger-Müller (GM) counters, the value of the gas amplification (approximately $M = 10^8$, see Table 6) causes such an increase of the total charge, that all output voltage signals have the same amplitude.

In both types, the charge response elicits a potential response $U(t)$. If gradually changing detector bias U_d under otherwise constant conditions of the measurement, and if measurement is taken at each value of the counting rate *N*, elicited by the constant disintegration rate, we obtain a dependence called a counting curve (Fig. 6). The region of an insignificant slope is called the counting plateau and it represents the area of operation. Usually the working voltage U_w of the counter is chosen at approximately one third of the plateau. A good counter is characterized by a long plateau, usually 200-400 V and a small slope (e.g., less than 1% per 100 V) which contributes distinctly to the stability of the conditions of the measurement.

At present, proportional and GM counters are used predominantly in radiation protection, relatively seldom for precision analysis. For gamma emitters they were always not useful with regard to their low counting efficiency. In measuring of alpha and beta emitters in solid samples [Winkler et al. (1989)], e.g., filters, chromatographic papers, dry residues etc., they were mostly replaced by computer controlled liquid scintillation spectrometers with highly effective non-cooled photomultipliers (see 1.5.3). Only internal gas counters are still used in a relatively greater extent especially in radiocarbon dating (Chap. 4, 2 and 4) although liquid scintillation spectrometry is also often applied for these purposes.

The internal gas detector is evacuated and filled with a measured amount of gas, the activity of which is determined. The measurement geometry 4π sr and the negligible self-absorption causes high detection efficiency for low-energy beta emitters or also soft photons. The 100 % counting efficiency in the effective volume is diminished only by a relatively small number of particles emitted near the walls or ends of the effective volume and escaping without generation of the charge response exceeding the equivalent noise charge: so-called wall or end effect (see 1.3). Usually the internal detector has a relatively large volume up to several dm^3 and it is filled at a pressure higher then an atmospheric one, so that it enables – in view of its high counting efficiency – the determination of low specific activity. A specific feature of the lowering of the background is that when using a higher filling pressure and a filling consisting of a hydrocarbon with a larger number of hydrogen atoms in the molecule (e.g., propane), an effective shielding of neutrons is required in view of the increase of the probability of a false response by reflected protons.

The sample is obtained either directly in the form of a gas (e.g., separated krypton), or, more commonly, it is converted to a form of suitable gaseous fillings of the detector from the solid phase (see 1.4). When measuring low activity the sample in the gaseous phase usually forms a total filling of the detector; in some instances it is mixed with another, radionuclide-free gas to obtain the necessary properties of the detector filling. Internal detectors are used both in the GM and – more frequently – in the proportional region of a gas amplification. Preference is given to the latter for the following three reasons:

– In view of the lower operating voltage on the detector in the proportional region, it is possible to use a higher filling pressure at the same voltage,

– in view of the lower gas amplification, a higher concentration of electronegative admixtures is admissible, and thus an easier preparation of the sample,

– in view of the proportional regime, some components of the background may be decreased by simple amplitude discrimination (e.g., by contamination of construction materials of the detector itself by alpha emitters).

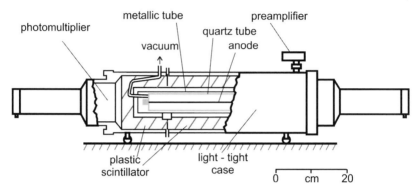

Figure 7: Scheme of an internal gas counter with plastic scintillator in anticoincidence, after [Šilar et al. (1977)].

The simplest type of internal gas detector is formed from a cylindrical cathode with a wire-net anode in the axis, which is surrounded both by an independent shielding detector and by mechanical shielding. The shielding detector, connected in anticoincidence, is formed usually by a plastic scintillator which replaced the formerly used screening wreath of GM counters with regard to its simpler long-term maintenance and easy formation according to the experimental conditions used. An example of such an arrangement is shown in Fig. 7 [Šilar et al. (1977)]. The outer part of the measuring detector was made from a copper pipe from the time before the Second World War, protected from the influence of radioactive fallout.

The cathode is made of the vacuum golded inner surface of a pure quartz tube, which is inserted into the copper tube. The molybdenum wire for the anode was repeatedly annealed before assembly in a radionuclide-free medium. The surfaces of the insulating materials were defatted in a non-contaminating medium and the assembling was done in a radionuclide-free atmosphere. The photomultipliers in the anticoincidence arrangement were made of glass with a very low content of potassium. All materials of further parts were also controlled carefully for radionuclide contents. The solder used did not contain lead or any other active contaminants.

Other types of large-volume internal gas detectors [Houtermans et al. (1958)], [Moljk et al. (1957)], [Povinec (1980)] are less widespread, primarily because of their difficult production and demanding maintenance.

To special types of internal gas counters belong also small counters (with volume of the order of magnitude of units of cm^3) used mainly for relatively simple radiocarbon dating (Chap. 4, 4) of milligram-sized archaeological and environmental samples [Sayre et al. (1981)]. However, for measurements of such very low specific ^{14}C-activity in small samples, even more accelerator-based mass spectrometry has been used (see 2.).

1.5.3.   Scintillation Detectors

If the ionizing radiation transmits a part of its entire energy or its whole energy to the scintillator during its passage through it, it emits scintillation photons. This emission of photons is an intramolecular property, which has its origin in the electronic structure of the scintillating molecules. Through excitation of their electrons and subsequent return of them to the ground electronic state, emission of photons with a characteristic spectrum takes place, i.e., luminescence. General findings on scintillation measurements and the physical explanation of individual phenomena are surveyed in detail in the literature [Birks (1964)].

The operation of a scintillation arrangement may be divided into the following stages:
 – Absorption of the incident radiation by the scintillator.
 – The scintillation process, i.e., the energy lost of the radiation in the scintillator resulting from the emission of scintillation photons.
 – Transfer of the scintillation photons onto the cathode and absorption by the cathode of the photomultiplier.
 – Emission of photoelectrons by the cathode and collection on the first dynode of the photomultiplier.
 – Multiplying process of the photoelectrons in the photomultiplier.
 – Collection of the total charge response on the anode.
 – The accumulated charge is converted to a corresponding voltage or charge response which is fed into the electronic circuits to be analyzed and recorded.

The linear response between the energy lost by the registered particle in the scintillator and the amplitude of the voltage at the photomultiplier (see 2.2.3) outlet is jeopardized by selfabsorption of scintillation photons in the scintillator, in consequence of their reflexion at interfaces and also by scattering. For spectrometry and/or measurement the effect of selfabsorption should be considered most important, because it can be so high that it significantly decreases the efficiency and prevents thus the measurement of a lower activity, mainly in liquid scintillators, where it is called quenching.

The majority of inorganic scintillators is created by activation with low concentrations of certain admixtures. The most widely used inorganic scintillator is NaI(Tl) with about 0.1 mol % of Tl-activator per mol of NaI. In view of its high density, this scintillator is mainly used in the detection and spectrometry of X- and gamma-radiation. As it was shown already at the beginning of this part (Table 6), spectrometric measurements are possible only for simple mixtures of gamma emitters. For complex gamma spectra, germanium detectors have to be used, sometimes using NaI(Tl) as screening detectors (see 1.5.4). On the contrary, for detection of a single radionuclide emitting gamma radiation, a NaI(Tl) detector has a high counting efficiency (Fig. 8) in relation to Ge(Li)-detectors (see 1.5.4). Therefore, for resolving of low levels in a mixture of gamma radionuclides, NaI(Tl) measurements are carried out on the individual components after their radiochemical separation. Another possibility for such purposes offer the high purity germanium detectors (see 1.5.4).

Organic scintillators may be divided into three types: *unitary* (primarily pure crystals, e.g., anthracene), *binary* (two-component solutions in liquid or solid form, e.g., p-terphenyl in polystyrene) and *tertiary,* primarily liquid scintillators in which the third component ensures the shift of the emission spectrum of photons into the region of maximum spectral sensitivity of the photomultiplier, e.g., 1,4-di-[2-(5-phenyloxazol)]-benzene (POPOP) in a solution of 2,5-diphenyloxazol (PPO) in toluene (Fig. 9). In comparison to inorganic scintillators they are characterized by low density and a low proton number and their existence in liquid or plastic forms. Organic scintillators are suitable for detection of low levels of alpha or beta emitters. They are also utilizable for spectrometry, but in view of the low energy resolution they do not permit analyses of more complex mixtures.

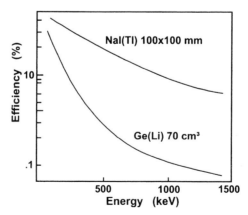

Figure 8: Comparison of detection efficiency of NaI(Tl) and Ge(Li) detectors for photons of different energy.

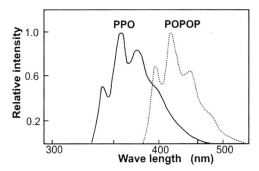

Figure 9: The emission spectra of two components of a tertiary liquid scintillator.

In organic scintillators, liquid scintillation counting (LSC) is most widespread. It is treated from various aspects in a number of specialized publications as, e.g., [Noujaim et al.(1976)]. The extended use of this detection procedure – especially in measurements of the samples of environmental and/or biological character – is caused by the high detection efficiency for low-energy beta emitters due to elimination of selfabsorption and the counting geometry of practically 4π sr. Moreover, the high performance of PC controlled spectrometers and the scintillation cocktails, which permit the introduction of samples with different characters, contribute to application of LSC. Recently, a portable LSC spectrometer has been commercially available for direct outdoor analyses of environmental radionuclides. Liquid scintillation spectrometers are also applicable for alpha spectrometry in environmental samples as, e.g., in mineral water [Schönhofer et al. (1987)]. Nevertheless, analogously to NaI(Tl) and Ge detectors, for a complex alpha spectrum silicon detectors are preferable with respect to their higher energy resolution (Table 6).

As we mentioned already previously, an important role is played by quenching effects which appear with the decrease of the measured counting rate in quenched samples in comparisons with non-quenched. In individual samples it differs both in quality and quantity. The following types of quenching are known: ionization quenching, quenching by dilution and concentration, chemical quenching, colour quenching, phase quenching and photon quenching. The first five types quenching, given by the sample character, take place in the scintillation solution, the photon quenching includes events which decrease the number of photons on the way between the scintillation solution and the photocathode of a photomultiplier. Quenching should be considered by users of liquid scintillation spectrometers especially in preparation of samples for measurement, although corrections of the measured values on quenching are automatically carried out in modern equipment. The approaches to sample preparation decreasing quenching are very different according to a sample character.

Other important effects in liquid scintillation spectrometry are represented by phosphorescence and chemiluminescence. The phosphorescence effect is the emission of photons owing to the excitation of molecules of some scintillation media – mainly on the basis of dioxane – or some types of measuring vials by solar or some other type of light. It is mainly observed during the detection of low-energy parts of the spectrum. The fluorescence is followed by the emission of photons at approximately 1 ns after excitation of the molecules, while in phosphorescence this time is longer, from 10 ms to several days. The occurrence of phosphorescence can be decreased to a minimum by preventing the access of light to the sample before its measurement. Chemiluminescence is the emission of photons elicited by a chemical reaction taking place in a mixture of sample and the scintillation medium. It is frequently observed, e.g., in cases when a basic solvent is used or also peroxide for the suppression of the colour quenching. The duration of chemiluminescence depends on the rate of the chemical reaction and the life time of the molecules in the excited state. Bioluminescence is a special case of chemiluminescence where the emission of photons is usually caused by bacteria, moulds or exothermic reaction catalyzed by proteins, enzymes etc. In modern equipment all these phenomena are automatically recorded.

The detection efficiency of each individual radionuclide is usually determined from the self-made calibration curve, on the basis of the value of the indication parameter of quenching. For the construction of this curve a set of reference samples may be used, which have equal activity but different quenching and which are usually supplied by the producer of the apparatus. Another methods could be also used [McQuarrie et al. (1980)], e.g., H-number supplied direct in modern spectrometers.

LSC spectrometers find also application for measurements of very low specific activity in ^{14}C or ^{3}H dating using synthesis of primary (usually ^{14}C$_6$H$_6$ or C$_6$^{3}H$_6$, respectively) or secondary solvent [Polach (1992)]. Another possibility lies in special commercially available spectrometers reaching, e.g., background of 0.60-0.65 cpm in ^{3}H-window (4 ml water sample+16 ml scintillator in a polyethylene vial), with which the measurement of 1.0±0.1 T.U. (see 1.4) is possible after electrolytic enrichment (see 1.4). The background of 0.35-0.40 cpm in ^{14}C-window is reached. In such arrangements it is sometimes necessary to consider the concentration of radon in the gaseous space above the scintillator during ^{3}H measurements [Murase et al. (1989)]. However, the decisive influence is the radionuclide content of the vial material (see 1.2). Also the vial holder should be radionuclide free.

1.5.4. Semiconductor Detectors

The results achieved in research and applications, both in technological and diagnostic processes for semiconductor structures create starting conditions for the development of semiconductor structures for analyses of activity of environmental radionuclides. Therefore the present-day development of detection procedures and their application is substantially broader in comparison with scintillation and mainly ionization detectors. For low levels of environmental radionuclides the use of semiconductor material is of great advantage because the detector dimensions can be kept much smaller than the equivalent gas-filled or scintillation detector and the energy resolution is considerably higher even in comparison with scintillation detectors (Table 6.).

The basic charge response of a semiconductor detector is formed by electron-hole pairs created along the path of the detected particle through the detector. Analogously to ionization chambers, the motion of the generated charge carriers in an applied electric field enables charge collection and the corresponding voltage response. The basic materials for detectors are silicon and germanium, other materials (e.g., CdTe, HgI_2, InSb or GaAs) have practically not been used in environmental radionuclide analyses till now. Tetravalent silicon in the normal crystalline structure forms covalent bonds with the four nearest silicon neighbour atoms. When a pentavalent admixture is present in small concentration (the order of magnitude of 10^{-6} or less), the admixture atoms will occupy a substitutional site within the silicon lattice, taking the place of a normal silicon atom. After all corresponding covalent bonds have been formed, there is one valence electron left over. Admixtures of this type are referred to as donors and such material as n-type silicon, with electrons called the majority carriers and holes the minority carriers. Correspondingly, after the addition of a trivalent admixture (acceptor) p-type silicon is created with holes as the majority carriers. If donors and acceptors are present in equal concentration, the semiconductor is said to be compensated. Analogous relations apply to germanium detectors.

The function of the semiconductor detector is enabled by forming a junction. At present, the introduction of doping admixtures at the semiconductor surface is carried out mainly by ion implantation using a beam of appropriately accelerated ions: either donor (e.g., phosphorus) on p-type or acceptors (e.g., boron) on n-type. To a smaller extent surface-barrier technology is applied based on a contact metal-silicon (e.g., gold on n-type silicon) prepared by special technological operations. Both types can be applied in environmental radionuclide analyses. The electron-hole pairs generated during the interaction of the detected particle can be collected only from the depleted layer of the detector where all free carriers are removed by working voltage. According to the ratio of this effective detection layer to the whole detector thickness, the detectors are divided into partially- or totally-depleted detectors.

The basic characteristics of silicon ($Z=14$, density 2.328 g·cm^{-3} and the mean energy per electron-hole pair 3.62 eV at 300 K) afford very good conditions for the use of silicon detectors for detection and spectrometry of alpha and beta emitters at room temperature [Tykva (1995)]. Furthermore, silicon detectors achieve a low background, mainly due to their low efficiency of the background particles with a lower specific ionization and the thin detector layer (only planar geometry is used). This layer is determined in spectrometry by the range of the detected particles with maximum energy, e.g., for alpha particles of 6 MeV energy a silicon layer of 34 μm is sufficient. The great advantage of silicon detectors for radionuclide analyses is thus the possibility with a high detection efficiency to tailor the thickness of the depleted layer and, in this way, to minimize the background. Taking into consideration the low efficiency of detection of the high-energy component of the background, it is frequently possible to measure the low activity without any shielding. The recently measured characteristics of silicon detectors are summarized in Table 7.

For measurement of extremely low activity levels it is useful to avoid contamination of silicon with ^{32}Si [Plaga (1991)]. At the same time, the background can be limited by choosing a corresponding counting channel. So, e.g., the background of 200 mm^2 planar surface-barrier detector in a ^{14}C-counting channel can be decreased to 0.19 cpm. Simultaneously, the ^{14}C-counting efficiency of 25 % can be obtained for samples without selfabsorption [Tykva (1977)]. Except analyses of beta emitting radionuclides, the silicon detector is used predominantly for detection and spectroscopy of alpha emitters in solid planar samples at 2π counting geometry [Matyjek et al. (1988)].

At present, two types of germanium detectors are used: In the older types lithium ions have to be drifted to compensate impurities while the modern technology of Ge crystals makes possible to use crystals having high purity (Ge HP) for detector preparation directly. In this way the low detection efficiency of Ge(Li)-detectors (Fig. 8) is considerably increased. Germanium detectors today afford the best conditions of all detection systems for analysis of complex gamma spectra. From the point of view of achieving a high sensitivity for the measurement (Eq. 1), it is important to pay great attention to the suppression of the detector background (see 1.2). In the analysis of gamma-spectra it is necessary to consider also the Compton and back-scattering. Photopeaks used in gamma or also X-ray spectrometry are formed by a photoelectric effect where a photon transfers all of its energy to an orbital electron. In the Compton scattering a photon undergoes elastic collision with a loosely bound orbital electron (essentially regarded as free) resulting in a transfer of some energy to the electron and the photon deflection with a lower energy (Fig. 10).

For the measurements of lower activities of environmental radionuclides the Compton continuum has to be suppressed. This is made possible by placing another detector(-s) around the measuring detector to create anticompton shielding. The reduction factor of "Compton background" for selected radionuclides reached by using an anticompton shielding of a germanium gamma-ray spectrometer is presented in Table 8.

A certain disadvantage of a germanium detector also consists in the relatively low energy per hole-electron pair (2.96 eV at 77 K). Moreover, due to thermal excitation of electrons at room temperature, an unacceptable increase of noise generated in the detector volume occurs. Therefore, germanium detectors have to be cooled with liquid nitrogen. Earlier production technologies afforded germanium single crystals with high concentrations of impurities. Therefore Ge(Li)-detectors were used exclusively in which the acceptors in p-type germanium were compensated by drifted lithium ions. These detectors are now mostly replaced by products of the present production technology of high purity germanium single crystals HPGe.

Table 7: Basic characteristics of silicon detectors for detection and spectrometry, after [Tykva (2000)].

Wafer Diameter (mm)	Min. Depleted Layer (μm)	Energy Resolution[1] (keV)	Disc. Level of Noise (keV)
3	50	3.4	4.7
3	150	3.9	6.2
6	150	4.7	8.3
22	150	6.9	12.8

[1] values obtained by α particles from ^{241}Am (5486 keV)

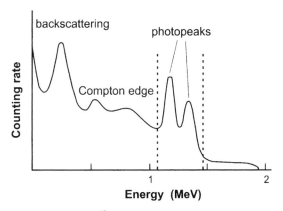

Figure 10: Pluse-height spectrum of ^{60}Co.

Table 8: Reduction factor R_p of the interfering peak after application of an anticompton arrangement, based on [Das (1987)].

Radio-nuclide	Half-Live	E_γ (keV)	R_p	Radio-nuclide	Half-Live	E_γ (keV)	R_p
^{46}Sc	84 d	889	7.3±0.5	^{134}Cs	2.1 y	563	17 ± 2
		1121	9.1±0.9			605	13 ± 2
^{59}Fe	45 d	143	2.4±0.2			796	14 ± 2
		192	7.7±0.4	^{152}Eu	13.3 y	122	≈ 10^2
		1095	1.1±0.1			245	≈ 10^2
		1292	1.1±0.1			344	3.5 ± 1.5
^{75}Se	120 d	121	48±2			779	20 ± 4
		136	55±5			965	3.5 ± 1
		265	17±1			1087	1.6 ± 0.2
		401	1.0±0.1	^{154}Eu	8.8 y	1004	3.0 ± 0.5
^{124}Sb	60 d	603	11±1			1278	2.8 ± 0.3
		1692	> 10^2			1408	5 ± 1
^{181}Hf	42 d	133	10±2	^{182}Ta	144 d	1189	2.2 ± 0.4
		482	2.5±0.5			1222	2.5 ± 0.5
						1231	2.5 ± 0.5

The counting efficiency can be influenced by the counting geometry. For environmental radionuclides often classical well-type coaxial detectors of different types are used, introducing the sample into a hole in the detector [Kemmer (1968)]. A planar type [Hedvall et al. (1987)] is used only exceptionally. For environmental samples with very large volume (e.g., 0.5 L), the Marinelli beaker around the detector is applied [Walford et al. (1976)].

1.5.5. Other Types of Detectors

Other types of detectors (e.g., autoradiographic emulsions [Rogers (1979)] track detectors [Nikezic et al. (1999)]) are only rarely applied for analyses of environmental radionuclides. Detailed descriptions of their principles and application are given in the instrumental monographs [Knoll (1979)], [Tsoulfanides (1983)].

1.6. Special Detection Arrangements

For environmental radionuclide analyses some special arrangements are applied. Four of them, finding broader application for analyses of environmental radionuclides, are described: assessment of radon and its decay products, field and area monitors, whole body counters and biosensors.

1.6.1. Assessment of Radon and its Decay Products

The naturally radioactive noble gas radon (Chap. 2, 3.4.6) is present in air and can accumulate in closed or poorly ventilated spaces, including buildings. For assessment of radon and its decay products its two isotopes are significant, namely ^{222}Rn, the immediate decay product of ^{226}Ra, deriving from the uranium series of natural radionuclides, and ^{220}Rn, the immediate decay product of ^{224}Ra, coming from the thorium series (Chap. 2, 1.2). These radionuclides are commonly known as radon and thoron. Thoron has a short half-life of 55.6 seconds, and a low abundance relative to radon, whose half-life is 3.82 days. Because of their different characteristics and, especially, their different contribution to the total annual effective dose – radon and thoron and their decay products are responsible for 1.2 mSv and 0.07 mSv (world-wide average), respectively – radon and its decay products (progeny) are of primary interest. From the radiological point of view, radon decay products are more important than radon itself. Actually, the effective dose from the inhalation of radon represents only about 5 % of the total radon-related dose, i.e., 95% of this dose is delivered by the short-lived radon decay products ^{218}Po, ^{214}Pb, ^{214}Bi, and ^{214}Po. Consequently, for radiation protection purposes it is desirable to monitor preferably the presence and the concentration of radon decay products, while for the identification of the sources and origin of radon, the measurement of its concentrations in air or water, and sometimes its exhalation from soil and building materials, are more valuable.

There are many different methods of measuring radon in air; some are based on active, some on passive detectors which are in both cases specially modified. Most techniques rely on the detection and spectrometry of alpha particles emitted by radon or its progeny, but there are some methods based on the evaluation of gamma and also of beta radiations from decay products.

The precision and accuracy of the individual measuring techniques depend on several factors, such as the statistical character of radioactive decay, variations in detector response, interference with unmeasurable species, and unfulfilled assumptions concerning the atmosphere being sampled.

The instruments and methods used for the measurement and estimation of radon and radon progeny concentrations in air have been discussed and summarized in several references [(OECD (1985)], [NCRP 97 (1988)], [Harley (1992)].

There is growing concern over the health hazard associated with the presence of dissolved radon in public and private water supplies. High waterborne radon conircentrations may directly elevate radon concentrations in air inside residential and public buildings and therefore it is important to measure the content of radon in water in order to check whether its quality complies with the regulatory requirements. Several different methods for the measurement of radon in water have been developed and used. The following four techniques are most often applied. Extraction of radon into an organic phase and measuring by liquid scintillation spectrometry [Schönhofer (1992)], analysis of photons emitted by short-time decay products [Farai et al. (1992)], gas extraction method [Mathieu et al. (1988)], and measuring in the air above the water sample [Kotrappa et al. (1993)].

Concerning radon decay products, unlike radon, which is more or less homogeneously distributed in space, they are attached to dust particles or deposited on the surfaces of surrounding materials. Most methods of monitoring radon decay products are based on their collection on a filter which is then measured using a detector for simple counting. More preferably, a spectrometry system is used for the evaluation of alpha particles emitted by ^{218}Po and ^{214}Po. In this arrangement, the sampling air with a known rate is drawn through a suitable filter which can collect radon products with high efficiency. In addition to the filter, the decay products can also be collected electrostatically and then detected. Besides active detectors for the registration of particles from decay products, some passive detectors can be used as well.

Methods for measuring radon decay products can be divided into three principal categories:
- Grab-sampling or semi-instantaneous method is based on the measurement of the limited volume of air sampled during a relatively short time period,
- semi-continuous method usually relies on a modified grab-sample technique with successive repetition of sampling. In principle, a continuously moving filter that collects decay products while simultaneously measuring them is another possibility. In any case, however, the response of a semi-continuous monitor will always be delayed,
- integrating method consists of the measurement of a time-averaged concentration of radon decay products. The integration may be carried out either on a filter using an integrating detector, or the results from semi-continuous monitors may be evaluated to obtain the average concentration of radon decay products.

Most of these methods can be used for measurements which can be interpreted in terms of the potential alpha-energy concentration (PAEC) or the equilibrium-equivalent decay-product concentration (EEDC). In order to determine these concentrations it is necessary to find the concentration of individual decay products. Since the half-life of ^{214}Po is too short (164 μs), its contribution to both PAEC and EEDC is negligible and it may not be considered. Ideally, it would be highly desirable to know the individual concentrations of ^{218}Po, ^{214}Pb, and ^{214}Bi with half-lives of 3.11 minutes, of 26.8 minutes, and of 19.9 minutes, respectively. In practice, however, a direct measurement of each concentration is very complicated and difficult. Instead of measuring simultaneously both alpha and beta radiation or also gamma, practically all monitors of decay products rely only on the detection of alpha particles because they can be measured selectively. Older monitors usually measured the total gross alpha particles using, for example, a ZnS(Ag) scintillation detector, but nowadays most instruments are based on alpha spectrometry so that they can distinguish between alpha particles emitted by ^{218}Po and ^{214}Po. Their emission rate from ^{214}Po actually represents the activity of ^{214}Bi.

Several types of universal computer-based instruments are commercially available for measurements of both radon and thoron decay products.

1.6.2. Field and Area Monitors

Field and area monitors are used for environmental measurements *in situ* or *on site*, respectively, including primarily the measurement of external gamma radiation. These monitors are used for the measurement of the dose rates or of the doses from terrestrial radionuclides, radionuclides deposited on surfaces and radionuclides dispersed in the air. The response of these instruments usually also includes the contribution from cosmic rays. The natural gamma background must be respected carefully so that any changes and anomalies in this background should be detected with an adequate exactness. To obtain a sufficient measuring range, starting from the background dose rate to the dose rates envisaged during emergency situations, sometimes two or more similar or different detectors are employed in one system.

In the event of any increase of radionuclides in the atmosphere and their deposition in the environment, measurement of the gamma dose rate can provide a rapid and straightforward indication of even very small changes in the radioactive contamination. An assessment of the need for further, more detailed, measurements or other actions depends many times on the results of external gamma radiation monitoring. The gamma monitors usually measure air dose rate or exposure rate, and most of them can integrate their response so that results also in the form of total dose or exposure are available. These measurable quantities can be, in principle, converted into the main radiation protection quantities, namely the effective dose equivalent or effective dose (see 1.5). This can be done, however, only when we know or assume at least some information as to the energy and direction of gamma radiation.

Essentially, two types of gamma monitors may be considered:

- – Monitors are based either on non-spectrometric or spectrometric detectors, respectively, but don't use energy spectrum evaluation. These instruments register gross gamma radiation without relying on the information about its energy. Preferably their response should be energy independent; this is achieved simply by making use of suitable properties of a radiation sensor (e.g., ionization chambers), by appropriate modification of the detector response or a combination of two detection media (scintillation detectors), or using a special type and shape of detector shielding to modify its energy response (Geiger-Müller detectors).

- – Gamma monitors based on spectrometric methods using scintillation or semiconductor detectors (e.g., NaI(Tl) or Ge) in a portable version. First after unfolding of the pulse-height spectrum the final quantity is calculated. These instruments have a built-in microprocessor-controlled multichannel analyzer, or they use an analyzer combined with a small portable computer.

In addition to portable gamma monitors, there are a number of various stand-alone continuous monitoring systems or measuring stations that are remotely controlled and programmed to deliver required information including alarm signals in the case of accident or emergency. These monitors are usually designed for outdoor applications and they are largely self-contained. They have batteries to be operable in case of power loss and are equipped for communication of data to a central computer. All components are housed in weatherproof enclosures. The minimum sensitivity is usually better than 10 nSv·h^{-1}.

Since radioactive material released into the atmosphere is transported very rapidly through the air, special attention should always be paid to monitoring. These stations are equipped with a lot of different monitors, which measure – continuously or at certain selected time intervals - the concentration of some important radionuclides in the air, especially iodine, caesium, plutonium, and noble gases, but also other radionuclides, including tritium and radon/thoron and their decay products.

Special stationary universal units have been developed for the measurement of radioactive aerosols emitting α, β or γ radiation. Radionuclides are collected on one or more fibreglass or other special filters whose activity is then spectrometrically determined. Usually, an internal air flow measurement capability is provided to correct the filter data for actual flow rates through the filter; other necessary corrections, some of which would be difficult to introduce in small portable instruments, are also made. These monitors are characterized by a very high air-flow. The measurement sensitvity of the environmental alpha emitters can be demonstrated by ^{239}Pu, which should be detected on levels below 40 mBq·m^{-3}. This concentration corresponds to the value of the derived activity concentration for occupationally exposed persons. One has to realize that these low concentrations are supposed to be measured in the presence of background concentrations of radon decay products of several tens of Bq·m^{-3}. To reduce the background due to radon when measuring other alpha radionuclides, sometimes a pseudocoincidence between the beta particle from the radon decay product ^{214}Bi and the subsequent alpha particle ^{214}Po is used.

Other monitors are designed in order to measure gross alpha and gross beta particulate concentrations in air with the highest possible sensitivity required for the early detection of airborne radioactivity. These instruments can achieve gross alpha and beta detection limits ≥ 0.2 Bq·m^{-3} and ≥ 0.5 Bq·m^{-3} respectively.

The monitors for iodine are equipped with a disposable or a refillable charcoal cartridge to trap the radioiodine component. Aerosol pre-filtering, cartridge pre-heating to remove moisture for more efficient capture of organic iodine compounds, and automatic gain stabilization ensure higher accuracy than in conventional monitors. It is possible to obtain detection limits better than 100 mBq·m^{-3}.

For selective measurements of concentrations of some radionuclides, high-resolution gamma spectrometry is used. In a two-hour measurement cycle, detection limits as low as 100 mBq·m^{-3} for ^{131}I and ^{137}Cs in air can be achieved. For ^{60}Co the limit is around 30 mBq·m^{-3} obtained from one hour of measuring. Current automatic analyzers are capable of evaluating the spectra of up to 100 different radionuclides. On-line radionuclide identification is important for a reliable assessment of the actual situation in the case of an accident, which on the basic of such results can be better evaluated as to its origin and character.

Nuclear facilities (Chap. 3, 1.5) release some amount of various radionuclides through their chimney stacks. These releases have to be supervised in order to keep the effluent activities below the allowed emission values set by national or international prescriptions. In most cases, the licence for the operation of a nuclear power plant stipulates the amounts of radioactive materials that can be discharged into the environment. Some special monitors with various fixed as well as moving filters have been developed for this particular purpose.

The operation of reactors and critical facilities (see Chap. 3, 2.2) results in the production of fission products in the fuel and activation products in all materials and elements irradiated by high neutron fluences. Since water is a common coolant and moderator in most reactors, its exposure to neutrons results in the transformation of hydrogen and oxygen into tritium and ^{16}N. Neutron irradiation of boron and lithium also produces tritium. Air is often used as a coolant for reactors and critical facilities and is always trapped or dissolved in water in the reactor. The activation of oxygen in air also gives ^{16}N and the activation of noble gases leads to the generation of such radionuclides as ^{41}Ar, ^{82}Kr, ^{89}Kr, ^{135}Xe, and ^{137}Xe. These radioactive noble gases are essentially a source of external exposure. Their impact as internal contaminants is relatively unimportant. The most sensitive method for assaying the concentration of radioactive noble gases is the method based on an internal gas counting. Continuous monitoring of radioactive noble gas effluents released through stack ventilation systems is often required at nuclear power plants. Airborne radioactivity is, in this case, often measured continuously by a high-efficiency solid-state detector.

In some nuclear installations dedicated tritium monitors are required. Tritium air monitors are often based on ionization chambers, but much higher sensitivity (about 100 times) can be achieved by instruments using proportional flow-through gas detectors with rise-time discrimination. This technique makes it possible to distinguish the responses of tritium from those of radioactive noble gases. In this way, tritium can be measured simultaneously and separately with some other radioactive gases. The measuring range covers the interval from about 200 Bq·m^{-3} to more than 10 kBq·m^{-3}. Such tritium monitors are used also near accelerators, in fusion research, and nuclear materials handling facilities.

1.6.3. Whole-Body Counter

In the beginning, arrangements of whole body counting were developed to determine the activity of ^{226}Ra in dial painters in the 1930s. The actual whole-body counters are used in radiation protection, for monitoring of internal contamination, in human medicine, mainly in the diagnostics, and in the research of laboratory animals after application of radiotracers [Matsusaka et al. (1988)]. The retention in the body and the excretion are determined. Besides, a human body contains natural radioisotopes of the body elements as, e.g., about 4 kBq of ^{40}K.

Usually, the activity of the measured body is very low so that the background has to be reduced by large and expensive shielding (see 1.2). Different detectors have been applied for whole-body counting, mainly liquid, plastic and NaI(Tl) scintillators (see 1.5.3), in the latest constructions also germanium detectors (see 1. and 4). Several arrangements were described in detail by [Berg et al. (1985a)]. A whole-body counter applied for people and big animals (as, e.g., a cow) is demonstrated in Fig. 11 (A) including a measured spectrum (B).

Applying a simple whole body counter for the determination of artificial contamination of the human body [Berg et al. (1990)], several relatively large sodium iodide crystals, a bed-like geometry and a large shielding box

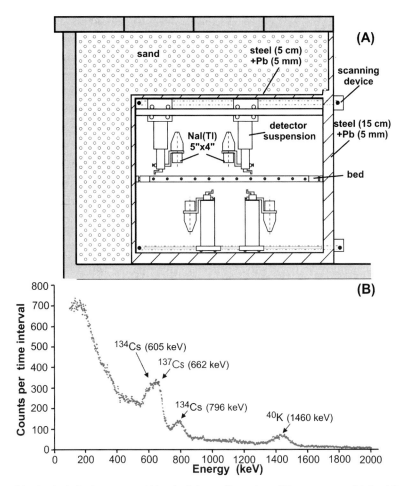

Figure 11: A whole body counter: (A) principle configuration, (B) a spectrum obtained for a contaminated person (counting time 1500 s), based on [Berg et al. (1985a)].

composed of lead, steel and quartz sand were used. For measurement with a time interval of 25 minutes, the detection limit of ^{137}Cs for an adult person was 70 Bq. The error of determination of this level is also affected by poor reproducibility of the positioning of the measured person and it achieves up to 10-20 %. The counter described was also used in the area of Munich in Germany [Berg et al. (1987a)] for whole body radioanalyses of children and adults after the Chernobyl accident (Chap. 3, 2.3 and Fig 11).

Another illustrative example of the whole body counting application is the radiation protection of children by monitoring of ^{137}Cs transfer into mother's milk [Gall et al. (1991)]. After the Chernobyl accident, between May 1987 and December 1988, duplicates were found in daily food intakes as well as in corresponding mother's milk samples. The milk was collected from 12 nursing mothers for two to four weeks in order to measure the ^{137}Cs activity. Once during the collection period, the total-body activity of each of the mothers involved was measured. Based on these results, ^{137}Cs was transferred approximately 19 % of the ^{137}Cs activity from the daily food intake into 1 litre of the mother's milk.

1.6.4.   Biomonitoring

In addition to detection arrangements based on physical principles, different cells and tissues can be used as biosensors of environmental radionuclides in soils and/or waters. For such purposes, e.g., different terrestrial and/or aquatic plants are applied to intake radionuclides through their root systems [Wolterbeek et al. (1996)]. After translocation of radionuclide(-s), the activity is analyzed in the appropriate tissues. In relation to classical field monitoring (see 1.6.2), such biosensors are environment friendly cost – effective *in situ* and *on site*, large – scale and long-term applicable. Moreover, analogously to some other environmental pollutants (e.g., heavy metals), such a biotechnological process can be applied not only for monitoring but bioremediation as well [Terry et al. (2000)]. On the contrary, its dependence on the year period and local conditions on one side, and, on the other, a lower accuracy may be the disadvantages of biomonitoring for some purposes. Nevertheless, radiophytomonitoring as well as radiophytoremediation should be taken for perspective methods in which improvements could be made by further studies (e.g., selection of biological species, evaluation of optimum time intervals etc.). Such studies are in progress as for surface waters [Soudek et al.(2000)] as in the neighbourhood of a nuclear power plant or a uranium mill [Tykva et al. (2002)].

1.6.5.   Airborne Measurements

Airborne radioactivity analyses are carried out using helicopters for terrestrial gamma emitting radionuclides [Winkelmann et al. (1995)] and aircrafts for radiocontaminated air masses [Dyck et al. (1993)].

For a rapid evaluation of gamma radiation sources in the topmost layer of the ground with a complete area for coverage, helicopter surveys are used. This approach represents sometimes the only possibility coverage as, e.g., in hilly regions. Such airborne monitoring is also useful in the case of nuclear accidents (Chap. 3, 2.3) because its coverage is about 2500 times larger than that of a comparable ground system. Therefore, the resulting cost per a large surveyed area are lower than for a comparable terrestrial survey. Another important application of helicopter surveys is making of dose rate maps enabling analysis of the effects of low doses of terrestrial environmental radiation on human health. During measurements, the helicopter flight altitude is about 100 m and its flight velocity about 25 m·s^{-1}. The relevant instrumentation may be different, e.g., four NaI(Tl) crystals (total volume 16.8 L) [Schwarz et al. (1993)] or a combination of three NaI(Tl) total volume 12 L) and one Ge (HP) crystals [Winkelmann et al. (1995)], respectively. As an example of the attainable measuring sensitivity ^{137}Cs can be given, for which units of kBq·m^{-2} are detectable.

Aircraft surveys are used for estimations of location, extent and radionuclide composition of air masses up to about 10000 meters. The gamma dose rate and activities of aerosol radionuclides are measured. The detection limit for ^{137}Cs can be 2.2 mBq·m^{-3} [Dyck et al. (1993)].

1.7. Accuracy of Activity Measurements

1.7.1. Statistical Errors

The radioactive decay is a random, stochastic process. Radioactive nuclei undergo disintegrations occurring with a certain probability depending on the decay constant of the radionuclide under consideration. The number of radioactive atoms $N(0)$, initially present at time $t=0$, due to decay will be reduced to $N(t)$ at time t (Chap. 1, Eq. 3), $N(t)$ and $N(0)$ are considered to be the expectation values of the relevant parameters. In any actual individual observation, however, the observed number of atoms may differ from the corresponding expectation value. The observed value will in fact fluctuate around the expectation value. Since $N(t)$ represents the number of radioactive atoms at time t, obviously the number of atoms which have decayed in the time interval $(0,t)$ will be equal to $N(0)-N(t)$. Taking into account Eq. 3 (Chap. 1), the probability that a radioactive atom will not decay within time t can be expressed as the ratio

$$\frac{N(t)}{N(0)} = e^{-\lambda \cdot t} \qquad (4)$$

and, similarly, the probability p that a single atom will decay in time t is given by:

$$p = \frac{N(0)-N(t)}{N(0)} = e^{-\lambda \cdot t} \qquad (5)$$

As long as the half-life of a radionuclide is sufficiently long (much longer than the observation interval t, then the number of decayed atoms $M=N(0)-N(t)$ will be very small in comparison to a large number $N(0)$. Under these conditions, the probability $P(M)$ of observing exactly M atoms decaying in time t can be expressed by the binomial distribution:

$$P(M) = \frac{N(0)!}{M! \cdot (N(0)-M)!} p^{n} \cdot (1-p)^{N(0)-M} \qquad (6)$$

Its derivation can be found in many books dealing with radioactivity and counting statistics as, e.g., [NCRP 58 (1978)], [Berg et al. (1985b)].

Since the number $N(0)$ is usually very large (about 10^{15} or more), the binomial distribution is computationally cumbersome and in nuclear counting applications it is more practical to use either the Poisson distribution or the normal distribution which can approximate the binomial distribution. The Poisson distribution can be written in the form

$$P(M) = \frac{(p \cdot N(0))^M}{M!} e^{-p \cdot N(0)} \tag{7}$$

where $P(M)$ is the probability of measuring, M decays in the time interval t during which the expected number of decays is equal to the expectation value $p \cdot N(0)$.

The Poisson distribution may be presented in the more appropriate form

$$P(M) = \frac{(n \cdot t)^M}{M!} e^{-n \cdot t} \tag{8}$$

where n is the expectation value of decays in a unit of time. In practice, it is approximated by a mean value.

The normal or, as it is sometimes called, Gaussian distribution, is often used to approximate the Poisson distribution under the condition that $p \ll 1$. This distribution is given as:

$$P(x) = \frac{1}{\sqrt{2\pi \cdot \sigma}} e^{\frac{-(x-\bar{x})^2}{2 \cdot \sigma^2}} \tag{9}$$

where x is a continuous variable (as distinct from the binomial or Poisson distributions defined only for integer values), $\bar{x}$ is its true (mean) value, and σ is the standard deviation which is related to the Poisson distribution parameters by the expression:

$$\sigma^2 = p \cdot N(0) = n \cdot t \tag{10}$$

The normal distribution is commonly used to interpret nuclear counting experiments and it is useful to summarize here some of the important properties of this distribution. The normal distribution is plotted in Fig. 12 which illustrates its various features.

The normal distribution is symmetric around $\bar{x}$, and defined uniquely by the two parameters σ and $\bar{x}$, extending x from $-\infty$ to $+\infty$. The probability of obtaining the value x from the interval (x_1, x_2) is equal to

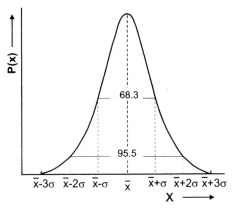

Figure 12: Normal distribution and some of its characteristic parameters.

$$P(x_1, x_2) = \int_{x_1}^{x_2} P(x) \cdot dx \qquad (11)$$

The infinite integral of the distribution satisfies the condition:

$$\int_{-\infty}^{+\infty} P(x) \cdot dx = 1 \qquad (12) \qquad \bar{x} = \int_{-\infty}^{+\infty} x \cdot P(x) \cdot dx = 1 \qquad (13)$$

The mean value $\bar{x}$, the variance $V(x)$, the standard deviation σ, and the full width at half maximum (*FWHM*, see Fig. 4) are given by

$$V(x) = \int_{-\infty}^{+\infty} (x - \bar{x})^2 \cdot P(x) \cdot dx = \sigma^2 \quad (14) \qquad \sigma = \sqrt{V(x)} \qquad (15)$$

$$FWHM = \left(2 \cdot \sqrt{2 \cdot \ln 2}\right) \cdot \sigma = 2.35 \cdot \sigma \qquad (16)$$

Pulses at the output of a detector follow essentially the same distribution as radioactive decay. Therefore, as long as the dead time of the detector and electronics can be neglected, pulses registered by a counter obey the Poisson law, which for this purpose can be written in the form:

$$P(M) = \frac{N^M}{M!} e^{-N} = \frac{(n \cdot t)^M}{M!} e^{-n \cdot t} \qquad (17)$$

where $P(N)$ is the probability of observing (measuring) M pulses during time t, N is the extended (true) number of pulses in the same interval, and n is the expected pulse rate. Both N and n are in practice replaced by their corresponding mean values $\bar{N}$ and $\bar{n}$:

$$\overline{N} = \frac{\sum_{i}^{K} N_i}{K} \qquad (18)$$

$$\overline{n} = \frac{\sum_{i}^{K} n_i}{K} = \frac{\overline{N}}{t} \qquad (19)$$

$$N = \lim_{K \to \infty} \left(\overline{N} \right) \qquad (20)$$

$$n = \lim_{K \to \infty} \left(\overline{n} \right) \qquad (21)$$

In some counting applications it is important to know distribution of time intervals between successive pulses. The relevant distribution can be derived using the probability that the next pulse will occur in the infinitesimal time interval dt following the time t which elapsed since the appearance of the previous pulse. Obviously, this probability, $P(t)$ dt, will be equal to the product of the probability $P(0)$ that no pulse appears in the interval $(0,t)$, and the probability $P_1(dt)$ that a pulse occurs during dt. These probabilities are given by

$$P(0) = e^{-n \cdot t} \qquad (22)$$

$$P_1 \cdot (dt) = n \cdot dt \qquad (23)$$

which after the substitution results in the equation representing the distribution function for time intervals between adjacent detector pulses

$$P(t) = n \cdot e^{-n \cdot t} \qquad (24)$$

The time interval distribution has a simple exponential shape with the most probable interval being zero. The mean or average time interval can be calculated using a standard formula:

$$\overline{t} = \frac{\int t \cdot P(t) \cdot dt}{\int P(t) \cdot dt} = \frac{\int t \cdot n \cdot e^{-n \cdot t} \cdot dt}{\int n \cdot e^{-n \cdot t} \cdot dt} = \frac{1}{n} \qquad (25)$$

which is what can intuitively be expected.

When the dead time τ of a counting system is taken into consideration the time interval probability density is be given by

$$P_\tau(t) = \begin{cases} 0 & \text{for } t < \tau \\ n \cdot e^{-n(t-\tau)} & \text{for } t \geq \tau \end{cases} \qquad (26)$$

and the main time interval in this case is given by:

$$\overline{t} = \tau + \frac{1}{n} \qquad (27)$$

The situation is illustrated in Fig. 13, which shows that the time interval distribution in the presence of a certain dead time is shifted (just by τ) in relation to the distribution of a system having zero dead time. This correction is especially serious for measuring of higher activities by ionization detectors.

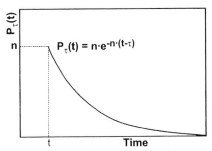

Figure 13: Time interval distribution of pulses from a counting system with a dead time.

In low-level counting, where measuring time is usually very long and each sample is measured only once, the simultaneous evaluation of the time interval distribution may give us further information which can be of use in eliminating false pulses carried out by, e.g., a power supply net.

There are, however, always some uncertainties and fluctuations which are caused by the statistical nature of radioactive decay. Statistical errors differ principally from instrumental errors; the first can never be fully eliminated, but they can be predicted, evaluated, and taken into account in the presentation and interpretation of final results.

The most important parameter used for the assessment of statistical fluctuations is the standard deviation σ, which is calculated from the equation

$$\sigma = \sqrt{\frac{\sum\limits_{i}^{K}(N_i - \overline{N})^2}{K-1}} \tag{28}$$

where i refers to a given observation, N_i is the value of a random variable referring to the ith observation, $\overline{N}$ is the mean value, and K is the number of measurements. The meaning of the standard deviation can be interpreted in terms of the probability that a single value N_i will lie inside the range $(N-k\cdot\sigma, N+k\cdot\sigma)$ where N is the true value and k is a coefficient defining the so-called confidence limits or intervals. The probability or confidence level that a value N_i will fall within these limits is often expressed as percentage. Some limits are specified by special names of relevant errors. Values of a confidence level as a function of the confidence interval together with error names commonly used are presented in Table 9.

The interpretation of the confidence levels related to a chosen confidence interval is self-explanatory: for example, the probable error-interval ($\overline{x} \pm 0.675\cdot\sigma$) is equally likely to be exceeded or not, since the confidence level is just 50 %. This means that 50 % of the observations will fall within the limits ($\overline{x} \pm 0.675\cdot\sigma$). Similarly, the standard deviation represents the 68.3 % probability of finding a result x within the interval from ($\overline{x}-\sigma$) to ($\overline{x}+\sigma$) where $\overline{x}$ is the true mean value.

Table 9: Confidence intervals $\pm k \cdot \sigma$ (limits) and corresponding confidence levels *CL* including commonly used error terminology (the confidence level represents the probability that values will lie within confidence interval limits).

k	*CL* (%)	**Name of Error**
0.675	50	Probable error
0.841	60	-
1.000	68.3	Standard deviation
1.038	70	-
1.150	75	-
1.281	80	-
1.645	90	9/10 error
1.960	95	95 % error
2.000	95.4	-
2.240	97.5	-
2.575	99	99 % error
3.000	99.73	-
3.291	99.90	-
4.000	99.99	-

The confidence level *CL* for any confidence interval $(\pm k \cdot \sigma)$ can be calculated as

$$CL = P(\bar{x} - k \cdot \sigma, \bar{x} + k \cdot \sigma) = \int_{\bar{x}-k\cdot\sigma}^{\bar{x}+k\cdot\sigma} P(x) \cdot dx \qquad (29)$$

where $P(x)$ is the distribution given by Eq. 7.

It is important to realize that the standard deviation σ represents an error of a single measurement. For such a standard deviation, it may be better to use the symbol $\sigma(N_i)$ for it can be calculated on the basis of experimental values N_i using Eq. 25. In that measurement we obtain *K* values of N_i from which the mean value is determined. Now suppose that we repeated this measurement *L* times, getting *L* values of $\bar{N}_j$ $(j = 2,3,4,...,L)$, and we are interested in finding the standard deviation $\sigma \cdot \bar{N}_j$ of these mean values $\bar{N}_j$. Generally, for this standard deviation the following expression can be derived:

$$\sigma(\bar{N}_j) = \sqrt{\frac{\sum_{i}^{K}(\bar{N}_i - \bar{N})^2}{L \cdot (K - 1)}} \qquad (30)$$

In practice, it is often necessary to use and combine results of various measurements which have different statistical errors. In this case the final mean value should be calculated by weighting the results of individual measurements with respect to their standard deviations. The weighted mean value can be found using the equation [Bevington (1969)]

$$\overline{N} = \frac{\sum\limits_{i}^{K} \frac{N_i}{\sigma_i^2}}{\sum\limits_{i-1}^{K} \frac{1}{\sigma_i^2}} \tag{31}$$

Many quantities and parameters involving radiation or radiation sources rely on the counting of pulses. Often only one such measurement is performed, giving the result in the form of a single value, say M_1. In this case we have no other option but to consider the value obtained as the best estimate of the true mean value. Consequently, its standard deviation will be $\sqrt{M_1}$, and the result will be given as $M_1 \pm \sqrt{M_1}$.

Results of the measurement are sometimes reported in terms of the relative standard error δ. Assuming a single value M_1 in a counting experiment, the relative error can be expressed as

$$\delta = \frac{\sigma(M_1)}{M_1} = \frac{\sqrt{M_1}}{M_1} = \frac{1}{\sqrt{M_1}} \tag{32}$$

Using the above equation one can assess how many counts have to be accumulated in order to keep the relative standard error below the selected level. The number of counts required for a given value of the relative error are listed in Table 10.

Table 10: The minimum number of counts needed to keep the relative standard error at the stated level.

Number of Counts M_1	Corresponding Relative Error δ $\left(\delta = \dfrac{100\%}{\sqrt{M_1}} \right)$
100	10
1000	3.16
10000	1.0
100000	0.316
1000000	0.1

1.7.2. Standardization

The accuracy of radionuclide analyses is further based on internationally agreed upon and maintained standards as well as measuring procedures [IAEA (2000d)], assuring world-wide conformity of radioactivity measurements.In order to improve the accuracy and uniformity of measurements throughout the world, international comparisons of radionuclides and other radiation sources are periodically organized. In these intercomparisons many national laboratories and institutions take part. In addition to such national laboratories as NIST (USA), PTB (Germany), or NPL (UK), an active role in radioactivity standardization is played by international organizations, agencies, and commissions, especially, the International Atomic Energy Agency, the World Health Organization, the International Commission on Radiation Units and Measurements, and the International Committee on Radionuclide Metrology.

There are many different types of radioactivity standards (reference sources), some of them represent individual radionuclides, usually of the highest purity and known impurities, or more complex sources such as natural-matrix standard reference materials used for the calibration of equipments measuring environmental radioactivity. Ideally, these standards should be traceable to the relevant international standards within the hierarchical international reference system, [Mann et al. (1991)].

Traceability in radioactivity must be based primarily on periodic demonstrations of the reliability of radiochemical procedures, the correct calibration of the measuring instruments, and the performance of the laboratory with test samples. To achieve these goals, NIST, for example, prepares a number of classes of test samples to check the calibrations of instruments, test sources with chemical and radioactivity interferences to check the adequacy of radiochemical procedures and the competence of the technicians, and natural-matrix materials to test sample-handling and dissolution techniques [Inn et al. (1984)].

Usually, the first calibration of any detection system for the measurement of activity consists in the conversion of pulse-heights into the energy scale. In the majority of cases a response is a linear function of the energy deposited in the detector and, in principle, two sufficiently different energy points are usually enough to draw a calibration line. For measuring systems with non-linear detector responses and also for the more precise calibration of systems with a linear or rather a quasi-linear energy-response relationship, more energy calibration sources are required. The preparation of these sources should be done in such a way as to preserve the energy of emitted particles as much as possible. Consequently, gamma sources tend to have a "point" form to avoid attenuation and scattering within the source, while alpha and beta sources are usually prepared as disks on which a very thin layer of active material is deposited.

Table 11: Radionuclides suitable for the preparation of reference gamma point sources used for the energy calibration of photon detectors, based on [Liden et al (1985)].

Radionuclide	Half-Life	Energy (keV)	Number of Photons per Decay
^{241}Am	433 ± 2 y	26.35	0.0258 ± 0.0022
		59.54	0.363 ± 0.004
^{109}Cd	453 ± 2 d	88.04	0.0373 ± 0.0006
^{57}Co	270.9 ± 0.6 d	122.06	0.8559 ± 0.0019
^{139}Ce	137.65 ± 0.05 d	165.85	0.8006 ± 0.0013
^{203}Hg	46.59 ± 0.05 d	279.19	0.815 ± 0.008
^{51}Cr	27.704 ± 0.002d	320.08	0.0980 ± 0.0010
^{113}Sn	114.9 ± 0.1 d	391.69	0.6490 ± 0.0020
^{85}Sr	64.85 ± 0.03 d	513.99	0.980 ± 0.010
^{207}Bi	38 ± 0.3 y	569.67	0.978 ± 0.005
		1063.61	0.74 ± 0.03
		1770.22	0.073 ± 0.004
^{137}Cs	30.0 ± 0.2 y	661.65	0.899 ± 0.004
^{94}Nb	(2.3 ± 0.16) $\cdot 10^4$ y	702.63	1
		871.10	1
^{54}Mn	312.5 ± 0.5 d	834.83	0.999760 ± 0.000002
^{88}Y	107 ± 1 d	898.83	0.934 ± 0.007
		1836.04	0.9935 ± 0.0003
^{65}Zn	244.1 ± 0.2 d	1115.52	0.5075 ± 0.0010
^{60}Co	5.271 ± 0.001 y	1173.21	0.99900 ± 0.00020
		1332.46	1
^{22}Na	2.602 ± 0.002 y	1274.54	0.99940 ± 0.00020

Recommended reference sources, including energy values, for the calibration of photon detectors can be found in various references. A good information source for such data is NCRP Report [NCRP 58 (1978)], which contains a comprehensive table of gamma rays as reference standards for energy calibration. Details of some gamma reference sources are given in Table 11.

Some data relevant to such calibration standards for beta detection systems are summarized in Table 12.

The energy calibration of detection and spectrometry systems for measuring alpha-emitting radionuclides requires specially prepared alpha sources with minimum distortion of their monoenergetic properties. The most often used alpha calibration sources are ^{241}Am, and also ^{210}Po, ^{239}Pu, and ^{244}Cm. In addition to energies, radionuclide impurity, chemical stability, and accompanying radiation should also be considered. Parameters of some alpha sources suitable for calibration are listed in Table 13.

Table 12: Radionuclides emitting electrons for calibration, based on [Mann et al. (1991)].

Source	Energy (keV)	Electrons per Decay (%)
^{125}I	3.7	79.3
^{55}Fe	4.9-5.2	48.5
	5.7-5.9	11.0
	6.3-6.5	0.9
^{133}Ba	17.2	10.6
^{125}I	21.8-23.0	13.1
	25.8-27.4	6.0
	29.8-31.7	0.8
	30.5-31.2	10.7
^{133}Ba	45.0	45.2
^{131}I	45.6	3.5
^{199}Au	125.1	5.5
^{203}Hg	193.6	13.5
^{131}I	329.9	1.5
^{207}Bi	481.7	1.6
^{137}Cs	624.1	8.1

Table 13: Parameters of some alpha sources used in energy calibration (ΔE_α - uncertainly), based on [Mann et al. (1991)].

Source	Half-Life	E_α (MeV)	ΔE_α (keV)	Alpha per Decay (%)
^{232}Th	$1.2 \cdot 10^{10}$ y	4.012	± 5	77
		3.953	± 8	23
^{238}U	$4.5 \cdot 10^{9}$ y	4.196	± 5	77
		4.149	± 5	23
^{230}Th	$7.7 \cdot 10^{4}$ y	4.6875	± 1.5	76.3
		4.6210	± 1.5	23.5
^{239}Pu	$2.4 \cdot 10^{4}$ y	5.1554	± 0.7	73.3
		5.1429	± 0.8	15.1
		5.1046	± 0.8	11.5
^{210}Po	138 d	5.30451	± 0.07	> 99
^{241}Am	433 y	5.48574	± 0.12	85.2
		5.44298	± 0.13	12.8
^{238}Pu	88 y	5.49921	± 0.20	71.1
		5.4565	± 0.4	28.7
^{244}Cm	18 y	5.80496	± 0.05	76.4
		5.762835	± 0.03	23.6

Table 14: Some standard reference materials for environmental analyses.

Material	Radionuclides
Marine sediment	^{60}Co, ^{90}Sr, ^{137}Cs, ^{238}Pu, ^{239,240}Pu, ^{241}Am
Marine algae	^{40}K, ^{54}Mn, ^{60}Co, ^{90}Sr, ^{99}Tc, ^{137}Cs, ^{226}Ra, ^{239,240}Pu
Homogenized fish	^{90}Sr, ^{134}Cs, ^{137}Cs, transuranics
Calcined animal bone	^{90}Sr, ^{266}Ra
Milk powder	^{90}Sr, ^{137}Cs
Peruvian soil	^{137}Cs, ^{230}Th, ^{232}Th, ^{239,240}Pu, ^{241}Am
River sediment	^{60}Co, ^{137}Cs, ^{252}Eu, ^{254}Eu, ^{226}Ra, ^{238}Pu, ^{239,240}Pu, ^{241}Am
Rocky Flats soil	^{40}K, ^{90}Sr, ^{137}Cs, ^{226}Ra, ^{228}Ac, ^{230}Th, ^{232}Th, ^{234}U, ^{238}U, ^{238}Pu, ^{239,240}Pu, ^{241}Am
Human lung	^{232}Th, ^{234}U, ^{238}U, ^{239,240}Pu, ^{238}Pu/(^{239,240}Pu)
Human liver	^{238}Pu, ^{239,240}Pu, ^{241}Am

The measurement of the environmental radioactivity for use in analytical intercomparison and as standard reference materials, requires very large homogeneous samples of a variety of matrices, each naturally contaminated by a number of long-lived radionuclides, at several different ranges of concentrations [Bowen (1978)].

A number of standards environmentally quasi-equilibrated natural-matrix materials to test sample preparation and analytical techniques for radionuclides in soil, sediments, animal and human organs have been prepared and produced. These include such matrices as fresh water, sea water, river sediment, lake sediment, ocean sediment, milk powder, soils of various origin, animal bones, human lung, human liver, and other materials.

It is important that the natural-matrix materials contain "useful" activities of all the artificial radionuclides having half-lives of about one year and longer, and which are now known or expected to be of environmental concern [Bowen (1978)]. Some examples are listed in Table 14.

Some manufacturers and producers of calibration sources offer solid reference sources in various shapes and sizes [Debertin (1991)]. These calibration sources are filled with a plastic material containing the selected radionuclides. The density of the plastic varies from 0.7 to 2.5 $g \cdot cm^{-3}$.

The containers used are usually Marinelli beakers and polyethylene or glass bottles. In addition, containers supplied by the customers can be filled according to special requirements. It is assumed that the attenuation properties of the plastic material are similar to those of a sample material of the same density.

It is also possible to use readily available standard solutions of accurately known radionuclide concentrations to prepare one's own calibration sources. In this case the material under study is spiked with known amounts of the calibrated solution. Using such a procedure one can produce a calibration source in the same matrix as the actual sample.

Some manufacturers supply custom-made radionuclide calibration standards where there is an increasing demand, especially for mixed gamma sources. These sources are usually traceable to recognized national laboratories such as NIST and they are tailored to meet the user's individual needs (custom geometries, activities and matrices, etc.).

Some special evaluations of the measured values as, e.g., in radiocarbon dating, are given in the corresponding parts of the book.

2.   ANALYTICAL TECHNIQUES IN RADIOGENIC DATING

Jan Košler

The principal tool for radiogenic dating of geological samples is a mass spectrometer – an analytical device that is capable of separating and detecting ions or charged particles of different masses. A typical mass spectrometer usually consists of four parts: an ion source , an accelerator, a mass filter and a detector. The most commonly used type of instrument for radiogenic dating is a mass spectrometer with a thermal ionization source (TIMS) and magnetic sector as a mass filter, although secondary ion mass spectrometry (SIMS), gas-source mass spectrometry (GSMS), inductively coupled plasma mass spectrometry (ICP-MS) and negative-ion thermal ionization mass spectrometry (N-TIMS) techniques are also employed (see [Platzner (1997)], [Montaser (1998)]).

For classical TIMS, the elements of interest must be chemically separated prior to analysis in order to avoid interference of identical masses from different elements or molecules. In most dating techniques (e.g., ^{87}Rb-^{87}Sr, ^{147}Sm-^{143}Nd, ^{176}Lu-^{176}Hf), the age of the rock is derived from the slope of the regression through the progeny element isotopic compositions of several cogenetic samples with different parent element contents, e.g., different parent/progeny isotope ratios. While the progeny element isotopic composition of a sample can be directly measured using TIMS, the parent/progeny element ratio (i.e., concentrations of parent and progeny elements) is commonly determined using the isotope dilution technique (see 2.3).

A similar scheme is also employed in radiocarbon dating, although it is based primarily on activity measurements (see 1.). Compared to ionization- and LSC-measurements, ^{14}C accelerator mass spectrometry (^{14}C AMS) employs smaller sample sizes (less than 1 mg) and shorter measuring times, but requires considerable investment [Bertsche et al. (1991)]. For application of AMS, see Chapt. 4, 4.2 and 4.3.

2.1. Sample Dissolution and Chemical Separation

Samples have to be dissolved for chemical separation. As geological samples (rocks, minerals) often consist of silicates, the most common technique involves dissolution in hydrofluoric acid (HF) or in a mixture of acids containing HF. HF is the only inorganic acid that effectively dissolves silicate materials, mainly due to the increased solubility of SiF_6^{2-} in acidic environment. However, as less soluble salts (e.g., salts of Ca and K) may also be present in the sample and because numerous elements occur in more than one oxidation state, a mixture of acids (e.g., HF, HNO_3, $HClO_4$) is used for silicate dissolution [Šulcek et al., (1989)]. Some minerals (e.g., galena or carbonates) may be dissolved in HCl or in acetic acid (carbonates). Some chemical and physical properties of acids commonly used in silicate analysis are given in Table 15.

Dissolution of a weighed amount of sample is carried out in digestion vessel (often made of fluorinated carbons known as Teflon - tetrafluoroethylene TFE, fluorinated ethylene propylene FEP), either closed or open to the air. Acid-resistant samples (e.g., zircon) are dissolved in pressurised vessels made of stainless steel with Teflon lining [Krogh (1973)] or special capsules that allow transfer of HF vapour [Parrish (1987)]. After dissolution, samples are usually dried to evaporate SiF_4 and the soluble part is again dissolved in

Table 15: Chemical and physical properties of some acids used in silicate rock and mineral analysis.

Acid	Formula	Concentration		Specific mass	Boiling point
		(%)	(M)	$(g \cdot cm^{-3})$	(°C)
Hydrofluoric	HF	48	29	1.15	112
Perchloric	$HClO_4$	70	12	1.67	203
Nitric	HNO_3	70	16	1.42	83
Hydrochloric	HCl	36	6.8	1.18	110
Sulphuric	H_2SO_4	98	18	1.84	338
Phosphoric	H_3PO_4	85	15	1.70	213
Acetic	CH_3COOH	99.5	17.5	1.05	118

dilute acid; the type of acid and its concentration depend on the subsequent separation procedure. If the concentrations of elements are to be determined by isotope dilution, the sample solution is split (aliquot) by weight; one portion is used for isotopic ratio measurement and the other is mixed with a known amount of an internal isotopic standard (spike).

To prevent isobaric interference, elements to be analysed on TIMS or ICP-MS must be separated from the matrix. The separation is usually carried out using chromatographic techniques, most commonly ion exchange or high-pressure liquid chromatography [Stray (1992)]. Ion-exchange techniques are based on the reversible equilibrium between the stationary and mobile phases.

The stationary phase consists in an ion-exchange resin – usually an organic compound (polymerising styrene or methacrylic acid with some proportion of divinyl benzene), to which functional groups are attached. The functional groups are either acidic (e.g., $-SO_3H$, $-PO(OH)_2$, $-COOH$) or basic (e.g., $-N(CH_3)_3OH-$, $-NH_2$, $=NH$) and the corresponding ion exchangers are referred to as cation and anion exchangers, respectively. The proportion of divinyl benzene (DVB) in the resin is directly linked to the degree of cross-linking. A high degree of cross-linking means that the resin has a stronger skeleton, which is less susceptible to swelling and has more functional groups, but also that the material is less porous with smaller exchange capacity, especially for larger ions. The exchange resins are usually manufactured in the form of spherical particles, whose size varies according to the specific application. In geochronological applications, ion-exchange resin is usually packed in glass, quartz or plastic columns, through which the mobile phase, consisting in the sample solution, is passed.

The reversible reaction between the stationary (sulphonic acid resin) and the mobile phase containing ion A^+ in aqueous solution A^+_{aq} can be written as

$$Resin\ SO_3^-H^+ + A^+_{aq} = Resin\ SO_3^-A^+ + H^+_{aq} \qquad (33)$$

for which the concentration-based ([ ]) selectivity constant of equilibrium K can be expressed in the form:

$$K = \frac{[Resin\cdot SO_3\cdot A^+]\cdot[H^+_{aq}]}{[Resin\cdot SO_3\cdot H^+]\cdot[A^+_{aq}]} \qquad (34)$$

For $K<1$, the resin exhibits an affinity for H^+, for $K>1$ the resin preferentially retains ion A^+. The selectivity constants represent a simple and convenient way of expressing the ion-exchange properties of the resins. Ion selectivity depends on a variety of factors, especially: the ion charge, the swelling of the resin, chemical interactions between the ion and the resin, the pore size of the resin, the temperature, the pressure and the ionic strength of solutions. In general, the affinity of ions for a strongly acidic resin increases in the order [Marhol (1982)], [Patterson (1970)], [Pecsok et al. (1976)], [Potts (1987)]:

$Li^+<Na^+<NH4^+<K^+<Rb^+<Cs^+<Tl^+<Ag^+$,

$Mg^{2+}<Ca^{2+}<Sr^{2+}<Ba^{2+}<Ra^{2+}$,

$Fe^{2+}<Co^{2+}<Ni^{2+}<Cu^{2+}<Zn^{2+}$,

$Al^{3+}<Sc^{3+}<Lu^{3+}<Yb^{3+}<Tm^{3+}<Er^{3+}<Ho^{3+}<Y^{3+}<Dy^{3+}<Tb^{3+}<Gd^{3+}<Eu^{3+}<$
$<Sm^{3+}<Pm^{3+}<Nd^{3+}<Pr^{3+}<Ce^{3+}<La^{3+}$,

$Th^{4+}<La^{3+}<Ca^{2+}<Na^+$

The affinity of ions to a strongly basic ion exchanger is in the order

$F^-<Cl^-<Br^-<I^-$,

$SO_4^{2-}<AsO_4^{3-}<MoO_4^{2-}<CrO_4^{2-}$,

$I^-<NO_3^-<Br^-$.

The ion exchange separation procedures differ in relation to the sample size, matrix composition, amount of resin and size of the columns. The columns must be calibrated so that the elements of interest can be effectively separated one from another.

2.2.   Mass Spectrometry with Thermal Ionization Source

The developments in thermal ionization mass spectrometry (TIMS) have largely occurred through its applications in isotope geology and geochronology and it still represents the most precise and accurate technique for isotopic determinations in geological samples. Following chemical separation, a sample solution is loaded onto the metal filament of a sample holder, placed in the source chamber of a mass spectrometer and heated under vacuum by passage of a current until the formation of an amorphous residue. Further heating causes atomisation and ionization of the sample. Ions are then extracted from the source chamber by a potential gradient (6-8 kV), accelerated and focused into the curved flight tube, which passes through the magnetic sector (mass filter). The ion beam is dispersed in the magnetic field into several fractions according to the masses of the ions and is focused into a detector (Faraday cup, secondary electron multiplier, Daly detector). Different masses can be focused into the detector by changing the magnet settings or several detectors can be placed at the end of the flight tube and ions of different masses can be collected simultaneously without changing the magnet setting. A typical configuration of TIMS is given in Fig. 14. Data are collected in blocks with each block containing several scans across a chosen part of the mass spectrum and the measurement is usually terminated after the required precision is reached.

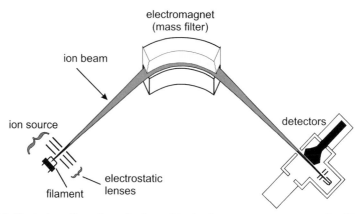

Figure 14: Typical configuration of a thermal ionization mass spectrometer, after [Potts (1987)].

2.2.1.   Ion Source and Production of Ions:

The ratio of positively charged n^+ to neutral n^o ions emitted from a heated metal filament is given by the Langmuir-Saha equation

$$\frac{n^+}{n^0} = e^{\frac{e \cdot (W - I_P)}{k \cdot T}} = e^{\frac{11600 \cdot (W - I_P)}{T}} \qquad (35)$$

where W is a working function of the filament, I_P is the first ionization potential, T is temperature, e is the charge of an electron ($1.60 \cdot 10^{-19}$ C) and k refers to the Boltzmann constant ($1.38 \cdot 10^{-23}$ J/K).

The working function reflects the amount of energy (in volts, see Table 16) necessary for emission of electrons from the filament. Several analytical parameters can be estimated using Eq. 35. For the formation of positively charged ions, the expression *(W-I$_P$)* must be positive. Accordingly, materials with a high working function are usually used for the filaments in TIMS. If the working function exceeds the value of the first ionization potential of the measured element, formation of positively charged ions is greater at lower temperatures and vice versa. Elements with first ionization potentials greater than 8 eV (e.g., $I_{p,As}=9.81$, $I_{p,Au}=9.22$, $I_{p,Hg}=10.43$, $I_{p,S}=10.36$, $I_{p,Pt}=9.0$) are difficult to measure on TIMS because the production of positively charged ions from atoms of these elements is very low.

As the second ionization potential of almost all elements exceeds 10 - 15 eV, interference between 1^+ and 2^+ ions in TIMS is rare. The working functions of metals which can be used as filaments in TIMS are given in Table 16; however, usually only Re, Ta and W are used, especially because of their high working function values, high melting points and chemical resistance.

Table 16: Working functions and melting points of selected metals, after [Potts (1987)].

Metal	Working function [V]	Melting point [°C]
Ni	5.03	1453
Nb	4.00	2468
Rh	4.80	1966
Pd	4.99	1522
Ta	4.19	2996
W	4.52	3410
Re	5.10	3180
Pt	5.32	1772

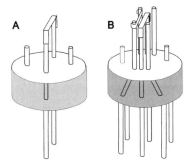

Figure 15: Sample holder of a thermal ionization mass spectrometer; (A) single filament assembly, (B) double filament assembly, after [Dickin (1995)].

One, two or three filaments (metal strips 15 x 1 mm, ca 10 μm thick) are welded to each sample holder (see Fig. 15). In the single filament arrangement, the sample is evaporated and ionised at the same time and the affinity of the filament for electrons from the sample exceeds the electron affinity of the sample itself. A single filament arrangement is used for elements with low ionization potential (Rb, Sr or oxide form of the rare earth, e.g., NdO^+). In a double or triple filament arrangement, the sample is loaded onto a side filament and its evaporation and ionization are controlled independently. Elements with high ionization potential (REE^+) are measured from the double or triple filament assembly. The evaporation/ionization ratio must be optimised in order to maintain a sufficiently strong and stable ion beam. Accordingly, the chemical form of the sample (e.g., chloride, nitrate) must allow for its effective evaporation and ionization at temperatures corresponding to its first ionization potential and the working function of the filament. The technique of sample loading can also affect the evaporation/ionization ratio. For instance, for Pb isotopic measurements, the sample is loaded onto a single Re filament and then onto a layer of colloidal Si-gel together with phosphoric acid.

Thermal atomisation of a sample in the source of a mass spectrometer results in mass fractionation of the isotopes. Lighter masses are more volatile and these isotopes are preferentially emitted during the initial part of the measurement. This effect (which is temperature-dependent) must be corrected for either by measurement of standards under the same conditions as used for the samples or, where possible, the measured results must be normalised to a known natural isotopic ratio (e.g., $^{86}Sr/^{88}Sr=0.1194$).

Thermal ionization has several advantages over other ionization techniques:

- the kinetic energy dispersion of the ions is small (ca 0.2 eV),
- a simple magnetic sector is sufficient for dispersion of an ion beam; an electrostatic sector is used only for high-precision measurements and for increasing the abundance sensitivity and
- the mass spectrum is free of interference and increased background level caused by residual gases in the vacuum system of the mass spectrometer.

2.2.2. Mass and Energy Ion Dispersion

As charged particles (ions) move in the magnetic field, their trajectory is curved by a force that is perpendicular to both their original trajectory and the magnetic field intensity. The radius of the trajectory curvature r depends on the kinetic energy of the ions and the magnetic field intensity H. An ion with mass m and charge e, which has been accelerated through a potential gradient U to velocity v has kinetic energy that is given by

$$e \cdot U = \frac{m \cdot v^2}{2} \qquad (36)$$

Accordingly, ions with different masses, which have been accelerated by the same potential gradient, must have different velocities. The magnitudes of the centrifugal force ($F_1 = m \cdot v^2/r$) and the force which curves the ion trajectory ($F_2 = H \cdot e \cdot v$) must be equal. Therefore,

$$R = \frac{\sqrt{2 \cdot U \cdot m/e}}{H} \qquad (37)$$

Assuming that the magnetic field intensity is expressed in gauss, potential gradient in volts, trajectory radius in centimetres and ion mass in ^{12}C units, then

$$r = 143.95 \frac{\sqrt{U \cdot m/e}}{H} \qquad (38)$$

The ion trajectory is, however, restricted by the curvature and inner diameter of the flight tube of the particular mass spectrometer. The heavier ions will be less deflected from their original trajectory and will have a larger trajectory radius compared to lighter ions of identical charge. In order to change the ion trajectory in the magnetic field and focus a particular mass into the detector, one must either change the kinetic energy of the ions by changing the potential gradient U that causes ion acceleration, or change the intensity of the magnetic field H. The latter is more common as it enables relatively simple magnet operation.

The ion source-magnet-detector geometry is important for discrimination between neighbouring masses in the mass spectrum. Assuming that the trajectories of individual ions in an ion beam are slightly divergent, the beam can be focused into the detector by setting the source-magnet and magnet-detector distances equal and by arranging the curvature centre, the source and the detector so that they lie on a straight line. Alternatively, beam focus into the detector and mass spectrum spreading can be achieved if the ion beam enters the magnetic sector at an angle of $26.5°$ and the source-magnet and magnet-detector distances are twice the curvature radius (see Fig. 16).

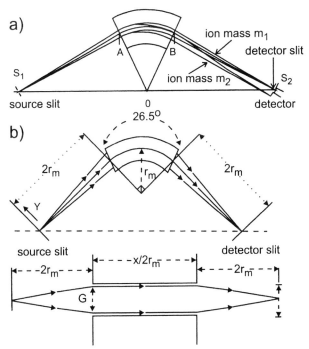

Figure 16: Possible arrangements of the source, magnet and detector in a thermal ionization mass spectrometer; a) Nier's arrangement, b) Cross' arrangement, top and side view, after [Potts (1987)].

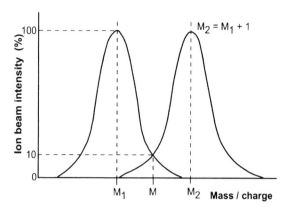

Figure 17: Resolution of a mass analyser of a thermal ionization mass spectrometer, the peak
tails overlap at 10 % intensity, after [Potts (1987)].

The mass discrimination is a measure of the ability to distinguish
between two neighbouring peaks in a mass spectrum. For two peaks with the
same intensity, the mass discrimination equals $M/\Delta M$ if the peak difference
ΔM is such that the peaks overlap at 10 % of their relative intensities (see
Fig. 17). Mass discrimination $M/\Delta M=500$ is sufficient for most geological
applications but it can be further improved if an electrostatic sector is
coupled to the magnetic sector in the mass spectrometer.

Similar to the magnetic sector, in the electrostatic sector (filter) two
forces affect a moving charged particle. The centrifugal force ($F_1=m{\cdot}v^2/r$)
must compensate for the force which causes electrostatic deflection of the
ion ($F_2=E{\cdot}e/d$, where E is the potential difference in volts between the
electrodes of the filter, d is the distance between the electrodes and e is the
ion charge). Accordingly,

$$r = \frac{E \cdot e}{d \cdot m \cdot v^2} \tag{39}$$

The kinetic energy of an ion that enters the electrostatic field is $m{\cdot}v^2/2$
and the radius of the curved trajectory depends only on the kinetic energy of
the particular ion and the electrostatic field intensity (E/d), but does not depend on the mass of the ion. As the curvature of the trajectory is given by the
shape of the flight tube, the intensity of the electrostatic field alone determines which ions can pass through the system.

2.2.3. Ion Detection

The most widely used ion detectors in mass spectrometry are the Faraday
cup, secondary electron multiplier, Daly detector and photographic plate;
only the first three are used in TIMS.

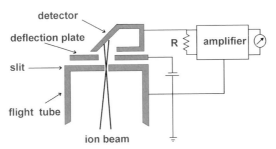

Figure 18: Principle of the Faraday cup detector, after [Potts (1987)].

The Faraday cup is a simple electronic device consisting of a slit, an electrode and a plate, which deflects the secondary electrons (see Fig. 18). Every ion that enters the detector carries a charge e which is transferred to the electrode. For a number of ions n per second, this charge equals:

$$n \cdot e = 1.602 \cdot 10^{-19} \cdot n \quad (C/s) \tag{40}$$

The charge is transferred through a resistor (10^8-10^{12} Ω) and the number of detected ions is proportional to the potential difference measured on the two sides of the resistor. The loss of signal due to the emission of secondary electrons is minimized by placing a plate with negative potential (50-90 V) in front of the electrode. Both the deflection plate and the shape of the electrode minimise the loss of charge. Faraday cups are used as single collectors and also in multicollector systems.

The principle of a secondary electron multiplier (SEM) is very similar to that of a photomultiplier, but the first set of electrodes (made of Be-Cu or Ag-Mg alloys) is designed to produce a large number of electrons when bombarded with ions (see Fig. 19). Secondary electrons are then accelerated in a set of dynodes (concave electrodes arranged in increasing negative potential). On an average, every secondary ion causes emission of two electrons from the surface of a dynode; the total electron yield is 2^n, where n is the number of dynodes in the multiplier. Finally, the electron cloud can be

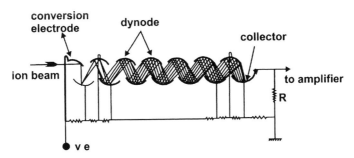

Figure 19: Principle of the secondary electron multiplier.

detected on a Faraday cup or an ion-counting device. SEM is much more sensitive compared to a Faraday cup and it allows detection of very small signals (ca 200 ions/s). Alternatively, in the channel electron multiplier, the set of electrodes used in the SEM is replaced by a tube, whose inner surface is covered by a semiconductor material.

The Daly detector (see Fig. 20) is the most sensitive type of detector used in TIMS. It consists of a conversion electrode with high negative potential, scintillator and photomultiplier. Ions are accelerated onto the electrode by the negative potential (-30 kV). On an average, each ion causes emission of eight secondary electrons, which are detected on the scintillator. The signal of the photons is then amplified in the photomultiplier.

Modern mass spectrometers allow simultaneous detection of different masses, sometimes on different types of detectors. For instance, radiogenic lead consists of masses 208, 207 and 206, but a small amount of non-radiogenic ^{204}Pb is usually also present in the sample. Detection of mass 204 is important for discrimination between the radiogenic and non-radiogenic portions of lead. The measurement of large lead signals (208, 207 and 206) can be carried out using Faraday cups but, for precise age dating, mass 204 must be measured simultaneously either on an SEM- or Daly-detector.

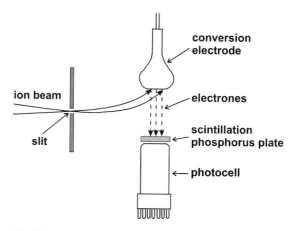

Figure 20: Principle of the Daly detector, after [Potts (1987)].

2.3.   Accelerator Mass Spectrometry

Extension of dating of minute amounts of sample is a new venture in radiocarbon dating and efforts are being made to perfect this dating technique. Amongst several methods, accelerator mass spectrometry (AMS) has been the most successful [Wölfli et al. (1984)] and developments in the

AMS, combined with enrichment of the ^{14}C content in milligram-size samples using laser technology, may in the future permit routine use of the ^{14}C time frame beyond 50 000 years [Taylor (1987)].

Accelerator mass spectrometry was developed in the early seventies and has been routinely used since 1977 [Gove (1992)]. Unlike decay counting, AMS enables direct detection of long-lived cosmogenic radioisotopes in the presence of much larger quantities of their stable nuclides in very small (milligram-size) samples. The introduction of AMS contributed greatly not only to ^{14}C dating, but also to measurements of other cosmogenic radioisotopes, such as ^{10}Be, ^{26}Al, ^{36}Cl, ^{41}Ca, and ^{129}I. The high cost of the instrumentation is still prohibitive for the routine use of the AMS technique in radiocarbon dating. However, mainly due to the very small amount of the sample required, AMS has become indispensable in dating of very rare samples [Gove (1989, 1992)] or samples in which carbon is present in very small quantities (e.g., dating of wrought iron [Igaki et al., (1994); Nakamura et al. (1995)] and dating of pottery using the smoke-derived carbon of potsherds [Delqué et al. (1995)]. AMS has also proven very effective also in dating marine and lake sediments, where only small samples are available from the drilling cores and where separate dating of thin layers is essential for determining the rate of sedimentation. Combined with paleontological analysis, measurements of ^{14}C in sediments enabled better definition of paleoclimatic events [Heier et al. (1995)].

AMS measurement of ^{14}C uses a negative ion beam (C^-), which is often generated from the sample by sputtering using a caesium (Cs^+) ion source. Most of the measurements are performed on carbon-rich organic matter, which is converted to graphite. Graphite can be made by catalytic reduction of CO_2 released from charcoal, bituminous coal, wood, cotton, sucrose, coconut charcoal or other carbon-bearing materials. Carbonates and CO_2 can be reduced at high pressure and temperature [Rubin et al. (1984)], [Gupta et al. (1985)]. Copper, iron, zeolite, montmorillonite or kaolinite are used as dilutants [Gupta et al. (1985)]. The sputtered ions are accelerated to ca 10 kV and the ion beam is cleaned in an electrostatic filter, similar to that used in TIMS, to obtain a narrow range of energies. The ions are then stripped of their charge by passing through an electron-stripping gas or just a lower vacuum chamber in the AMS instrument. Charge stripping is effective in removing isobaric interferences and in improving mass resolution without the need for a high-mass resolution magnetic filter, but it requires external calibration, e.g., by standards with known $^{14}C/^{12}C$ ratios. Finally, the ions are accelerated to the high-vacuum end of the AMS instrument where the isotopic ratios of carbon are detected either on a single detector or by a multi-collector array.

2.4. Isotope Dilution

Isotope dilution combined with TIMS is the most precise method for measuring the element concentration and is often employed in geochronology to measure the concentrations of Rb, Sr, REE's, U, Pb, Hf, Re, Os, etc. Most of the elements in nature have more than one isotope and the natural isotopic ratios for most elements are well known or can be measured using TIMS. If a sample of known isotopic composition is mixed with a tracer containing a known amount of an element that is synthetically enriched in one of the isotopes – internal isotopic standard (spike) – the concentration of the element in the sample can be calculated. The spike is usually added to the sample prior to or after sample dissolution and, if the sample-spike mixture is homogeneous, no further quantitative operations (e.g., ion exchange element separation) need be performed.

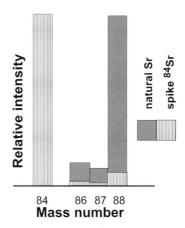

Figure 21: Isotope dilution analysis of Sr using ^{84}Sr spike (90 % enrichment).

The principles of isotope dilution can be demonstrated on Sr (see Fig. 21), which has four natural isotopes: ^{84}Sr with an abundance of 0.56 %, ^{86}Sr (9.86 %), ^{87}Sr (6.98 %) and ^{88}Sr (82.5 %). The spike is enriched in ^{84}Sr and usually contains over 99 % ^{84}Sr. The Sr concentration in the sample can be calculated as:

$$[Sr_{sam}] = [^{84}Sr_{sp}] \frac{W_{Sr} \cdot W_{sp}}{F_{84} \cdot W_{sam}} \cdot \frac{\dfrac{N_{88Sr_m}}{N_{84Sr_m}} - \dfrac{N_{88Sr_{sp}}}{N_{84Sr_{sp}}}}{\dfrac{N_{88Sr_{nat}}}{N_{84Sr_{nat}}} - \dfrac{N_{88Sr_m}}{N_{84Sr_m}}} \qquad (41)$$

where

$[Sr_{sam}]$	is the Sr concentration in the sample (ppm)
$[^{84}Sr_{sp}]$	is the ^{84}Sr concentration in the spike (ppm)
W_{Sr}	is the atomic weight of natural Sr (=87.6079 amu)
W_{sp}	is the weight of added spike (g)
F_{84}	is the fraction of ^{84}Sr in natural Sr (=0.0056)
W_{sam}	is the weight of sample (g)
N_{88Sr_m}/N_{84Sr_m}	is the measured isotopic ratio
$N_{88Sr_{sp}}/N_{84Sr_{sp}}$	is the known isotopic ratio in spike
$N_{88Sr_{nat}}/N_{84Sr_{nat}}$	is the known natural isotopic ratio (=148.01).

Measured isotopic ratios must be corrected for mass fractionation during the measurement, i.e., the measured Sr isotopic ratios are normalised to $N_{86Sr}/N_{88Sr}=0.1194$.

The amount of spike added to the sample influences the precision and accuracy of the age determination. Ideally the amount of added spike should be such that

$$\frac{N_{88Sr_m}}{N_{84Sr_m}} = \sqrt{\frac{N_{88Sr_{sam}}}{N_{84Sr_{sam}}} + \frac{N_{88Sr_{sp}}}{N_{84Sr_{sp}}}} \tag{42}$$

i.e., the measured isotopic ratio should be close to 1. In fact, a slight deviation (i.e., within one order of magnitude) from 1 has only a small effect on the precision of the concentration measurement. However, if the sample is significantly overspiked or underspiked, the error increases drastically [Colby et al. (1981)]. Accordingly, semiquantitative analysis of the sample prior to the isotope dilution is always useful as an indication of the concentration of the element that is to be determined by isotope dilution.

The isotope dilution technique has several limitations:
- Only the concentrations of elements which have more than one isotope in nature can be determined.
- Elements which have in nature two isotopes but the proportion of one of them is less than 1 % is always problematic.
- Correction for mass fractionation in samples where a spike has been added can only be employed for elements which have at least three isotopes (one isotopic ratio must be known).
- Elements with high ionization potential (>8 V) are difficult to measure using conventional TIMS.

On the other hand, the concentration of monoisotopic elements can be determined using isotope dilution if there is a suitable decay product with sufficiently long half-life, e.g., the half-life of ^{230}Th, a decay product of ^{238}U, is 75 200 years, and therefore the concentration of monoisotopic ^{232}Th can be determined by isotope dilution.

References

AARDSMA, G.E., (1995): Evidence for a lost millenium in biblical chronology. *Radiocarbon 37, 267-273*

AARKROG, A., (1968): Strontium 90 in the deciduous teeth collected in Denmark, the Faroers and Greenland from children born in 1950 -1958. *Health Phys. 15, 105 - 114*

AARKROG, A., (1971a): Prediction models for ^{90}Sr in shed deciduous teeth and infant bone, *Health Phys. 21, 803 - 809*

AARKROG, A., (1971b): Prediction models for strontium-90 and caesium-137 levels in the human food chain. *Health Phys. 20, 297 - 311*

AARKROG, A., (1971c): Radioecological investigations of plutonium in an artic marine environment. *Health Phys. 20, 31 - 47*

AARKROG, A., DAHLGAARD, H., FRISSEL, M., (1992): Sources of antropogenic radionuclides in the southern Urals. *J. Environ. Radioact. 15, 69 - 80*

AARKROG, A., DAHLGAARD, H., NILSSON, K., (1984): Further studies of plutonium and americium at Thule, Greenland. *Health Phys. 46, 29 - 44*

ADAMS, A., (1993): The origin and early development of the Belgian radium industry, *Environ. Int. 19, 491 - 501*

ADAR, E.M., LEIBUNDGUT, C., eds., (1995): Application of Tracers in Arid Zone Hydrology, Wallingford, *IAHS Publ. No. 232*

AEN - NEA, (1997): *Chernobyl. Ten years on radiological and health impact.* NEA Committee on Radiation Protection and Public Health. Paris, OECD

AEN - NEA, (2002): *Assesment of radiological and health impacts. 2002 update of Chernobyl: ten years.* NEA Committee on Radiation Protection and Public Health, Paris, OECD

AGLIETTA, M. et al., (1992): Study of the low energy background radiation and the effect of the [222]Rn in the LSD underground experiment. *Nucl. Phys. B., Proc. Suppl. 28A, 430 - 434*

AHRENS L.H., (1955): Implications of the Rhodesia age pattern. *Geochim. Cosmochim. Acta 8, 1 - 15*

AITCHESON, S.J., FORREST, A.H., (1994): Quantification of crustal contamination in open magmatic systems. *J. Petrol. 35, 461 - 488*

AITKEN, M.J., MICHAEL, H.N., BETANCOURT, P.P. (1988): The Thera eruption: continuing discussion of the dating. Resume of dating; Further arguments for an early date. *Archaeometry-UK 30, 1, 165 - 182*

AKLEYEV, A.V., LYUBCHANSKY, E.R., (1994): Environmental and medical effects of nuclear weapon production in the Southern Urals. *Sci. Tot. Envir. 142, 1 - 8*

ALGAZIN, A.I., et al., (1995): Radiation impact of nuclear weapon tests at the Semipalatinsk test site on the population of the Altai region. In: *Proc. IAEA Conference on environment impact of radioactive releases.* Vienna, International Atomic Energy Agency, *435 - 447*

ALLEN, S.E., ROWLAND, A.P., (1980): Iodine-129 in the terrestrical environment. *NRPB/NERC Contract EMR 4/79, ITE Project 667, Report June.*

ALLSOPP, H.L., (1961) : Rb - Sr measurements on total rock and separated mineral fractions from the Old Granite of the Central Transvaal. *J. Geophys. Res. 66, 1499 - 1508*

ALMEREY, R., BOUZO , R.,(1999) : Stochiometric study of phosphovanadomolybdate complex and its use to determine effect on uranium extraction by DEHPA / TOPO on the concentration of phosphoric acid using spectrophotometric method. *Afinidad 56, 290 - 294*

ANDRASI, A., et al., (1988): Measurements for the estimation of external doses in Hungary from Chernobyl releases. In: *IVth European Congress, XIIIth Regional Congress of IRPA, Twenty Years Experience in Radiation Protection, Salzburg, Sep. 15 - 19, 1986, 896 - 901*

ANDREWS, J.N., et al., (1984): Environmental isotope studies in two aquifer systems. In: *Isotope hydrology 1983.* Vienna, International Atomic Energy Agency, *535 – 576*

ANDREWS, J.N., FONTES, J.CH., (1992): Importance of the in situ production of ^{36}Cl, ^{36}Ar and ^{14}C in hydrology and hydrogeochemistry. In: *Isotope techniques in water resources development 1991.* Vienna, International Atomic Energy Agency, *245 - 269*

ANSDELL K.M., KYSER T.K., (1993): Textural and chemical changes undergone by zircon during the Pb-evaporation technique. *Am. Mineral. 78, 36 - 41*

ANSPAUGH, L.R., CHURCH, B.W., (1986): Historical estimates of external exposure and collective external exposure from testing at the Nevada test site. I. Test series through Hardtack II. *Health Phys. 51, 35 - 51*

ARELLANO, D.M., et al., (1987): Radiocarbon dating in a karstic coastal aquifer in tropical climatic conditions. In: *Isotope techniques in water resources development.* Vienna, International Atomic Energy Agency, *721 - 722*

ARELLANO, D.M., et al., (1989): Radiocarbon dating in a karstified coastal aquifer in Cuba. *Acta Universitatis Carolinae - Geologica, 3, Praha, 367 - 387*

ARNDT, N.T., GOLDSTEIN, S.L., (1987): Use and abuse of crust-formation ages. *Geology 15, 893 - 895*

ARTHUR, R.J., REEVES, J.H., MILEY, H.S.,(1987): Use of low - background germanium detectors to preselect high - radiopurity materials intended for constructing advanced ultra-low detectors. *Nuclear Science Symposium, October 21-23, San Francisco, 14-*

ARTINGER, R., et al., (1995): Influence of sedimentary organic matter on dissolved fulvic acids in groundwater. In: *Isotopes in water resources management, Vol. 1.* Vienna, International Atomic Energy Agency, *57 - 72*

ARTINGER, R., et al., (1998): Radiocarbon dating on aquatic fulvic acids in the presence of sedimentary organic carbon: A correction method for groundwater age. In: *Isotope techniques in the study of environmental change.* Vienna, International Atomic Energy Agency, *898 - 900*

AWSIUK, R., PAZDUR, M.F., (1986): Regional Suess effect in the Upper Silesia urban area. *Radiocarbon 28, 655 - 660*

BAIR, W.J. (1979): Metabolism and biological effects of alpha - emitting radionuclides. In: *Radiation Research.* OKADA, S., ed., Tokyo, Japan *Association of Radiation Research, 903 - 912*

BAIR, W.J., RICHMOND, C.R., WACHHOLZ, B.W., (1974a): A radiobiological assessment of the spatial distribution of radiation dose from inhaled plutonium. Washington D.C., United States Atomic Energy Commission September 1994, *WASH -1 320*

BAIR, W.J., THOMPSON, R.C., (1974b): Plutonium: Biomedical research. More is know about the toxicology of plutonium than about most other hazardous elements. *Science 183, 715 - 722*

BAKER, S.I., BULL, J.S., GOSS, D.L., (1997): Leaching of accelerator-produced radionuclids. *Health Phys. 73, 912 - 918*

BALEK, J., (1983): Identifikační metody pro bilanční výpočty v české křídě. In: *Hydrologické a hydraulické procesy v krajině, Proceedings, Bratislava, ÚHH SAV, 157 - 163*

BALLOU, J.E., PARK, J.F., (1972): Pacific Northwest Laboratory Annual Report for 1971. Richland (Washington), Battelle Northwest, *BNWL 1650, part 1, 146*

BARBECOT, F., et al., (1998): Geochemical evolution of a coastal aquifer to a Holocene seawater intrusion (Dogger aquifer, northern France). In: *Isotope techniques in the study of environmental change*. Vienna, International Atomic Energy Agency, *275 - 282*

BARD, E., et al., (1993): ^{230}Th-^{234}U and ^{14}C ages obtained by mass spectrometry on corals. *Radiocarbon 35, 191 - 199*

BARETT, E.W., HUEBNER, L., (1961): Atmospheric tritium analysis. USAEC, Univ. Chicago, *Technical progress report Nr. 3, Rep. Nr. TID-14425*

BARTHEL, F.H., (1993): Die Urangewinnung auf dem Gebiet der ehemaligen DDR von 1945 bis 1990. *Geol. Jb. A142, 335 - 346*

BATES, R.L., JACKSON, J.A., (1980): *Glossary of Geology*. 2nd ed., Falls Church, Va., Amer. Geol. Inst.,

BATH, A.H. (1983): Stable isotopic evidence for paleo-recharge conditions of groundwater. In: *Palaeoclimates and palaeowaters: A collection of environmental isotope studies*. Vienna, International Atomic Energy Agency, *169 - 186*

BAUMANN, W., MATTAUSCH, E., (1995), Nuklearkriminalität in der Bundesrepublik -Erfahrungen und Konsequenzen. *Strahlenschutz Praxis 1, 6 - 8*

BAVERSTOCK, K.F., VENNART, J., (1976): Emergency reference levels for reactor accidents: a re-examination of the Windscale reator accident. *Health Phys. 30, 339 - 344*

BAYER, A., et al., (1996): Kontamination und Strahlenexposition in Deutschland nach dem Unfall im Kernkraftwerk Tschernobyl. In: *Zehn Jahre nach Tschernobyl, eine Bilanz. Seminar des Bundesamtes für Strahlenschutz und der Strahlenschutzkommission, München 6. - 7. März (1996)*. Bayer, A., Kaul, A., Reiners, Chr., eds., Stuttgart, Gustav Fischer Verlag, *129 - 148*

BEARD, B.L., et al., (1995): Geochronology and geochemistry of eclogites from the Mariánské Lázně complex, Czech Republic : Implication for Variscan orogenesis. *Geol. Rundsch. 84. 552 – 567*

BEASLEY, T.M., HELD, E.E., CONARD, R.M., (1972): Iron-55 in Rongelap people, fish and soils. *Health Phys. 22, 245 - 250*

BECK, J.W., DRUFFEL, E.R.M., McNICHOL, A.P., eds., (1996): ^{14}C cycling and the oceans (in tribute to Reidar Nydal). Tucson, Arizona *Radiocarbon 38, 386 – 642*

BECKER, B., (1992): The history of dendrochronology and radiocarbon calibration. In: *Radiocarbon after four decades*. Taylor, R.E., Long, A., Kra, R.S., eds., New York, Springer-Verlag, *34 - 49*

BECKER, D., MERTENS, D., (1995), Handel mit Kernsprengstoffen in Europa, - poltische und soziale Hintergründe. *Strahlenschutz Praxis 1, 4 - 5*

BEDFORD, J., et al., (1960):The metabolism of strontium in children. *Brit. Med. J. 1, 589 - 591*

BENETT, B.G., (1976): Transuranic element pathways to man. In: *Transuranic nuclides in the Environment*. Vienna, International Atomic Energy Agency

BERG, D. et al., (1987a): Radioactive iodine and cesium in Bavarian citizens after the nuclear reactor accident in Chernobyl. In: *Trace Substances in Environmetal Health-XXI*. Hemphill, D.D., ed., Columbia, University of Missouri, *219 - 225*

BERG, D., (1976): Beeinflussung des Jodstoffwechsels der Ratte durch Jodmangel und Narkotika. *Dissertation*, Saarbrücken, Mathematisch - Naturwissenschaftliche Fakultät der Universität des Saarlandes

BERG, D., et al., (1987b): Ganzkörpermessungen an Kindern und Erwachsenen in München von Mai bis Dezember 1986. *Der Nuklearmediziner 5 (10), 353 - 358*

BERG, D., et al., (1991): Radioaktivitätsmessung in der Russischen Republik. In: *11. Arbeitstagung für Mengen- und Spurenelemente 12./13. Dezember 1991*. Anke, M., ed., Jena, Friedrich Schiller Universität Jena, *317 - 326*

BERG, D., HENRICHS, K., (1986): Ganzkörpermessungen an Kindern und Erwachsenen aus dem Raum München seit Mai 1986. In: *Radioaktivitätsmessungen in der Schweiz nach Tschernobyl und ihre wissenschaftliche Interpretation*. André, L., Born, E.J., Fischer, G., eds., Bern, Bundesamt für Gesundheitswesen, *Vol. 1, 529 - 535*

BERG, D., KOLLMER, W.E., KRIEGEL, H., (1987c): Ganzkörpermessungen nach dem Reaktorunfall von Tschernobyl an Kindern und Erwachsenen aus dem Raum München. *Der Nuklearmediziner 1(10), 87 - 92*

BERG, D., KRETSCHKO, J., (1985a): Meßgeräte in Radioisotopenlaboratorien und in der Nuklearmedizin. In: *Grundlagen der Nuklearmedizin*. Kriegel, H., ed., Stuttgart, Gustav Fischer, *139 - 189*

BERG, D., KRETSCHKO, J., (1985b): Statistik bei Messungen von Radionukliden. In: *Grundlagen der Nuklearmedizin*. Kriegel, H., ed., Stuttgart, Gustav Fischer, *125 - 137*

BERG, D., OBERHAUSEN, E., MUTH, H., (1973): Untersuchungen zur Wechselwirkung von ^{47}Ca, ^{85}Sr, ^{133}Ba und ^{226}Ra mit Proteinen im Rinderserum. *Biophysik 10, 309 - 319*

BERG, D.et al; (1990): Whole body content and turnover of Cs and K. In: *Biological trace element Research*. Schrauzer, G.N., Totowa, N.J., eds., The Humana Press Inc., *249 - 256*

BERGER, G.W., YORK D., (1981): Geothermometry from ^{40}Ar/^{39}Ar dating experiments. *Geochim. Cosmochim. Acta 45, 795 - 811*

BERGER, R., JOHNSON, R.M., HOLMES, J.R., (1983): Radiocarbon measurements of particulates in smog. *Radiocarbon 25, 615 – 620*

BERGER, R., McJUNKIN, D., JOHNSON, R., (1986): Radiocarbon concentration in California aerosols. *Radiocarbon 28, 661 - 667*

BERNHARDT, D.E., CARTER, M.W., BUCK, F.N., (1971): Protective actions for radioiodine in milk. *Health Phys. 21, 401 - 416*

BERTIN, M., (1982): Comparison of the risks of various energy forms. In: *Radionuclide metabolism and toxicity*. Galle, P. Masse, R., eds., Paris, Masson, *281 - 299*

BERTSCHE, K.J., KARADI ,C.A., MUELLER, R.A. (1991). Radiocarbon determination with a small low energy cyclotron. *Nucl. Instrum. Methods Phys. Res. A 301, 358 - 371*

BEVINGTON,P.R., (1969): *Data reduction and error analysis for the physical sciences*. New York, NY, Mc Graw-Hill

BIGN, J., (1992): Contamination of alpha-particle detectors by desorption of ^{222}Rn progeny from metal surfaces. *Appl. Radiat. Isot. 43, 747 – 751*

BIRKS, J.B., (1964): *Theory and practice of scintillation counting*. Oxford, Pergamon Press

BLACK, D., (1984): *Investigation of the possible increased incidence of cancer in West Cumbria*. Report of the Independent Advisory Group. London, Her Majesty's Stationery Office

BLACK, S.C., POTTER, G.D., (1986): Historical perspectives on selected health and safety aspects of nuclear weapons testing. *Health Phys. 51, 17 - 33*

BOBROW, M., (1986): *The implications of the new data on the releases from Sellafield in the 1950s for conclusions of the report on the investigation of the possible increased incidence of cancer in West Cumbria.* Committee on Medical Aspects of Radiation on the Environment (COMARE). London, Her Majesty's Stationery Office

BÖCK, H., et al., (1988): Der Reaktorunfall von Tschernobyl und seine radiologischen Folgen für Österreich. In: *IVth European Congress, XIIIth Regional Congress of IRPA, Twenty Years Experience in Radiation Protection, Sep. 15 - 19, 1986, Salzburg.* Fachverband für Strahlenschutz of Switzerland and the Federal Republic of Germany, *712 - 728*

BOGNER, L., et al., (1987): Ganzkörpermessungen im Raum München nach den Reaktorunfall von Tschernobyl bis zum Juli 1987. In: *Tagungsbericht der Deutschen Gesellschaft für. Medizin. Physik und der Österreich Gesellschaft f. Medizin. Physik, 9. - 12. Sept. 1987, Innsbruck.*

BØHMER, N., NILSEN, T., (1995): Reprocessing plants in Sibiria. *Bellona Report Nr. 4*

BELLONA, (2003): *http:www.bellona.no*

BOJANOVSKI, R., KNAPINSKA-SKIBA, D., (1990): Determination of low-level ^{90}Sr in environmental materials: a novel approach to the classical method. *J. Radioanal. Nucl. Chem. Articles 138, 207 - 218*

BOLIN, B., et al., eds., (1979): *The global carbon cycle.* Chichester, John Wiley & Sons

BOLIN, L.L.,(1964): *Research in geophysics, Vol. 2.* Cambridge (Mass.), MIT Press, *479 -*

BONKA, H., (1982): *Strahlenexposition durch radioaktive Emissionen aus kerntechnischen Anlagen im Normalbetrieb.* Köln, Verlag TÜV Rheinland

BOOIJ, E., GALLAHAN, W.E., STAUDIGEL, H., (1995): Ion - exchange experiments and Rb/Sr dating on celadonites from the Troodos ophiolite, Cyprus. *Chem. Geol. 126, 155 - 167*

BOONE, F.W., et al., (1985): Residence half-times of I-129 in undisturbed surface soils based on measured soil concentration profiles. *Health Phys. 48, 401 - 413*

BORAK, T.B., et al., (1972): The underground migration of radionuclides produced in soil near high energy proton accelerators. *Health Phys. 23, 679 - 687*

BOTTOMLEY, D.J., VEIZER J., (1992): The nature of groundwater flow in fractured rock: Evidence from the isotopic and chemical evolution of recrystallized fracture calcites from the Canadian Precambrian Shield. *Geochim. Cosmochim. Acta 56, 369 - 388*

BOWEN, V.T., (1978): Natural Matrix Standards. *Environ. Int. 1, 35 - 39*

BREITENSTEIN, B.D., NEWTON, C.E., MORRIS, H.T., (1985): The U.S. transuranium registry report on the ^{241}Am content of a whole body, Part I: Introduction and history of the case. *Health Phys. 49, 565 - 567*

BRIAN, G.P., MICHALAK, J., McGEE, M.K., (1985): Adult human thyroid weight. *Health Phys. 49, 1097 - 1103*

BRINKMANN, R., MÜNNICH, K.O., VOGEL J.C., (1960): Anwendung der ^{14}C-Methode auf Bodenbildung und Grundwasserkreislauf. *Geol. Rdsch. 49, 244 - 253*

BRODZINSKI, R.J., et al., (1988): Achieving ultralow background in a germanium spectrometer. *J. Radioanal. Nucl. Chem. Articles 124, 513 - 521*

BRODZINSKI, R.L., MILEY, H.S., REEVES, J.H., (1990): Further reduction of radioactive backgrounds in ultrascusitive germanium spectrometers. *Nucl. Instrum. Methods Phys. Res. A 292, 337 - 342*

BROECKER, W.S., OLSON E.A., (1959): Lamont radiocarbon measurements VI. *Am. J. Sci. Radiocarbon Suppl. 1, 111 - 132*

BROOKS, C., HART, S.R., WENDT, I., (1972) : Realistic use of two – error regression treatment as applied to Rb–Sr data. *Rev. Geoph. Space Phys. 10, 551 – 577*

BROWN, K.W., McFARLANE, J.C., (1973): Deposition and retention of ^{131}I on Grayia spinosa following Baneberry. *Health Phys. 24, 680 - 682*

BRUNNER, H., GENSICKE, F., (1995): Möglichkeiten und Maßnahmen zur Dekorporation und Dekontamination von Plutonium. *Strahlenschutz Praxis 1, 23 - 28*

BRUNS, M., et al., (1980): Regional sources of volcanic carbon dioxide and their influence on ^{14}C content of present-day plant material. *Radiocarbon 22, 532 - 536*

BUBNER, M., SCHMIDT, H., (1966): *Die Synthese Kohlenstoff - [14]C - markierter organischer Verbindungen.* Leipzig , Georg Thieme Verlag

BUCHAN K.L., et al., (1977): Thermal overprinting of natural remanent magnetization and K/Ar ages in metamorphic rocks. *J. Geomagn. G. 29, 401 - 410*

BUCHANAN, J.R., (1963): SL-1 final report. *Nucl. Saf. 4, 83 - 86*

BUGLOVA, E.E., KENIGSBERG, J.E., SERGEEVA, N.V., (1996): Cancer risk estimation in Belarussian children due to thyroid irradiation as a consequences of the chernobyl nuclear accident. *Health Phys. 71, 45 - 49*

BULDAKOV, L.A., DEMIN, S.N., KOSTYUCHENKO, V.A., (1990): Medical consequences of the radiation accident in the southern Urals in 1957. In: *Proc. of a symposium on recovery operations in the event of a nuclear accident or radiological emergency.* Vienna, International Atomic Energy Agency, *STI/PUB/826, 419-431*

BULL, J.S., et al., (1997): Groundwater activation at the superconducting super collider: a new design model. *Health Phys. 73, 800 - 807*

BUNZL, K., (1991): The migration of radionuclides in the soil. In: *Low-level measurements of man-made radionuclides in the environment.* Garcia-Leon, M., Madurga, G., eds., World Scientific Singapore, *328 - 353*

BUNZL, K., KRACKE, W., (1983): [238]Pu, [239+240]Pu, [241]Pu und ^{137}Cs in Lebensmittel aus der Umgebung eines Kernkraftwerkes und anderen Teilen der Bundesrepublik Deutschland. *Arch. Leb. Mitt. Hyg. 34, 113 - 116*

BUNZL, K., KRACKE, W., (1994): Fate of fall-out plutonium and americium in the environment: selected examples. *J. Alloys and Compounds 213/214, 212 - 218*

BURCHARDT, B., FRITZ, P., (1980): Environmental isotopes as environmental and climatological indicators. In: *Handbook of environmental isotope geochemistry. The terrestrial environment.* Fritz,P., Fontes, J.C., eds., Amsterdam, Elsevier Scientific Publishing Company, *Vol. 1, 473 - 504*

BURKE, et al., (1982): Variations of seawater [87]Sr/[86]Sr throughout Phanerozoic time. *Geology 10, 516 - 519*

BURLEIGH, R., BAYNES-COPE, A.D., (1983): Possibilities in the dating of writing materials and textiles. *Radiocarbon 25, 669 - 674*

BURNAZYAN, A.I., (1990): *Results and studies and experience in elimination of the consequences of accidental contamination of a territory with the products of uranium fission.* Moscow, Energoatomized

BURT, D.M., (1989): Compositional and phase relations among rare earth element minerals. *Rev. Mineral. 21, 259 - 302*

BUTLER, H.L., LeROY, J.H., (1965): Observations of biological half life of tritium. *Health Phys. 11, 283 - 285*

BUTTLAR, H., LIBBY, W.F, (1955): Natural distribution of cosmic-ray-produced tritium. Part II. *J. Inorg. Nuc. Chem. 1, 75 - 91*

CAIN, W.F., SUESS, H.E., (1976): Carbon 14 in tree rings. *J. Geophys. Res. 81, 3688 - 3694*

CAMBRAY, R.S., et al., (1983): Radioactive fallout in air and rain: results to the end of 1982. *AERE-R10859*

CAMERON A.E., SMITH D.H., WALKER R.L., (1969): Mass spectrometry of nanogram-size samples of lead. *Anal. Chem. 41, 525 - 526*

CATALDO, D.A., et al., (1980): Foliar absorption of transuranic elements: influence of physiochemical form and environmental factors. *J. Envir. qual. 9, 364 - 369*

CATE, S., RUTTENBER, A.J., CONKLIN, A.E., (1990): Feasibility of an epidememiologic study of thyroid neoplasia in persons exposed to radionuclides from the Hanford nuclear facility between 1944 and 1956. *Health Phys. 59, 169 - 178*

CEC (1979): *Methodology for evaluating the radiological consequences of radioactive effluents released in normal operations.* Joint Report by the National Radiological Protection Board and the Commisariat à l'Energie Atomique. Commission of the European Communities

CHAMBERLAIN, A.C., (1986): Environmental impact of particles emitted from Windscale piles 1954 - 7. Environmental and medical Sciences Devision Harwell Lab., Oxfordshire, *AERE R 12163*

CHANG, J., (1999): Table of nuclides. Korea Atomic Energy Research Institute, Brookhaven National Laboratory, *http://www.dne.bnl.gov/CoN/*

CHAPMAN H.J., RODDICK J.C., (1994): Kinetics of Pb release during the zircon evaporation technique. *Earth Planet. Sci. Lett. 121, 601 - 611*

CHAPPELL, B.W., WHITE A.J.R., (1992): I- and S-type granites in the Lachlan Fold Belt. *Trans. Roy. Soc. Edinburgh: Earth Sci. 83, 1 - 26*

CHATTERS, R.M., CROSBY, J.W.III, Engstrand, L.G., (1969): Fumarole gaseous emanations. Their influence on carbon-14 dates. Coll. of Engineering, Washington State Univ., Pullman, Wa., *Circular 32, 1 - 5*

CHEN, J.H., WASSERBURG G.J., (1983): The least radiogenic lead in iron meteorites. In: *14th Lunar and Planetary Science Conference Abstracts, Part 1, 103 - 104*

CHERONIS, N.D., (1954) : Micro and Semimicro Methods. In: *Technique of organic chemisty, Vol. VI.* Weissberger, A., ed., New York , Interscience Publ., *1 - 409*

CHERTOK, R.J., LAKE, S., (1971): Availability in the peccary pig of radionuclides in nuclear debris from the plowshare excavation buggy. *Health Phys. 20, 313 - 316*

CLARK, A.M., (1984): Mineralogy of the rare earth elements. In: *Rare earth element geochemistry.* Henderson, P., ed., Amsterdam, Elsevier, *33 - 62*

CLARK, I., FRITZ, P., (1997): *Environmental isotopes in hydrogeology.* Boca Raton, Lewis Publishers

CLARK, R.H., (1974): Analysis of the 1957 Windscale accident using the reactor safety code WEERIE. *Ann. Nucl. Sci. Eng. 1, 73 - 82*

CLARK, R.H., (1989): Current radiation risk estimates and implications for health consequences of Windscale, TMI and Chernobyl accidents. In: *Medical response to effects to ionizing Radiation.* Crosbie, W.A., Gittus, J.H., eds., Elsevier *Applied Sciences, 102 - 118*

CLEMENTE, G.F., MARIANI, A., SANTARONI, G.P., (1971): Sex differences in Cs metabolism in man. *Health Phys. 21, 709 - 711*

CLIFF, R.A., et al., (1985): Isotopic dating in metamorphic belts. *J. Geol. Soc. London 142, 97 - 110*

CLIFF, R.A., GRAY, C.M., HUHMA, H., (1983): A Sm - Nd isotopic study of the South Harris igneous complex, the Outer Hebrides. *Contrib. Mineral. Petrol. 82, 91 - 98*

CLINE, J.F., RICKARD, W.H., (1972): Radioactive strontium and cesium in cultivated and abandoned field plots. *Health Phys. 23, 317 - 324*

COHEN, B.L., (1981): The role of radon in comparison of environmental effects of nuclear energy, coal burning, and phosphate mining. *Health Phys. 40, 19 - 25*

COHEN, B.L., (1982): Health effects of radon from coal burning. *Health Phys. 42, 725 - 727*

COHEN, B.L., (1983): Long term waste problems In. *Electricity production: Nucl. & Chem. Waste Mngmt. 4, 219 - 229*

COHEN, B.L., (1985a): Transport of elements from soil to human diet: an alternativ approach to pathway analysis. *Health Phys. 49, 239 - 245*

COHEN, B.L., (1985b): Survey of one - year average Rn levels in Pittsburgh area homes. *Health Phys. 49, 1053 - 1059*

COHEN, B.L., (1985c): Comparison of radiological risks from coal burning and nuclear power. *Health Phys. 48, 342 - 343*

COHEN, N., RAHON, T.E., HIRSFIELD, H., (1985d): A comparison of radionuclide concentrations in 1956 and 1973 Enewetak beach material. *Health Phys. 48, 228 - 230*

COLBY, B.N., ROSECRANCE, A. E., COLBY, M.L., (1981): Measurement parameter selection of quantitative isotope dilution gas chromatography/ mass spectrometry. *Anal. Chem. 53, 1907 – 1911*

COMAR, C.L., WASSERMANN, R.H., NOLD, M.M., (1956): Strontium - calcium discrimination factors in the rat. *Proc. Soc. Exp. Med. 92, 859 -*

COMPSTON, W., WILLIAMS, I.S., BLACK, L.P., (1983): *Use of the ion microprobe in geological dating. BMR 82.* Canberra, Yearb. Bureau Mineral. Res., Geol. Geophys.

COOK, G.T., et al., eds., (1995): Proceedings of the 15th International Radiocarbon Conference, Glasgow, Scotland, 15-19 August 1994. Austin., Arizona, *Radiocarbon 37, No.2, 844 -*

CORFU, F., NOBLE, S.R., (1992): Genesis of the southern Abitibi greenstone belt, Superior Province, Canada: evidence from zircon Hf isotope analyses using a single filament technique. *Geochim. Cosmochim. Acta 56, 2081 - 2097*

COTTON, F.A., WILKINSON, G., (1962): *Advance inorganic chemistry.* London, Interscience Publishers - John Wiley & Sons

COWAN, G.A., (1976): A natural fission reactor. *Sci. Am. 235, 36 - 47*

COX, K.G., BELL J.D., PANKHURST, R.J., (1979): *The interpretation of igneous rocks.* London, George Allen and Unwin, *450 -*

CRAIG, H., (1953): The geochemistry of the stable carbon isotopes. *Geochim. Cosmochim. Acta 3, 53 - 92*

CRAIG, H., (1954): Carbon 13 in plants and the relationships between carbon 13 and carbon 14 variations in nature. *J. Geol. 62, 115 - 149*

CREASER, R.A., PAPANASTASSIOU, D.A., WASSERBURG, G.J., (1991): Negative thermal ion mass spectrometry of osmium, rhenium and iridium. *Geochim. Cosmochim. Acta 55, 397 - 401*

CRICK, M.J., LINSLEY, G.S., (1982): An assessment of radiological impact of the 1957 Windscale reactor fire. *NRPB-R 135*

CRICK, M.J., LINSLEY, G.S., (1984): An assessment of radiological impact of the 1957 Windscale reactor fire, October 1957. *Int. J. Radiat. Biol. 46, 479 - 506*

CROSBY, J.W.III, CHATTERS, R.M., (1965): New techniques of water sampling for carbon 14 analysis. *J. Geophys. Res. 70, 2839 - 2844*

CROSS, F.T., HARLEY, N.H., HOMANN, W., (1985): Health effects and risks from ^{222}Rn in drinking water. *Health Phys. 48, 649 - 670*

CURIE, I., JOLIOT, F., (1934): Physique nucleaire. Un nouveau type de radioacitvité. *C.R. Acad. Sci. Paris 198, 254 -256*

CURRIE, L.A., et al., (1983): On the origin of carbonaceous particles in American cities: Results of radiocarbon „dating„ and chemical characterization. *Radiocarbon 25, 603 - 614*

CURRIE, L.A., et al., (1986): The power of ^{14}C measurements combined with chemical characterization for tracing urban aerosol in Norway. *Radiocarbon 28, 673 - 680*

CURRIE, L.A., POLACH, H.A., (1980): Exploratory analysis of the international radiocarbon cross-calibration data: consensus values and interlaboratory error, preliminary note. *Radiocarbon 22, 933 - 935*

CZARNECKI, J., et al., (1986): Bodenverstrahlung in der Schweiz aufgrund des Reaktorunfalls in Chernobyl. In: *Radioaktivitätsmessungen in der Schweiz nach Tschernobyl und ihre wissenschaftliche Interpretation.* André, L., Born, E.J., Fischer, G., eds., Bern, Bundesamt für Gesundheitswesen, *Vol. 1, 93 - 109*

CZOSNOWSKA, W., PIETRZAK - FLIS, Z., GRABOWSKI, D., (1972): Long - term study of the urinary strontium-90 and calcium in children. *Health Phys. 23, 215 - 221*

DALE, W.S.A., (1987): The Shroud of Turin: Relic or icon? – In: *Nucl. Phys. Res. Section B 29, 187 - 192*

D'AMICO, A., MAZZETTI, F., (1986): *Noise in Physical Systems and 1/f Noise.* Amsterdam, North-Holland Publ.

DAMON, P.E., (1992): The natural carbon cycle, Preface. In: *Radiocarbon after four decades.* Taylor, R.E., Long, A., Kra, R.S., eds., New York, Springer-Verlag, *17 - 18*

DAMON, P.E., et al., (1989): Radiocarbon dating of the Shroud of Turin. *Nature London 337, 611 - 615*

DAMON, P.E., LONG, A., WALLICK, E.I., (1972): Dendrochronologic calibration of the Carbon-14 time scale. In: *Proceedings of the 8th International Conference on Radiocarbon Dating 1.* Lower Hutt, New Zealand, *44 - 59*

DAS, H.A., (1987): The advantage of anti - Compton counting in the measurement of low-level radioactivity by gamma-ray spectrometry. *J. Radioanal. Nucl. Chem. Articles 115, 159 - 173*

DAS, H.A., (1992): Release of ^{137}Cs from anoxic lacustrine sediment measurement and formulation. *J. Radioanal. Nucl. Chem. Articles 156, 129 - 149*

De CORT, M., et al., (1990): Radioactive measurements in Europe after the Chernobyl Accident. Part II: Fallout and deposition., Brüssel, CEC, *EUR 12800 EN*

De PAOLO, D.J., (1981): Trace element and isotopic effects of combined wallrock assimilation and fractional crystallization. *Earth Planet. Sci. Lett. 53, 189 - 202*

De PAOLO, D.J., (1988): *Neodymium isotope geochemistry.* Berlin, Springer Verlag

De PAOLO, D.J., LINN, A.M., SCHUBERT, G., (1991): The continental crustal age distribution: Methods of determining mantle separation ages from Sm-Nd isotopic data and application to the southwestern United States. *J. Geophys. Res. 96(B2), 2071-2088*

De PAOLO, D.J., WASSERBURG, G.J., (1976a): Nd isotopic variations and petrogenetic models. *Geophys. Res. Lett. 3, 249 - 252*

De PAOLO, D.J., WASSERBURG, G.J., (1976b) : Inferences about magma sources and mantle structure from variations of $^{143}Nd/^{144}Nd$. *Geophys, Res. Lett. 3, 743 - 746*

De SANTIS, V., LONGO, I., (1984): Coal energy vs nuclear energy:a comparison of the radiological risk. *Health Phys. 46, 73 - 84*

De VRIES, H.,(1958): Variations in concentration of radiocarbon with time and location on Earth. *Proceedings,* Nederlandsche Akademie van Wetenschappen, *Series B 61, 1*

DEBERTIN, K., (1991): Efficiency calibration in gamma-ray spectrometry with germanium detectors. In: *Low-level measurements of man-made radionuclides in the environment.* Garcia-León, M., Madurga, G., eds., Singapore , World Scientific, *3 - 14*

DEGTEVA, M.O., et al., (1996): An approach to dose reconstruction for the Urals population. *Health Phys. 71, 71 - 76*

DEGTEVA, M.O., KOZHEUROV, V.P., VOROBIOVA, M.I., (1994): General approach to dose reconstruction in the population exposed as a result of the release of radiaoctive wastes into the Techa river. *Sci. Total. Environ. 142, 49 - 62*

DEINES, A.C., et al., (1991): *Marshall Islands chronology 1944 to 1990.* Draft. Available from History Associates Incorporated. The Historic Montrose School, 5271 Radolph Road, Rockville MD20852

DEINES, P., (1980): The isotopic composition of reduced organic carbon. In: *Handbook of environmental isotope geochemistr. The terrestrial environment.* Fritz, P., Fontes, J.C., eds., Amsterdam, Elsevier Scientific Publishing Company, *Vol. 1, 329 - 406*

DELQUÉ KOLIČ, E., (1995): Direct radiocarbon dating of pottery: Selective heat treatment to retrieve smoke-derived carbon. *Radiocarbon 37, 275 - 284*

DEMOVIČ, R., HOEFS, J., WEDEPOHL, K.H.,(1972): Geochemische Untersuchungen an Travertinen in der Slowakei. *Contrib. Mineral. Petrol. 37, 15 - 28*

DICKIN A.P., (1995) : *Radiogenic isotope geology.* Cambridge Univ. Press

DIETRICH, P.G., HEBERT, D., (1986): Methodische Aspekte isotopenhydrogeologischer und hydrochemischer Untersuchungen an einem Karstgrundwasserleiter der DDR. *Freiberg. Forsch.-H., Reihe C, 76 - 92*

DODSON M.H., (1973): Closure temperatures in cooling geological and petrological systems. *Contrib. Mineral. Petrol. 40, 259 - 274*

DOE, (1987):*Task group on health and environmental aspects of the Soviet nuclear accident: Health and environmental consequences of the Chernobyl nuclear power plant accident. Report to the US Departement of Energy, Committee on the Assessment of Health Consequences in Exposed Population, DOE/ER-0332 UC-41 and 48.* Springfield, Virginia 22161, National Technical Information Service, US Dep. of Commerce

DOERFEL, H., (1987): 25 Jahre Referenzmessungen im Karlsruher Ganzkörperzähler. In: *Jahresbericht 1986 der Hauptabteilung Sicherheit.* Kernforschungszentrum Karlsruhe, *Report KfK-4207, 193-*

DOMINGO, J.L., (1994):Metal - induced developmental toxicity in mammals: a review. *J. Toxicol. Env. Health 42, 123 - 141*

DRAY, M., FERHI, A.A., JUSSERAND, C., OLIVE, P., (1998): Paleoclimatic indicators deduced from isotopic data in the main French deep aquifers. In: *Isotope Techniques in the Study of Environmental Change.* Vienna, International Atomic Energy Agency, *683 - 692*

DRUFFEL, E.M., LINICK T.W., (1978): Radiocarbon in annual coral rings of Florida. *Geophys. Res. Lett. 5, 913 – 916*

DRUFFEL, *E.M.*, SUESS, *H.M.*, *(1983)*: On the radiocarbon record in banded corals: Exchange parameters and net transport of $^{14}CO_2$ between atmosphere and surface ocean. *J. Geophys. Res. 88, C2, 1271 - 1280*

D'SOUZA, T.J., KIRCHMANN, R., LEHR, J.J., (1971): Distribution of radiostrontium and radiocesium in the organic and mineral fractions of pasture soils and their subsequent transfer to grasses. In: *Isotopes and radiation in soil - plant relationship including forestry, 13.-17.12.1971, Vienna, 595 - 603*

DUBASOV, Yu.V., et al., (1993): Radiation situation around the Semipalatinsk Test site. *Bulletin of Centre of Public Information in the Field of Nuclear Energy N9* (in Russian)

DUBASOV, Yu.V., et al., (1994a): Semipalatinsk and north test sites in the USSR: integrated programme of radiation and ecological research on evironmental consequences of nuclear tests. In: *Atmospheric nuclear tests, environmental and human consequences, 10. - 14. January 1994, Vienna*

DUBASOV, Yu.V., et al., (1994b): Underground explosions of nuclear installations in industrial centres in the territory of USSR in 1965 - 1988: Chronology and radiation consequences. *Bulletin of Centre of Public Information in the Field of Nuclear Energy N1, 18 - 28* (in Russian)

DUBASOV, Yu.V., et al., (1994c): Chronological list of the atmospheric nuclear tests at the Semipalatinsk test site and their radiological characteristics. *Вестник научнои программы „Семипалатинский полигон - Алтай" 4, 78 - 86* (in Russian)

DUNSTER, H.J., HOWELLS, H., TEMPLETON, W.L., (1958): Districts surveys following the Windscale incident, October 1957. In: *Proc. U.N. 2nd Int. Conf. Peaceful Uses Atomic Energy, Geneva*

DURBIN, P.W., (1975): Plutonium in mammals: influence of plutonium chemistry, route of administration and phsiological states of the animal on initial and long term metabolism. *Health Phys. 29, 495 - 510*

DURBIN, P.W., SCHMIDT, C.T., (1985): The U.S. transuranium registry report on the ^{241}Am content of a whole body, Part V: Implications for metabolic modelling. *Health Phys. 49, 623 - 661*

DYCK, W., BRUST, H., MUELLER, E. (1993). Airborne measurement of radioactivity. In: *Radioprotection, Spec. issue: Environmental Impact of Nuclear Installation.* Voelkle, H., Pretre, S., eds. *February, 375 - 379*

EDMUNDS, W.M., et al., (1998): Groundwater, palaeoclimate and palaeorecharge in the southwest Chad Basin, Borno State, Nigeria. In: *Isotope Techniques in the Study of Environmental Change.* Vienna, International Atomic Energy Agency, *693 - 707*

EDWARDS, R.R., (1962): I - 129, its occurrence in nature and its utility as a tracer. *Science 137, 851 - 853*

EDWARDS, R.R., REY, P., (1969): Terrestrical occurrence and distribution of iodine-129. *USAEC-Report, NYO-3624-3*

EICHINGER, L., RAUERT, W., WOLF, M., (1980): ^{14}C-Messungen an Weinen und Baumringen. In: *Traceruntersuchungen in Hydrogeologie und Hydrologie.* D-85764 Neuherberg, GSF - National Research Center for Environment and Health, *GSF-Bericht R 250, 105 - 118*

EISENBUD, M., (1981): The status of radioactive waste manegement: needs for reassessment. *Health Phys. 40, 429 - 437*

EISENBUD, M., (1987): *Environmental radioactivity from natural, industrial, and military sources. 3rd Edition.* San Diego, Academic Press

EISENBUD, M., (1990): *An environmental odyssey: People, pollution, and politics in the life of a practical scientist.* Seattle WA, University of Washington Press

EISENBUD, M., (1997):Monitoring distant fallout: the role of the Atomic Commission Health and Safety Laboratory during the Pacific tests with special attention to the events following BRAVO. *Health Phys. 73, 21 - 27*

ELLAM, R.M., CARLSON, R.W., SHIREY, S.B., (1992): Evidence from Re-Os isotopes for plume-lithospheric mixing in Karoo flood basalt genesis. *Nature 359, 718 - 721*

ELMORE, D., PHILLIPS, F.M., (1987): Accelerator mass spectrometry for measurements of long-lived radioisotopes. *Science 236, 543 - 550*

ERIKSSON, E., (1983): Stable isotopes and tritium in precipitation. In: *Guidebook on Nuclear Techniques in Hydrology, 1983 Edition.* Vienna, International Atomic Energy Agency, *Technical Report Series No. 91, 19 - 33*

ESSIEN, I.O., SANDOVAL, D.N., KURODA, P.K., (1985): Deposition of excess amount of natural U from the atmosphere. *Health Phys. 48, 325 - 331*

FABRIKANT, J.I., (1983a): Is nuclear energy an unacceptable hazard to health? *Health Phys. 45, 575 - 578*

FABRIKANT, J.I., (1983b): The effect of the accident at Three Mile Island on the mental health and behavioral responses of the general population and nuclear workers. *Health Phys. 45, 579 - 586*

FAIRES, R.A., PARKS, B.H., (1973): *Radioisotope laboratory techniques.* London, Butterworth.

FAO (1984): Food and agriculture organization. 1983 Fertilizer Yearbook. *FAO statistics series, Vol. 33, 56 -*

FARAI, I.P., SANNI, A.O., (1992): ^{222}Rn in groundwater in Nigeria: a survey, *Health Phys. 62, 96 - 98*

FARRIS, W.T., et al., (1996): Radiation dose from Hanford site release to the atmosphere and the Columbia River. *Health Phys. 71, 588 – 601*

FAURE, G., (1986): *Principles of isotope geology, 2nd ed.* New York., John Willey & Sons

FEHN, U., TENG, R., ELMORE, D., KUBIK, P. ,(1986) : Isotopic composition of osmium in terrestrial samples determined by accelerator mass spectrometry. *Nature 323, 707 – 710*

FERGUSON, C.W., HUBER, B., SUESS, H.E., (1966): Determination of the age of Swiss lake dwellings as an example of dendrochronologically-calibrated radiocarbon dating. *Z. Naturforsch. 21 (34), 1173 - 1177*

FERRONSKY, V.I., POLYAKOV, V.A., FERRONSKY, S.V., (1992): Isotope variations in the hydrological cycle as a tool in a climatic change mechanism study. In: *Isotope techniques in water resources development 1991.* Vienna, International Atomic Energy Agency, *567 - 586*

FISCHER, B., KIRCHNER, G., (2000): Tokaimura II: Vergleich mit früheren Kritikalitätsunfällen. *Stahlenschutz Praxis 6, 65 – 71*

FLORKOWSKI, T. (1992): Some aspects of the underground production of radionuclides used for dating groundwater. In: *Isotope techniques in water resources development 1991.* Vienna, International Atomic Energy Agency, *215 - 228*

FOLK, R.L., VALASTRO, S.Jr. (1979): Dating of lime mortar by ^{14}C. In: *Radiocarbon dating.* Berger, R., Suess, H.E., eds., Berkeley, University of California Press, *721 - 732*

FONTES, J.C., (1983): Dating of groundwater. In: *Guidebook on nuclear techniques in hydrology, 1983 Edition*. Vienna, International Atomic Energy Agency, *Technical Reports Series No. 91, 285 - 317*

FONTES, J.C., GARNIER,J.-M., (1979): Determination of the initial ^{14}C activity of the dissolved carbon. A review of the exixting models and a new approach. *Wat. Resour. Res. 15, 399 - 413*

FONTES,J.C., (1985): Some considerations on ground water dating using environmental isotopes. In: *Proceedings hydrogeology in the service of man, Part 1*. Cambridge, IAH, *118 - 154*

FREDRIKSSON, L., et al., (1958): Studies of soil - plant - animal interrelationship with respect to fission products: In*: 2. Int. Conf. on the Peaceful Uses of Atomic Energy, 1. -12. 9. 1958, Geneva, Vol. 58, 449 - 470*

FREUNDLICH, J.C., (1979): Fossil fuel exhaust-gas admixture with the atmosphere. In: *Radiocarbon dating*. Berger, R., Suess, H.E., eds., Berkeley, University of California Press, *388 - 393*

FRITZSCHE, A.F., (1981): Accident at the experimental nuclear power station in Lucens. *Nucl. Saf. 22, 87 - 102*

FRÖHLICH, K., (1992): Isotopenmethoden in der Hydrogeologie - Anwendungen in Entwicklungsländern. *Z. Dtsch. Geol. Gesell. 143, Hannover, 202 - 213*

FRÖHLICH, K., et al., (1987): Silicon-32 in different aquifer types and implications for groundwater dating. In: *Isotope techniques in water resources development*. Vienna, International Atomic Energy Agency, *149 - 163*

FRÖHLICH, K., JORDAN, H., HEBERT, D., (1977a): Radioaktive Isotope in der Hydrogeologie. *Freiberg. Forsch.-H., Reihe C 330, Leipzig, 1 - 92*

FRÖHLICH, K., KATER, R., MILDE, G., (1977b): Measurement of natural silicon-32. In: *Low-Radioactive Measurements and Applications, Proc. Symp., 1975, Bratislava, 371 - 374*

FRYER, B.J., et al., (1993): The application of laser ablation microprobe-inductive coupled plasma mass spectrometry (LAM-ICPMS) to in-situ (U)-Pb geochronology. *Chem. Geol. 109, 1 - 8*

FURCHNER, J.E., DRAKE, G.A., (1971a): Comparative metabolism of radionuclides in mammals - VI. Retention of ^{95}Nb in the mouse, rat, monkey and dog. *Health Phys. 21, 173 - 180*

FURCHNER, J.E., RICHMOND, C.R., DRAKE, G.A., (1971b): Comparative metabolism of radionuclides in mammals - VII. Retention of ^{106}Ru in the mouse, rat, monkey and dog. *Health Phys. 21, 355 - 365*

FURCHNER, J.E., RICHMOND, C.R., LONDON, J.E., (1973): Comparative metabolism of radionuclides in mammals. VIII Retention of beryllium in the mouse, rat, monkey and dog. *Health Phys. 24, 293 - 300*

GABBASOV, M.N., et al., (1995*): Evaluation of radioactive contamination levels at the Semipalatinsk test site*. Bulletin of Centre of Public Information in the Field of Nuclear Energy N5-6 (in Russian)

GABRIEL, T.A., ALSMILLER, R.G., BARISH, J., (1970): Calculation of the long-lived induced activity in the soil around high-energy accelerator target areas. U.S. AEC, *Report ORNL-TM-4599*

GALL, M.S., MAHLER, S., WIRTH, E., (1991): Transfer of ^{137}Cs into mother's milk. *J. Environ. Radioact. 14, 331 - 339*

GARRET, A.R., CUMMINGS, S.L., REGNIER, J.E, (1971): Accumulation of Cs-137 and Sr-85 by Florida's forages in uniform environment. *Health Phys. 21, 67 - 70*

GAVRILIN, Y.I., et al., (1999): Chernobyl accident: reconstruction of thyroid dose for inhabitants of the republic of Belarus. *Health Phys. 76, 105 - 119*

GENTRY, R.W., et al., (1982): Differential lead retention in zircons: implications for nuclear waste containment. *Science 216, 296 - 297*

GERA, F., (1975): Disposal of radioactive wastes in salt domes. *Health Phys. 29, 1 - 7*

GERUSKY, T.M., (1981): Three Mile Islands: assessment of radiation exposures and environmental contamination. In: *The Three Mile Islands nuclear accident: lessons and implications.* Moss, T.H., Sills, D.L., eds., New York, Academy of Sciences, *Ann. N.Y. Acad. Sci. 365, 54 - 62*

GEYER, S., et al., (1993): Isotope investigations on fractions of dissolved organic carbon for ^{14}C groundwater dating. In: *Isotope techniques in the study of past and current environmental changes in the hydrosphere and the atmosphere.* Vienna, International Atomic Energy Agency, *359 - 380*

GEYER, S., et al., (1994): Vergleich der ^{14}C-DOC- und ^{14}C-DIC-Datierungsmethode an Grundwässern aus dem Keupersandstein der Fränkischen Alb. In: *Proceedings, Isotopenkolloquium Freiberg 1994, Freiberg,* TU Bergakademie, *59 - 68*

GEYH, M. A., SCHLEICHER, H.,(1990): *Absolute age determination. Physical and chemical dating methods and their application.* Berlin, Springer Verlag

GEYH, M.A., (1971): *Die Anwendung der ^{14}C-Methode. Clausth. Tekt. H. 11.* Clausthal-Zellerfeld, Verlag Ellen Pilger, *1 - 118*

GEYH, M.A., (1972): Basic studies in hydrology and ^{14}C and ^{3}H measurements. In: *Int. Geol. Congr., 24th Sess., Sect. 11, Hydrogeology, Montreal, 227 - 234*

GEYH, M.A., (1980a): Kohlenstoff-14. In: *Isotopenmethoden in der Hydrologie.* Moser, H., Rauert, W., eds., Berlin, Gebrüder Borntraeger, *43 - 56*

GEYH, M.A., (1980b): Interpretation von Messungen des Tritium- und Kohlenstoff-14-Gehalts im Grundwasser.In: *Isotopenmethoden in der Hydrologie.* Moser, H., Rauert, W., eds., Berlin, Gebrüder Borntraeger, *213 - 224*

GEYH, M.A., (1980c): Interpretation of environmental isotopic groundwater data. In: *Arid-zone hydrology: Investigations with isotope techniques.* Vienna, International Atomic Energy Agency, *31 - 46*

GEYH, M.A., (1992): ^{14}C time scale of ground water - correction and linearity.In: *Isotope techniques in water resources development 1991.* Vienna, International Atomic Energy Agency, *167 - 177*

GEYH, M.A., (2000): An overview of ^{14}C analysis in the study of groundwater. *Radiocarbon 42, 99 - 114*

GEYH, M.A., BRÜHL, H., (1991): Versuche zur ^{14}C-Altersbestimmung von Grundwasser anhand gelöster organischer Stoffe. *Geol. Jb. E 48, Hannover, 385 - 397*

GEYH, M.A., et al., (1983): The unreliability of ^{14}C dates obtained from buried sandy podzols. *Radiocarbon 25, 409 - 416*

GEYH, M.A., et al., (1998): Isotope hydrological study on the origin and age of deep groundwater in the semi-arid Chaco Boreal, South America. In: *Isotope techniques in the study of environmental change.* Vienna, International Atomic Energy Agency, *283 - 291*

GEYH, M.A., MOSER, H., (1980): Weitere radioaktive Umweltisotope für hydrologische Untersuchungen. In: *Isotopenmethoden in der Hydrologie.* Moser,H., Rauert, W., eds., Berlin, Gebrüder Borntraeger, *56 - 61*

GEYH, M.A., MÜLLER, H., MERKT, J., (1970): ^{14}C-Datierung limnischer Sedimente und die Eichung der ^{14}C-Zeitskala. *Naturwiss. 57, 564 - 567*

GIFFORD, F.A., (1968): An outline of theories of diffusion in the lower layers of the atmosphere. In: *Metereology and Atomic Energy, Report TID-24190*. Sagan, D.H., ed., Washington DC, USAEC, *65 - 116*

GILET-BLEIN, N., MARIEN, G., EVIN, J., (1980): Unreliability of ^{14}C dates from organic matter of soils. *Radiocarbon 22, 919 - 929*

GILLESPIE, R., (1989): Fundamentals of bone degradation chemistry: Collagen is not "the way". *Radiocarbon 31, 239 - 246*

GLOSSARY of GEOLOGY, (1980): *Glossary of geology*. Bates, R.L., Jackson, J. A., eds., 2nd Ed., Amer. Geol Inst. Falss Church, Va.

GODWIN, H., (1962): Half-life of radiocarbon. *Nature 195, 984*

GOLDICH, S.S., MUDREY, M.G., (1972): Dilatancy model for discordant U-Pb zircon ages. In: *Contributions to recent geochemistry and analytical chemistry (Vinogradov volume)*, Moscow, Nauka, *415 - 418*

GÖMMEL, R. (1997): Endlagerung radioaktiver Abfälle. Heidelberg, *Spektrum der Wissenschaften, Dossier 1/97: Radioaktivität*

GORIN, V.V., et al., (1993): Chronology of underground nuclear explosions and their primary radiation effects (1961 - 1989). *Bulletin of Centre of Public Information in the Field of Nuclear Energy N9, 21 -33* (in Russian)

GOVE, H., (1989): Progress in radiocarbon dating the shroud of Turin. *Radiocarbon 31, 965 - 969*

GOVE, H.E., (1992): The history of AMS, its advantages over decay counting: Applications and prospects. In: *Radiocarbon after four decades*. Taylor,R.E., Long, A., Kra, R.S., eds., New York, Springer-Verlag, *214 - 229*

GRAN, F.C., (1960): Studies on calcium metabolism and Sr^{90} metabolism in rats. *Acta Physiol. Scand. 48 Suppl. 167*

GREENPEACE, (1996a): *Nuclear weapons testing countries, April 1996.* Washington DC, Greenpeace

GREENPEACE, (1996b): *Selected accidents involving nuclear weapons 1950 - 1993.* Washington DC, Greenpeace

GRINBERG, B., LE GALLIC, (1961): Basic characteristics of a laboratory designed for measuring very low activities. *Int. J. Appl. Radiat. Isotopes 12, 104 - 117*

GROMET, L.P., SILVER, L.T., (1983): Rare earth element distribution among minerals in a granodiorite and their petrogenetic implications. *Geochim. Cosmochim. Acta 47, 925 - 39*

GROOTES, P.M., et al., (1975): Enrichment of radiocarbon for dating samples up to 75,000 years. *Z. Naturforsch. 30A, 1 - 14*

GRUETER, H., (1971): Radioactive fission product ^{137}Cs in mushrooms in W. Germany during 1963 - 1970. *Health Phys. 20, 655 - 656*

GSF, (1986): Umweltradioaktivität und Strahlenexposition in Südbayern durch den Tschernobyl - Unfall. Bericht des Inst. f. Strahlenschutz. D-85764 Neuherberg, GSF - National Research Center for Environment and Health, *GSF-Bericht 16/86*

GSF, (1997): Wissenschaftler helfen Tschernobyl Kindern. Bericht der Phase I und Anhang zum Bericht der Phase I.. Reiners, et al., eds., D-85764 Neuherberg, GSF - National Research Center for Environment and Health, *GSF-Bericht 29/97*

GUDIKSEN, P.H., (1972), The activity of tungsten-181 as a function of time within a nuclear cratering cloud. *Health Phys. 23, 355 - 362*

GUENDOUZ, A., et al., (1998): Palaeoclimatic information contained in groundwaters of the Grand Erg Oriental, North Africa. In: *Isotope techniques in the study of environmental change, Vienna, 555 - 571*

GUPTA, S.K., POLACH, H.A., (1985): *Radiocarbon practices at ANU*. Canberra, ANU

HALLIDAY, A.N., (1984): Coupled Sm-Nd and U-Pb systematics in late Caledonian granites and the basement under northern Britain. *Nature 307, 229 - 33*

HANSHAW, B.B., BACK, W., RUBIN M., (1965): Radiocarbon determinations for estimating groundwater flow velocities in Central Florida. *Science 148, No. 3665, 494 – 495*

HARKNESS, D.D., (1981): Radiocarbon dating. *NERC News J. 2, No.10, Swindon, 10 - 12*

HARKNESS, D.D., BECKER-HEIDMANN, P., eds., (1996): ^{14}C and Soil Dynamics: Special Section. *Radiocarbon 38, 175 - 386*

HARKNESS, D.D., BURLEIGH, R., (1974): Possible carbon-14 enrichment in high-altitude wood. *Archaeometry 16, 121 - 127*

HARLAND, W.B., et al., (1982): *A geologic time scale*. Cambridge Univ. Press.

HARLEY, J.H., (1971): Worlwide plutonium fallout from weapons tests. In: *Proc. of Environmental Plutonium Symp*. Los Alamos Laboratory, *Report LA-4756, 13 - 19*

HARLEY, J.H., (1992): Measurement of ^{222}Rn. A brief history. *Radioat. Prot. Dosim. 54,13 - 18*

HARPER, W.R., (1961): *Basic principles of fission reactors*. New York, Interscience Wiley

HARRISON, J.D., (1982): Gut uptake factors for plutonium, americium and curium. Chilton, Didcot, Oxon ORQ, England, National Radiological Protection Board. *Publication R-129*

HARRISON, J.D., et al., (1989a): Biokinetics of plutonium-239 and americium in the rat after subcutaneous deposition of contaminated particles from the former nuclear weapon test site at Maralinga: Implications for human exposure. Chilton, National Radiological Protection Board, *NRPB-M198*

HARRISON, J.D., NAYLOR, G.P.L., STATHER, J.W., (1989b): Gastrointestinal absorption of plutonium and americium in rats and guinea-pigs after ingestion of dusts from former nuclear weapon test site at Maralinga: Implications for human exposure. Chilton, National Radiological Protection Board, *NRPB-M196*

HASHIMOTO, T., et al.,(1989): Measurement of radionuclides in sea water. In: *Proc. 15. Regional Congress of the International Radiation Protection Association (IRPA) on the Radioecology of Natural and Artificial Radionuclides*. Köln, Germany, Verlag TUEV Reinland, *480 - 483*

HASL-329, (1977): Final tabulation of monthly ^{90}Sr fallout data: 1954 - 1976. Health and Safety Laboratory, *HASL-329*

HAWKESWORTH, C.J., et al., (1991): Element fluxes associated with subduction related magmatism. In: *The behaviour and influence of fluids in subduction zones*. J. Tarney, ed., London, The Royal Society, *167 - 79*

HAYNES, C.V., (1967): Bone organic matter and radiocarbon dating.In: *Radioactive dating and methods of low-level counting*. Vienna, International Atomic Energy Agency, *163*

HAYWOOD, S.M., SMITH, J.G., (1990): Assessment of the radiological impact of residual radioactive contamination in the Maralinga and Emu areas. Chilton, National Radiological Protection Board, *NRPB-R237*

HAYWOOD, S.M., SMITH, J.G., (1992): Assessment of potential doses at the Maralinga and EMU test sites. *Health Phys. 63, 624 - 630*

HEALY, J.W., (ed.), (1975): Plutonium - health implications for man. In: Proc. of 2nd Los Alamos life sciences symposium, Los Alamos, New Mexiko 22 - 24 May 1974. *Health Phys. 29, 441 - 640*

HEAMAN, L.M., TARNEY, J., (1989): U-Pb baddeleyite ages for the Scourie dyke swarm, Scotland: evidence for two distinct intrusion events. *Nature 340, 705 - 708*

HEBERT, D., (1996): Methodische Aspekte der Grundwasserdatierung mit kosmogenen Radionukliden. In: *Proceedings of the Freiberger Isotopenkolloquium 1996, Freiberg.* T.U. Bergakademie, *111 - 119*

HEBERT, D., FRÖHLICH, K., (1983): Methodische Untersuchungen zur ^{14}C-Datierung an Kalksinter. *Z. Angew. Geol. 29, 123 - 128*

HEDGES, R.E.M., LEE-THORP, J.A., TUROSS, N.C., (1995): Is tooth enamel carbonate a suitable material for radiocarbon dating? *Radiocarbon 37, 285 - 290*

HEDVALL, R., PETTERSSON, H., ERLANDSON, M., (1987): Gamma-spectrometric determination of uranium isotopes in biofuel ash. *J. Radioanal. Nucl. Chem., Articles 115, 216 - 221*

HEEP, C.M., et al., (1996): Reconstruction of radionuclide releases from the Hanford site 1944 -1972. *Health Phys. 71, 545 - 555*

HEID, K.R., ROBINSON, B., (1985): The U.S. transuranium registry report on the [241]Am content of a whole body, Part II: Estimate of the inital systemic burden. *Health Phys. 49, 569 - 575*

HEIER-NIELSEN, S., et al., (1995): Radiocarbon dating of shells and foraminifera from the Skagen core, Denmark: Evidence of reworking. *Radiocarbon 37, 119 - 130*

HEINEMANN, K., (1991): Meßprogramm der Bundesrepublik Deutschland. Ergebnisse der Umweltmessungen in Rußland in der Zeit vom 21 Mai bis 11 Juni 1991. Jülich, Forschungszentrum Jülich, *Report JÜL-2531*

HEINEMANN, K., HILLE, R.,(1993): Meßprogramm der Bundesrepublik Deutschland. Ergebnisse der Umweltmessungen in Rußland, Weißrußland und der Ukraine in der Zeit vom 12. Mai bis 26. September 1992. Jülich, Forschungszentrum Jülich, *Report JÜL-2760*

HEINEMANN, K., HILLE, R.,(1994): Meßprogramm der Bundesrepublik Deutschland. Ergebnisse der Umweltmessungen in Rußland, Weißrußland und der Ukraine in der Zeit vom 17. Mai bis 2. September 1993 und vom 8. Oktober bis 1. November 1993. Jülich, Forschungszentrum Jülich, *Report JÜL-2925*

HELD, J., SCHUHBECK, S., RAUERT, W., (1992): A simplified method of [85]Kr measurement for dating young groundwaters. *Appl. Radiat. Isot. 43, 939 - 942*

HEMPELMANN, L.H., et al., (1973): Manhatten project plutonium workers:a twenty-seven year follow - up study of selected cases, *Health Phys. 25, 461 - 479*

HENDERSON, P.,(1982) : *Inorganic geochemistry.* Oxford, Pergamon Press.

HENRICHS, K., BERG, D., BOGNER, L., (1992): Ganzkörpermessungen nach Tschernobyl. In: *Die Folgen von Tschernobyl für Deutschland und für die ehemalige Sowjetunion, Journalisten Seminar 19.03.-19.03.1991.* Haury, H.J., ed., D-85764 Neuherberg, GSF - National Research Center for Environment and Health, *GSF-Bericht 17/92, 27 - 34*

HENRICHS, K., et al., (1989): Measurements of Cs absorption and retention in man. *Health Phys. 57, 571 - 578*

HENRY, D. O., (1992): The impact of radiocarbon dating on Near Eastern prehistory. In: *Radiocarbon after four decades*. Taylor,R.E., Long,A., Kra, R.S., eds., New York, Springer-Verlag, *313 – 323, 324 – 334*

HERR, W., et al., (1967): Development and recent application of the Re/Os dating method. In: *Radioactive dating and methods of low-level counting*. Vienna, International Atomic Energy Agency, *499 - 508*

HERRMANN, A.G., (1983): *Radioaktive Abfälle. Probleme und Verantwortung.* Berlin-Heidelberg, Springer Verlag

HESS, C.T., et al., (1985): The occurrence of radioacitivity in public water supplies in the United States. *Health Phys. 48, 553 - 586*

HILL, P., et al., (1995): Radiological assessment of long - term effects at the Semipalatinsk test site, NATO-Semipalatinsk Project 1996/96. Jülich, Forschungszentrum Jülich, *Report JÜL-3325*

HILL, P., HILLE, R., (1992): Meßprogramm der Bundesrepublik Deutschland. Ergebnisse der Ganzkörpermessungen in Rußland in der Zeit vom 17. Juni bis 4. Oktober 1991. J ülich, Forschungszentrums Jülich, *Report JÜL-2610*

HILLE, R., et al., (1996): The impact of the Chernobyl accident - an evaluation from the German perspective. Jülich, Forschungszentrum Jülich, *Report JÜL-3186*

HILLER, A., FUHRMANN, R., (1991): Radiocarbondatierungen an koexistenten Kohlenstoffträgern aus Binnenwasservorkommen Sachsens und Thüringens. *Z. Geol. Wiss. 19, 569 - 584*

HIRATA, T., NESBITT, R.W., (1995): U-Pb isotope geochronology of zircon: evaluation of laser probe-inductively coupled plasma mass spectrometry technique. *Geochim. Cosmochim. Acta 59, 2491 - 2500*

HODGE,V.F., FOLSOM, T.R., YOUNG, D.R., (1973): Retention of fallout constituents in upper layers of the Pacific ocean as estimated from studies of a tuna population. Radioactive contamination of marine environment. Vienna, International Atomic Energy Agency, *STI/PUB 313, 435 - 445*

HOHL, R., ed., (1981): *Die Entwicklungsgeschichte der Erde.* Leipzig, F.A. Brockhaus Verlag

HOLLEMAN, D.F., LUICK, J.R., WHICKER, F.W., (1971): Transfer of radiocesium from lichen to reindeer. *Health Phys. 21, 657 - 666*

HOLMES A., (1946): An estimate of the age of the Earth. *Nature 157, 680 - 684*

HOLMES, A., (1954) : The oldest dated minerals of the Rhodesian Shield. *Nature 173, 612 – 617*

HOOKER, P.J., O'NIONS, R.K., PANKHURST, R.J., (1975): Determination of rare earth elements in USGS standard rocks by mixed solvent ion exchange and mass spectrometric isotope dilution. *Chem. Geol. 16. 189-196*

HORAN, J.R. GAMMILL, W.P., (1963): The health aspects of the SL-1 accident. *Health Phys. 9, 177 - 186*

HORN, I., RUDNICK, R.L., MCDONOUGH, W.F. (2000): Precise elemental and isotope ratio determination by simultaneous solution nebulization and laser ablation – ICP-MS: application to U-Pb geochronology. *Chem. Geol. 164, 3 - 4, 283*

HORVATINČIĆ, N., et al., (1989): Comparison of the ^{14}C activity of groundwater and recent tufa from karst areas in Yugoslavia and Czechoslovakia. *Radiocarbon 31, 884 – 892*

HÖTZL, H., WINKLER, R., (1984): Experiences with large area Frisch gird chambers in low-level alpha spectrometry. *Nucl. Instrum. Methods Phys. Res. A 223, 290 - 294*

HOUK, R.S., et al., (1980): Inductively coupled argon plasma as an ion source for mass spectrometric determination of trace elements. *Anal. Chem. 52, 2283 - 2289*

HOUTERMANS, F.G., (1946): The isotope ratios of natural lead and age of uranium. *Naturwissenschaft 33, 185 - 186*

HOUTERMANS, F.G., OESCHGER, H., (1958): Proportional counter for measurement of lower activities of weak beta emitters, (in German). *Helv. Phys. Acta 31, 117 - 126*

HOYER, F.W., (1968): Induced radioactivity in the earth shielding on top of high energy particle accelerators. European Organization for Nuclear Research, Geneva, *Report CERN 68-42*

HSK, (1986): Der Unfall Chernobyl. Ein Überblick über die Ursachen und Auswirkungen. Hauptabteilung für die Sicherheit der Kernanlagen, Bundesamt für Energiewirtschaft, Switzerland, *HSK-AN-1816*

HUBERT, P., et al., (1986): Alpha-rays induced background in ultra low-level counting with Ge spectrometers. *Nucl. Instrum. Methods Phys. Res. A 252, 87 - 90*

HULSE, S.E., et al., (1999): Comparison of ^{241}Am, ^{239,240}Pu and ^{137}Cs concentrations in soil around Rocky Flats. *Health Phys. 76, 275 - 287*

HUNT, G.J., (1980): Radioactivity in surface and coastal waters of the British isles, 1978. Ministry of Agriculture, Fisheries and Food, *Aquatic Environment Report Nr. 4*

IAEA, (1963): *Radioisotopes in hydrology.* Vienna, International Atomic Energy Agency

IAEA, (1967): *Isotopes in hydrology.* Vienna, International Atomic Energy Agency

IAEA, (1970): *Isotope hydrology 1970.* Vienna, International Atomic Energy Agency

IAEA, (1974): *Isotope techniques in groundwater hydrology 1974, Vol. I, Vol. II.* Vienna, International Atomic Energy Agency

IAEA, (1975): The Oklo phenomenon. Vienna, International Atomic Energy Agency, *Report STI/PUB/405*

IAEA, (1978): *Isotope hydrology 1978, Vol. I, Vol. II.* Vienna, International Atomic Energy Agency

IAEA, (1980a): Arid-zone hydrology: Investigations with Isotope techniques. Panel Proceedings Series. Vienna, International Atomic Energy Agency, *Report STI/PUB/547*

IAEA, (1980b): Underground disposal of radioactive wastes. Proceedings Series Proceedings of a symposium jointly organized with OECD/NEA, Otaniemi, Finland, 2-6 July 1979. Vienna, International Atomic Energy Agency, *Report STI/PUB/528*

IAEA, (1981a): Shallow ground disposal of radioactive wastes: A guidebook. Vienna, International Atomic Energy Agency, *Safety Series No. 53, STI/PUB/578*

IAEA, (1981b): Underground disposal of radioactive wastes: Basic Guidance. Vienna, International Atomic Energy Agency, *Safety Series No. 54., Report STI/PUB/579*

IAEA, (1982a): Site investigations for repositories for solid radioactive wastes in shallow ground. Vienna, International Atomic Energy Agency, *Technical Reports Series No. 216, STI/DOC/10/216*

IAEA, (1982b): Site investigations for repositories for solid radioactive wastes in deep continental geological formations. Vienna, International Atomic Energy Agency, *Technical Reports Series No 215, STI/DOC/10/215*

IAEA, (1983a): Concepts and examples of safety analyses for radioactive waste repositories in continental geological formations. Vienna, International Atomic Energy Agency, *Safety Series No. 58, STI/PUB/632*

IAEA, (1983b): Control of radioactive waste disposal into the marine environment. Vienna, International Atomic Energy Agency, *Safety Series No. 61, STI/PUB/609*

IAEA, (1983c): Criteria for underground disposal of Solid Radioactive Wastes. Vienna, International Atomic Energy Agency, *Safety Series No. 60, STI/PUB/612*

IAEA, (1983d): Disposal of low and intermediate level solid radioactive wastes in rock cavities. Vienna, International Atomic Energy Agency, *Safety Series No. 59, STI/PUB/610*

IAEA, (1983e): Disposal of radioactive grouts into hydraulically fractured shale. Vienna, International Atomic Energy Agency, *Technical Reports Series 232, STI/DOC/10/232*

IAEA, (1983f): Isotope techniques in the hydrogeological assessment of potential sites for the disposal of high-level wastes. Vienna, International Atomic Energy Agency, *Technical Report Series 228*

IAEA, (1984a): Design, construction, operation, shutdown and surveillance of repositories for solid radioactive wastes in shallow ground. Vienna, International Atomic Energy Agency, *Safety Series No. 63, STI/PUB/652*

IAEA, (1984b): Site investigations, design, construction, operation, shutdown and surveillance of repositories for low and intermediate level radioactive wastes in rock cavities. Vienna, International Atomic Energy Agency, *Safety Series No. 62, STI/PUB/659*

IAEA, (1984c): *Isotope hydrology 1983*. Vienna, International Atomic Energy Agency

IAEA, (1985a): Acceptance criteria for disposal of radioactive wastes in shallow ground and rock cavities. Vienna, International Atomic Energy Agency, *Safety Series No. 71, STI/PUB/710*

IAEA, (1985b): Deep underground disposal of radioactive wastes: Near-field effects. Vienna, International Atomic Energy Agency, *Technical Reports Series No. 251, STI/DOC/10/251*

IAEA, (1985c): Operational experience in shallow ground disposal of radioactive wastes. Vienna, International Atomic Energy Agency, *Technical Reports Series No. 253, STI/DOC/10/253*

IAEA, (1986a): An oceanographic model for the dispersion of wastes disposed of in the deep sea. Vienna, International Atomic Energy Agency, *Technical Reports Series No. 263, STI/DOC/10/263*

IAEA, (1986b):Siting, design and construction of underground repositories for radioactive wastes. In: *Proceedings Series Proceedings of a symposium, Hannover, Germany, 3-7 March 1986.* Vienna, International Atomic Energy Agency, *Report STI/PUB/715*

IAEA, (1986c): Mathematical models for interpretation of tracer data in groundwater hydrology. Vienna, International Atomic Energy Agency, *IAEA-TECDOC-381*

IAEA, (1987): *Isotope techniques in water resources development*. Vienna, International Atomic Energy Agency

IAEA, (1988a): Assessing the impact of deep sea disposal of low level radioactive waste on living marine resources. Vienna, International Atomic Energy Agency, *Technical Reports Series No. 288, STI/DOC/10/288*

IAEA, (1988b): The radiological accident in Goiânia. International Atomic Energy Agency, Vienna, *Report STI/PUB/815*

IAEA, (1989a): Guidance for regulation of underground repositories for disposal of radioactive wastes. Vienna, International Atomic Energy Agency, *Safety Series No. 96, STI/PUB/774*

IAEA, (1989b): Safety principles and technical criteria for the underground disposal of high level radioactive wastes. Vienna, International Atomic Energy Agency, *Safety Series No. 99, STI/PUB/854*

IAEA, (1990): Sealing of underground repositories for radioactive wastes. Vienna, International Atomic Energy Agency, *Technical Reports Series No. 319, STI/DOC/10/319*

IAEA, (1991): *The international Chernobyl project surface contamination maps.* Vienna, International Atomic Energy Agency

IAEA, (1992a): Design and operation of high level waste vitrification and storage facilities. Vienna, International Atomic Energy Agency, *Technical Reports Series No. 339, STI/DOC/10/339*

IAEA, (1992b): Design and operation of radioactive waste incineration facilities. Vienna, International Atomic Energy Agency, *Safety Series No. 108, STI/PUB/921*

IAEA, (1992c): *Isotope techniques in water resources development 1991.* Vienna, International Atomic Energy Agency

IAEA, (1993a): Containers for packaging of solid and intermediate level radioactive wastes. Vienna, International Atomic Energy Agency, *Technical Reports Series No. 355, STI/DOC/10/355*

IAEA, (1993b): Geological disposal of spent fuel and high level and alpha bearing wastes. In: *Proceedings Series Proceedings of a symposium organized jointly with CEC and OECD/NEA, Antwerp, Belgium, 19-23 October 1992.* Vienna, International Atomic Energy Agency, *Report STI/PUB/907*

IAEA, (1993c): *Isotope techniques in the study of past and current environmental changes in the hydrosphere and the atmoshere.* Vienna, International Atomic Energy Agency

IAEA, (1994): Mathematical models and their applications to isotope studies in groundwater hydrology. Vienna, International Atomic Energy Agency, *IAEA-TECDOC-777*

IAEA, (1995): The principles of radioactive waste management. Vienna, International Atomic Energy Agency Safety, *Series No 111-F.*

IAEA, (1996a): One decade after Chernobyl.- Summing up the Consequences of the Accident. In: *Proceedings Series. Proc. Internat. Conference 8-12 April 1996 Vienna.* Vienna, International Atomic Energy Agency, *Report STI/PUB/1001*

IAEA, (1996b): Emergency planning and preparedness for the re-entry of a nuclear powered satellite. Vienna, International Atomic Energy Agency, *Safety Series No. 119, STI/PUB/1014*

IAEA, (1996c): Regulations for the safe transport of radioactive material. Safety Standards Series No. ST-1/Requirements. Vienna, International Atomic Energy Agency, *Report STI/PUB/998*

IAEA, (1996d): *Isotopes in water resources management, Vol. I, Vol. II.* Vienna, International Atomic Energy Agency

IAEA, (1997a): Characterization of radioactive waste forms and packages. Vienna, International Atomic Energy Agency, *Technical Reports Series No. 383, STI/DOC/010/383*

IAEA, (1997b): Planning and operation of low level waste disposal facilities. In: *Proceedings Series Proceedings of a symposium on Experience in the Planning and Operation of Low Level Waste Disposal Facilities, Vienna, 17-21 June 1996.* Vienna, International Atomic Energy Agency, *Report STI/PUB/1002*

IAEA, (1998a): Radiological conditions at Bikini-Atoll: Prospects for resettlement. Vienna, International Atomic Energy Agency, *Radiological Assessment Series, STI/PUB/1054*

IAEA, (1998b): Radiological conditions at the Semipalatinsk test site, Kazakhstan: Preliminary assessment and recommendations for further study. Vienna, International Atomic Energy Agency, *Radiological Assessment Series, STI/PUB/1063*

IAEA, (1998c): Radiological conditions at the atolls of Mururoa and Fangataufa: main report. Reports by an international advisory committee. Vienna, International Atomic Energy Agency, *Radiological Assessment Series, STI/PUB/1028*

IAEA, (1998d): Radiological conditions at the atolls of Mururoa and Fangataufa: summary report. Reports by an international advisory committee. Vienna, International Atomic Energy Agency, *Radiological Assessment Series, STI/PUB/1029*

IAEA, (1998e):The radiological accident in the reprocessing plant at Tomsk. Vienna, International Atomic Energy Agency, *Report STI/PUB/1060*

IAEA, (1998f): Interim storage of radioactive waste packages. Vienna, International Atomic Energy Agency, *Technical Reports Series No. 390, STI/DOC/010/390*

IAEA, (1998g):Radiological conditions of the Western Kara Sea. Vienna, International Atomic Energy Agency, *Radiological Assessment Reports Series, STI/PUB/1068*

IAEA, (1998h): The radiological accident in Tammiku. International Atomic Energy Agency, Vienna, *Report STI/PUB/1053*

IAEA, (1998i): *Isotope techniques in the study of environmental change.* Vienna, International Atomic Energy Agency

IAEA, (1999): Report on the preliminary fact finding mission following the accident at the nuclear fuel processig facility in Tokaimura, Japan. International Atomic Energy Agency, Vienna, *http://www.iaea.org/worldatom/Press/P_release/1999/jap_report.shtm*

IAEA, (2000a): The radiological accident in Istanbul. International Atomic Energy Agency, Vienna, *Report STI/PUB/1102*

IAEA, (2000b): The radiological accident in Lilo. Vienna, International Atomic Energy Agency, *Report STI/PUB/1097*

IAEA, (2000c): The radiological accident in Yanango. Vienna, International Atomic Energy Agency, *Report STI/PUB/1101*

IAEA, (2000d): Calibration of radiation protection monitoring instruments. Vienna, International Atomic Energy Agency, *Safety Rep.Ser. No.16*

IAEA, (2001): Upgrading the safety ans security of radioactive sources in the Republic of Georgia. International Atomic Energy Agency, http://www.iaea.org/worldatom/Press/News/gorgia_radsources.shtml

ICRP 10, (1968): *Evaluation of radiation doses to body tissues from internal contamination due to occupational exposure.* International Commission on Radiological Protection, Oxford, Pergamon Press

ICRP 20, (1973): Alkaline earth metabolism in adult man. *Health Phys. 24, 125 - 331*

ICRP 23, (1975): *Report of the task group on reference man.* International Commission on Radiological Protection, Oxford, Pergamon Press

ICRP 30, (1979): *Limits of intakes of radionuclides by workers.* International Commission on Radiological Protection, Oxford, Pergamon Press

ICRP 56, (1989): *Age-dependent doses to members of the public from intake of radionuclides.* International Commission on Radiological Protection, Oxford, Pergamon Press

ICRP 60, (1991): *1990 Recommendations of the international commission on radiological protection.* International Commission on Radiological Protection,Oxford, Pergamon Press

ICRP 66, (1994): *Human respiratory tract model for radiological protection.* International Commission on Radiological Protection,Oxford, Pergamon Press

ICRP 67, (1993): *Age-dependent doses to members of the public from intake of radionuclides, Part 2: ingestion dose coeffients*. International Commission on Radiological Protection, Oxford, Pergamon Press

ICRP 69, (1995): *Age-dependent doses to members of the public from intake of radionuclides, Part 3*. International Commission on Radiological Protection, Oxford, Pergamon Press

ICRP 81 (2000): *Radiation protection recommendation as applied in the disposal of long-lived solid radioactive waste*. Pergamon, London.

IGAKI, K., et al., (1994): Radiocarbon dating study of ancient iron artifacts with the accelerator mass spectrometry. *Proceedings of the Japan Academy, Series B, 70 (1), 4 - 6*

IGGY LITAOR, M., (1999): Plutonium contamination in soils in open space and residential areas near Rocky Flats, Colorado. *Health Phys. 76, 171 - 179*

INGERSON, E., PEARSON, F.J.,Jr. (1964): Estimation of age and rate of motion of groundwater by the ^{14}C method. In: *Recent research in the fields of hydrosphere, atmosphere, and nuclear geochemistry, Tokyo, 263 - 283*

INN, K.G., MULLEN P.A., HUTCHINSON, J.M.R., (1984): Radioactivity standards for environmental monitoring II. *Environ. Int. 10, 91 - 97*

IPSN, (1999): Fiche relative à l' accident survenue le 30 septembre 1999 à Tokaimura (Japon) de 14.10.1999. Fontenay-aux-Roses, Institute de Protection et de Sureté Nucléaire, France, *IPSN/99 – 2729*

IRVING, W.N., HARRINGTON, C.R. (1973): Upper-Pleistocene radiocarbon-dated artifacts from the Northern Yukon. *Science 179, 335 - 340*

ISRAEL, H.; (1962): Die natürliche und künstliche Radioaktivität in der Atmosphäre. In: *Kernstrahlung in der Geophysik*. Israel, H., Krebs, A., eds, Springer Verlag

ISRAELI, M., (1985): Deposition rates of Rn progeny in houses. *Health Phys. 49, 1069 – 1083*

IZQUIERDO, L., SEYFFER, U., (1993): Results of the work - related personal dosimetry and the supervision of emission in a convoy plant. In: *Environmental impact of nuclear installations, Proc. of the Joint Seminary Sept. 15th - 18th, 1992 in Fribourg*. Völkle, H., et al., eds., *11 - 20*

IZRAEL, Yu.A., STUKIN, E.D., TSATUROV, Yu.S., (1994): On the possibility for identification of radioactive pattern from nuclear explosions and for reconstruction of population exposure doses using long - lived radionuclides analysis. *Metereology and Hydrology, N12* (in Russian)

JACOB, H., et al., (1998): Benefits and limits of isotope investigations in groundwater protection and remediation. In: *Isotope techniques in the study of environmental change*. Vienna, International Atomic Energy Agency, *67 - 68*

JACOBI, W., (1981): Umweltradioaktivität und Strahlenexposition durch radioaktive Emissionen von Kohlekraftwerken. D-85764 Neuherberg, GSF - National Research Center for Environment and Health, *GSF-Bericht S-760*

JACOBSEN, S.B., WASSERBURG, G.J., (1980): Sm - Nd evolution of chondrites. *Earth Planet. Sci. Lett. 50. 139 – 155*

JAFFEY, A.H., et al., (1971): Precision measurement of half lifes and specific activities of ^{235}U and ^{238}U. *Phys. Rev. C4*, 1889 - 1906.

JÄGER, E., HUNZIKER, J.C., eds., (1979): *Lectures in isotope geology*. Berlin, Springer

JAKEMAN, D., (1986): Notes on the level of radioactive contamination in the Sellafield area arising from discharges in the early 1950s. Dorchester Dorset, Atomic Energy Establishment Winfrith, *AEEW-R 2104*

JAMMET, H., DOUSSET, M., (1982): Nuclear toxicants: origin, presence or release in the environment, contribution to human exposure. In: *Radionuclide metabolism and toxicity.* Galle, P., Masse, R., eds., Paris, New York, Masson, *1 - 32*

JANOSSY, L., (1950): *Cosmic Rays. 2nd ed.* Oxford, Clarendon Press

JANOUŠEK, V., et al., (2000) :Modelling diverse processes in the petrogenesis composite batholith : the Central Bohemian pluton, Central European Hercynides. *Journ. of Petrol. 11, 511 – 543*

JANOUŠEK, V., ROGERS, G., BOWES, D.R., (1995): Sr-Nd isotopic constraints on the petrogenesis of the Central Bohemian Pluton, Czech Republic. *Geol. Rdsch. 84. XX - XX*

JEANMAIRE, L., (1982a): Metabolism and toxicity of tritium. In: *Radionuclide metabolism and toxicity.* Galle, P., Masse, R., eds., Paris, New York, Masson, *156 - 161*

JEANMAIRE, L., (1982b): Metabolism and toxicity of carbon 14. In: *Radionuclide metabolism and toxicity.* Galle, P., Masse, R., eds.,Paris, New York, Masson, *162 - 165*

JEE, W.S.S., (1976): *The health effects of plutonium and Radium.* Salt Lake City, J.W. Press

JELÍNEK, E., et al., (1984): Geochemistry of peridotites, gabbros and trondhjemites of the Ballantrae complex, SW Scotland. *Trans. Roy. Soc. Edinburgh: Earth Sciences 75, 193 - 209*

JÍLEK, P., et al., (1987): Dating of limnic Holocene carbonate sediments from Valča, Czechoslovakia. In: *Isotopes in Nature 1986.* Leipzig, Academy of Sciences of the GDR, Central Institute of Isotope and Radiation Research, *389 - 400*

JÍLEK, P., et al., (1995): Radiocarbon dating of Holocene sediments: Flood events and evolution of Labe (Elbe) River in central Bohemia (Czech Republic). *Radiocarbon 37, 131 - 137*

JISL, R., TYKVA, R., (1986):Correction of the continual scanning record of radioactivity distribution. II. Deconvolution. *Nucl. Instrum. Methods Phys. Res. A 251, 166 - 171*

JOHANSSON, L., (1983): Oral intake of radionuclides in the population. In : *Dosimetry, radionuclides, and technicology, Paper E5-05, Proc. of the 7th Intern. Congr. of Radiation Research.* Broerse, J.J., ed., Amsterdam, Martinus Nijhoff Publisher

JOHNSON, F., (1955): Reflections upon the significance of radiocarbon dates. In: *Radiocarbon Dating, (1952), 2nd ed. 1955, Fifth Impression 1965.* Libby, W.F., Chicago, University of Chicago Press, *141 - 161*

JOHNSTON, P.N., LOKAN, H.K., WILLIAMS, G. A., (1992): Inhalation doses for aboriginal people reoccupying former nuclear weapons testing ranges in South Australia. *Health Phys. 63, 631 - 640*

KAMIKUBOTA, N., et al., (1989): Low-level radioactive isotopes contained in materials used for beta-ray and gamma-ray detectors. In: *Radiation detectors and their uses.* Miyajima, M., Sasaki ,S., Doke, T., eds., Ibaraki (Japan);Tsukuba, *19 - 23*

KAUFMAN, A.J., JACOBSEN S.B., KNOLL A.H., (1993) : The Vendian record of Sr and C isotopic variations in seawater : implications for tectonic and paleoclimate. *Earth Planet. Sci. Lett. 120, 409 - 430*

KAUFMAN, S., LIBBY,W.F, (1954): The natural distribution of tritium. *Phys. Rev. 93, 1337 - 1344*

KEISCH, B., KOCH, R.C., LEVINE, A.S., (1965): *Modern trends in acitivation analysis: determination of biospheric levels of ^{129}I by neutron - activation analysis.* TX. Texas A and M University 284, College Station

KELLER, C., (1990): Cäsium - Fixierung in Pilzen. *GIT Fachz. Lab. 7/90, 888 - 889*

KELLER, G., (1993): Radiological aspects of former mining activities in the Saxon Erzgebirge, Germany. *Environ. Int. 19, 449 - 454*

KELLEY, S.P. , WARTHO, J.A. (2000): Rapid kimberlite ascent and he significance of Ar-Ar ages in xenoliths phlogopites. *Science 289, 609 – 611*

KEMENY, J.G., (1979): *President's commissions on the accident at Three Mile Islands.* Washington DC, Report Force on Public Health and Safety USGPO

KEMMER, J., (1968): Ge (Li) gamma spectrometer of low level gamma activities. *Nucl. Instrum. Methods 64, 268*

KENT, J. T., WATSON,G. S., ONSTOTT, T. C., (1990): Fitting straight lines and planes with an application to radiometric dating. *Earth Planet. Sci. Lett. 97, 1 – 17*

KIGOSHI, K., SUZUKI, N., SHIRAKI, M., (1980): Soil dating by fractional extraction of humic acid. *Radiocarbon 22, 853 - 857*

KING, L.J., Mc CHARLEY, W.T., (1961): Plutonium release accident of November 20. Oak Ridge, Tennessee, Oak Ridge National Lab., *Report ORNL-2989*

KINNE, O. (Ed.), (1970): *Marine ecology, Vol.1: Environmental factors.* New York, Interscience Wiley

KINNES, I., THORPE, I.J., (1986): Radiocarbon dating: use and abuse. *Antiquity, UK 60 (230), 221 -223*

KIRCHNER, T.B., et al., (1996): Estimating internal dose due to ingestion of radionuclides from Nevada test site fallout. *Health Phys. 71, 487 - 501*

KISELEV, V.I., LOBOREV, V.M., SHOIKHET, J.N., (1994): Problems of the Semipalatinsk test site impact upon Altai region population. *Вестник научнои программы „Семипалатинский полигон - Алтай" 1, 5 - 9* (engl. Summary)

KINZELMANN, T., (2003): 17 Jahre nach Tschenobyl: Wahrheit und Mythos. Gesundheitliche Folgen des Tschernobylunfalls. *Strahlenschutz Praxis 9,2 , 49 - 51*

KLEIN, J., et al., (1980): Radiocarbon concentration in the atmosphere: 8000-year record in tree rings: first results of a USA workshop. *Radiocarbon 22, 950 – 961*

KLENER, V., et al., (1996): Contamination of the territory of the Czechoslovakia; Internal and external exposure; Evaluation of the situation and response of the autorities. In: *Zehn Jahre nach Tschernobyl, eine Bilanz. Seminar des Bundesamtes für Strahlenschutz und der Strahlenschutzkommission, München 6. - 7. März (1996).* Bayer, A., Kaul, A., Reiners, Chr., eds., Stuttgart, Gustav Fischer

KLINE, J.R., COLON, J.A., BRAR, S.S., (1973): Distribution of ^{137}Cs in soils and vegetation on the islands of Puerto Rico. *Health Phys. 24, 469 - 475*

KLOTZ, D., MOSER, H., (1980): Messung der Ausbreitung von Tracern. In: *Isotopenmethoden in der Hydrologie.* Moser,H., Rauert,W., Berlin, Gebrüder Borntraeger, *252- 262*

KLOUDA, G.A., et al., (1986): Urban atmospheric ^{14}CO and ^{14}CH$_4$ measurements by accelerator mass spectrometry. *Radiocarbon 28, 2A, 625 - 633*

KNOLL, G.F., (1979): *Radiation detection and measurement.* New York, John Wiley and Sons

KOBER, B., (1986): Whole-grain evaporation for ^{207}Pb/^{206}Pb-age-investigations on single zircons using a double-filament thermal source. *Contrib. Mineral. Petrol. 93, 482 - 490*

KOBER, B., (1987): Single-zircon evaporation combined with Pb$^+$ emitter bedding for ^{207}Pb/^{206}Pb-age investigations using thermal ion mass spectrometry, and implications for zirconology. *Contrib. Mineral. Petrol. 96, 63 - 71*

KOCH, J., TADMOR, J., (1986): RADFOOD - a dynamic model for radioactivity transfer through the human food chain. *Health Phys. 50, 721 - 731*

KOCH, L., (1995): Nuklearer Fingerabdruck von Kernbrennstoffen. - Methoden zur Identifizierung von Herstellern und Anwendungen. *Strahlenschutz Praxis 1, 31 - 33*

KOCHER, D.C., RYAN, M.T., (1983): Animal data on GI - tract uptake of plutonium - implications for environmental dose assessments. *Radiat. Prot. Dosim. Vol. 5 No 1, 37 - 43*

KOCZY, F.F., (1960):The distribution of elements in the sea. In: *Disposal Radioact. Wastes, Proc. Sci. Conf. Monaco 1959*

KOLB, W, (1974): Radionuclide concentrations in ground level air from 1971 to 1973 in Brunswick and Tromsö. *PTB - Ra - 4*

KÖNIG, (1994): In-vivo Messungen von Strontium-90 und Dosisabschätzung. In: *Erste deutsche Aktivitäten zur Validierung der radiologischen Lage im Südural, BfS-ISH-166/94*, Burkhard, W., ed., Neuherberg, Bundesamt für Strahlenschutz, Inst. für Strahlenhygiene, *25 - 40*

KONSHIN, O.V., (1992a): Mathematical model of ^{137}Cs migration in soil: analysis of observations following the Chernobyl accident. *Health Phys. 63, 301 - 306*

KONSHIN, O.V., (1992b): Transfer of ^{137}Cs from soil to grass - analysis of possible sources of uncertainty. *Health Phys. 63, 307 - 315*

KONSHIN, O.V., (1992c): Applicability of the convection - diffusion mechanism for modeling migration of ^{137}Cs and ^{90}Sr in the soil. *Health Phys. 63, 291 - 300*

KOŠLER, J. et al., (1995): Temporal association of ductile deformation and granitic plutonism: Rb - Sr and 40Ar - 39Ar isotopic evidence from the roof pendants above the Central Bohemian Pluton, Czech Republic. *J. Geol., 103, 57 - 64*

KOŠLER, J. et al., (2001): Application of Laser Ablation ICPMS to U-Th-Pb Dating of Monazite. *Geost. Newslet., 25, 375 – 386*

KOŠLER, J. et al., (2002): U-Pb dating of detrital zircons for sediment provenance studies – a comparison of laser ablation ICPMS and SIMS techniques. *Chem. Geol., 182, 605 - 618*

KOSSENKO, M.M., (1996): Cancer mortality among Techa river residents and their offspring. *Health Phys. 71, 77 - 82*

KOSSENKO, M.M., DEGTEVA, M.O., (1994): Cancer mortality and radiation risk evaluation for Techa river population. *Sci. Total. Environ. 142, 73 - 90*

KOSTYUCHENKO, V.A., KRESTINA, L.Y. (1994): First results from the follow - up of persons exposed to the Kyshtym fallout. Long - term irradiation effects in the population evacuated from the East - Urals radioactive trace area. *Sci. Total. Environ. 142, 119 - 124*

KOTRAPPA, P., JESTER, W.A., (1993): Electret ion chamber radon monitors measure dissolved ^{222}Rn in water. *Health Phys. 64, 397 - 405*

KOVALCHUK, E.L., et al., (1982): U235, Th222, K40 along the main 3 km adit of the Baksan neutrino observatory. In: *Natural Radiation Environment*. Vohra W.G., et al., eds., New York, John Wiley and Sons, *201 - 205*

KOZHEUROV, V.P. (1994a): SICH - 91 - a unique whole - body counting system for measuring Sr-90 via bremsstrahlung:The main results from a long - term investigation of the Techa river population. *Sci. Total. Environ. 142, 37 - 48*

KOZHEUROV, V.P., DEGETEVA, M.O., (1994b): Dietary intake evaluation and dosimetric modelling for the Techa river residents based on in vivo measurements of strontium-90 in teeth ans skeleton. *Sci. Total. Environ. 142, 63 - 72*

KRA, R., (1986): Standardizing procedures for collecting, submitting, recording, and reporting radiocarbon samples. *Radiocarbon 28, 2A, 765 - 775*

KRAMER, L., SPENCER, H., (1973): Dietary strontium-90 intake in Chicago. *Health Phys. 25, 445 - 448*

KREY, P.W. (1967a): Atmospheric burnup of a plutonium-238 generator. *Science 158, 769 - 771*

KREY, P.W., (1967b): Stratospheric trends on inventory of SNAP-9A debris. In: Fallout Program Quarterly Summary Report. Washington DC, Health and Safety Laboratory, *HASL-181, 12-115*

KREY, P.W., et al., (1979): Atmospheric burn-up of the Cosmos - 954 reactor: *Science 205, 583 - 585*

KREY, P.W., KRAJEWSKY, B., (1970): Comparison of atmospheric transport model calculations with observations of radioacitve debris. *J. Geophys. Res. 75, 2901 - 2908*

KRIEGEL, H., KOLLMER, W.E., (1985): Nuklearbiologie. In: *Grundlagen der Nuklearmedizin - Fundamentals of Nuclear Medicine, Vol. 1.* Kriegel, H., ed., *311 - 327*

KROGH, T.E., (1973): A low contamination method for the hydrothermal decomposition of zircon and extraction of U and Pb for isotopic age determination. *Geochim. Cosmochim. Acta 37, 485 - 494*

KROGH, T.E., (1982a): Improved accuracy of U-Pb zircon dating by selection of more concordant fractions using a high gradient magnetic separation technique. *Geochim. Cosmochim. Acta 46, 631 - 635*

KROGH, T.E., (1982b): Improved accuracy of U-Pb zircon ages by the creation of more concordant systems using an air abrasion technique. *Geochim. Cosmochim. Acta 46, 637 - 649*

KRÜGER, F.W., et al., (1996): Der Ablauf des Reaktorunfalls Tschernobyl 4 und die weiträumige Verteilung des freigesetzten Materials: Neuere Erkenntnisse und ihre Bewertung. In: *Zehn Jahre nach Tschernobyl, eine Bilanz, 4 - 22. Seminar des Bundesamtes für Strahlenschutz und der Strahlenschutzkommission, München 6.-7-März (1996).* Bayer, A., Kaul, A., Reiners, Chr., eds., Stuttgart, Gustav Fischer Verlag

KRYSHEV, I.I., et al., (1998): Environmental contamination and assessement of doses from radiation releases in the Southern Urals. *Health Phys. 74, 687 - 697*

KUC, T., (1986): Carbon isotopes in atmospheric CO_2 of the Krakow-Region: A two-year record. New Haven, Connecticut. *Radiocarbon 28, 2A, 649 - 654*

KULKARNI, K.M., et al., (1998): Drinking water salinity problem in coastal Orissa, India. - Identification of past transgressions of sea water by isotope investigation. In: *Isotope techniques in the study of environmental change.* Vienna, International Atomic Energy Agency, *293 – 306*

LAGERQUIST, C.R., et al., (1973): Distribution of plutonium and americium in occupationally exposed humans as found from autopsy samples. *Health Phys. 25, 581 - 584*

LAL, D., (1992): Cosmogenic *in situ* radiocarbon on Earth. In: *Radiocarbon after four decades.* Taylor,R.E., Long,A., Kra, R.S., eds., New York, Springer-Verlag, *146 - 161*

LAL, D., NIJAMPURKAR, V.N., RAMA S., (1970): Silicon-32 hydrology. In: *Isotope hydrology.* Vienna, International Atomic Energy Agency, *847 - 863*

LAL, D., PETERS, B, (1967): Cosmic-ray-produced radioactivity on the earth. In: *Encyclopedia of Physics Vol. XLVI/2 (Cosmic Rays).* Sitte, K., ed., New York, Springer Verlag

LALIT, B.Y., SUBBA RAO, M., HINGORANI, S.B., (1972): Strontium-90 contamination of milk and food samples in India. *Health Phys. 23, 47 - 54*

LAMBERT, D.D., et al., (1989): Rhenium - osmium and samarium - neodymium isotopic systematics of the Stillwater Complex. *Science 224, 1169 - 1174*

LAMONT, L., (1965): *"Day of Trinity".* New York, Atheneum

LANDA, R.E., (1993): A brief history of the american radium industry and its ties to the scientific community of its early twentieth century. *Environ. Int. 19, 503 - 508*

LANGE, G., et al., (1991): Der Uranerzbergbau in Thüringen und Sachsen - ein geologisch-bergmännischer Überblick. *Erzmet. 44,3 162 - 171*

LANGHAM, W.M., (1959): Physiology and toxicology of plutonium-239 and its industrial medical control. *Health Phys. 2, 172 - 185*

LANGHAM, W.M., et al., (1950): Distribution and excretion of plutonium administered to man. Los Alamos, New Mexico, Los Alamos Scientific Laboratory, *Report LA 1151*

LANGMUIR, C.H., et al., (1978): A general mixing equation with application to Icelandic basalts. *Earth Planet. Sci. Lett. 37, 380 - 392*

LARSEN, R.J., (1983): Worldwide deposition of Sr-90 through 1981. *EML - 415*

LARSEN, R.J., (1985): Worldwide deposition of Sr-90 through 1983. *EML - 444*

LARSEN, R.P., et al., (1981): Plutonium retention in mice and rats after gastronintestinal absorption. *Radiat. Res. 87, 37 - 49*

LAYER, P.W., HALL C.M., YORK D., (1987): The derivation of $^{40}Ar/^{39}Ar$ age spectra of single grains of hornblende and biotite by laser step-heating. *Geophys. Res. Lett. 14, 757 - 760*

LCI, (1979): *Sicherheitsbericht über den Zwischenfall in Versuchsatomkraftwerk Lucens am 21. Januar 1969. Final Report. Lucens Investigation Committee.* Bern, Eidgenössische Drucksachen- und Materialzentrale

LE ROUX, L.J., GLENDENIN, L.E., (1963): Half-life of ^{232}Th. *Proc. Nat. Meeting Nucl. Energy, Pretoria, S. Africa, 83 - 94*

LEDERER, C.M., HOLLANDER, J.M., PERLMAN, I., (1967): *Table of isotopes.* 6th ed. New York, John Wiley & Sons

LEE. J.K.W., et al., (1991): Incremental heating of hornblende in vacuo implications for $^{40}Ar/^{39}Ar$ geochronology and the interpretation of thermal histories. *Geology 19, 872 - 876*

LEGGETT, R.W. (1985a): An upper - bound estimate of the gastrointestinal absorption fraction for Pu in adult humans. *Health Phys. 49, 1299 - 1301*

LEGGETT, R.W., (1985b): A model of the retention, translocation and excretion of systemic Pu. *Health Phys. 49, 1115 - 1137*

LEGGETT, R.W., (1986): Prediction the retention of cesium in individuals. *Health Phys. 50, 747 - 759*

LEGGETT, R.W., (1989):The behavior and chemical toxicity of U in the kidney: a reassessment. *Health Phys. 57, 365 - 383*

LENGEMANN, F.W., COMAR, C.L., WASSERMAN, R.H., (1957): Absorption of calcium and strontium from milk and nonmilk diets. *J. Nutr. 61, 571 - 583*

LENTSCH, J.W., et al., (1972): Stable manganese and Mn-54 distributions in the physical and biological components of the Hudson river estuary. In: *Proc. Nat. Symp. Radioecol., 3. Conf-710501-P2, USAEC, Washington D.C., 752 - 768*

LERMAN, J.C., (1972): Carbon-14 dating: Origin and correction of isotope fractionation errors in terrestrial living matter. In: *Proceedings of 8th International Conference on Radiocarbon Dating, Lower Hutt, New Zealand, Oct. 1972, 2, 613 - 624*

LÉVESQUE, B., et al., (1997): Radon in residences: influences of geological and housing characteristics. *Health Phys. 72, 907 - 914*

LEVIN, I., HESSHEIMER, V., (2000): Radiocarbon - a unique tracer of global carbon cycle dynamics. *Radiocarbon 42, 69 - 80*

LEVIN, I., MÜNNICH, K.O., WEISS, W., (1980): The effect of antropogenic CO_2 and C-14 sources on the distribution of C-14 in the athmosphere. *Radiocarbon 22, 379-391*

LIBBY, W.F. (1958): Radioactive fallout. *Proc. Nat. Acad. Sci (USA), 44, 800 - 819*

LIBBY, W.F., (1955): *Radiocarbon dating. 2nd ed.* Fifth Impression 1965, Chicago, The University of Chicago Press.

LIDEN, K., HOLM, E., (1985): Measurement and dosimetry of radioactivity in the environment. In *The dosimetry of ionizing radiation, Vol. 1.* Kase, K.R., Bräjngard, B.E., Attix, F.H., eds., Orlando, Academic Press

LIEW, T.C., HOFMANN, A.W., (1988): Precambrian crustal components, plutonic associations, plate environment of the Hercynian Fold Belt of central Europe: Indications from Nd and Sr isotopic study. *Contrib. Mineral. Petrol. 98, 129 - 38*

LIKHTAREV, I.A., et al., (1996): Internal exposure from investigation of foods contaminated by ^{137}Cs after the Chernobyl accident. Report 1. General model: ingestion doses and countermeasure effectiveness for the adults of Rovno oblast of Ukraine. *Health Phys. 70, 297 - 317*

LIKHTAREV, I.A., PEREVOZNIKOV, O.N., LITVINETS, L.A., (1992): The experience of using whole body counters for radiocesium body burden measurements of the general public after the Chernobyl power plant accident. In: *Radiation monitoring, clinical problems, socio-psychological aspects, demographic situation, and low-level exposure to ionizing radiation. Kiev; Information Bulletin, 2nd Ed., Vol 1.* Ukrainian Scientific Centre of Radiation Medicine, Ministry of Health, and Academy of Sciences, *219 - 224* (in Russian)

LINDSTROM, L.M., LANGLAND, J.K., (1990): A low-background gamma-ray assay laboratory for activation analysis. *Nucl. Instrum. Methods Phys. Res. A 299, 425 - 429*

LINGENFELTER, R.E., (1963): Production of carbon-14 by cosmic-ray neutrons. *Rev. Geophys. 1, 35 - 55*

LINIECKI, J., (1971): Kinetics of calcium, strontium, barium and radium in rabbits. *Health Phys. 21, 367 - 376*

LLOYD, R.D., (1973): Cesium-137 half times in humans. *Health Phys. 25, 605 - 610*

LOBOREV, V.M., et al., (1994a): Radiation impact of the Semipalatinsk test site upon Altai region and the problems of quantitative assessments of this impact. *Вестник научнои программы „Семипалатинский полигон - Алтай" 1, 10 - 26* (engl. Summary)

LOBOREV, V.M., et al., (1994b): The reconstruction of Altai region population irradiation doses due to the nuclear expolsion of August 29, 1949. *Вестник научнои программы „Семипалатинский полигон - Алтай" 1, 27 - 56* (engl. Summary)

LOGACHOV, V.A., et al., (1993): Analysis of data on medical biological research programme and inspection of health of critical population groups living in areas in the Altai and Gorny Altai region. *Bulletin of Centre of Public Information in the Field of Nuclear Energy, Special Edition, January 20.* (in Russian)

LONG, A., MURPHY, E.M., DAVIS, S.N. (1992): Natural radiocarbon in dissolved organic carbon in groundwater. In: *Radiocarbon after four decades.* Taylor,R.E., Long,A., Kra, R.S., eds., New York, Springer-Verlag, *288 - 308*

LOOSLI, H.H., OESCHGER, H., (1979): Argon-39, carbon-14 and krypton-85 measurements in groundwater samples. In: *Isotope hydrology 1978, Vol. II.* Vienna, International Atomic Energy Agency, *931 - 947*

LOOSNESKOVIC, C., et al., (1999): A copper hexacyanoferrate (polymer) silica composite as selective sorbent for the decontamination of radioactive caesium. *Radiochim. Acta 85, 143 - 148*

LÓPEZ-VERA, F., ŠILAR, J., SPANDRE, R., (1996): Evaluation of the sustained yield of groundwater resources in the aquifer of the detrital Tertiary of Madrid. In: *Isotopes in water resources management, Vol. I.* Vienna, International Atomic Energy Agency, *454 - 456*

LUCCI, F., MEROLLI, S., PELLICIONI, M., (1973), Measurements of induced radioactivity in dust around a 400 MeV linac. *Health Phys. 24, 411 - 415*

LUCK, J.M., ALLEGRE, C.J., (1983): [187]Re - [187]Os systematics in meteorites and cosmochemical consequences. *Nature 302, 130 - 132*

LUCK, J.M., BIRCK, J.L., ALLEGRE, C.J., (1980): [187]Re - [187]Os systematics in meteorites: early chronology of the solar system and the age of the galaxy. *Nature 283, 256 - 259*

MACFARLANE, R.D., KOHMAN, T.P., (1961): Natural alpha radioactivity in medium-heavy elements. *Phys. Rev. 121, 1758 - 1769*

MACHTA, L, (1974a): Globale scale atmospheric mixing. In: *Turbulent diffusion in environmental pollution, Vol. 18b.* Frankiel, F.N., Munn, R.E., eds., Advances in Geophysics Series, Landsberg, H.E., et al., eds., New York, J. Academic Press, *33*

MACHTA, L., FERBER, G.J., HEFFTER, J.L., (1974b): Regional and global scale dispersion of krypton-85 for population-dose calculations. In: *Physical behavior of radioactive contaminants in the atmosphere.* Vienna, International Atomic Energy Agency, *411*

MALONEY, B.K., McCORMAC, F.G., (1995): A 30,000 year pollen and radiocarbon record from highland Sumatra as evidence for climatic change. *Radiocarbon 37, 181 - 190*

MANGERUD, J., GULLIKSEN, S., (1975): Apparent radiocarbon ages of recent marine shells from Norway, Spitsbergen, and Arctic Canada. *Quat. Res. (New York-London) 5, 263 - 273*

MANN, W.B., (1983): An international reference material for radiocarbon dating. *Radiocarbon 25, 519 - 527*

MANN, W.B., RYTZ, A., SPERNOL, A., (1991): *Radioactivity measurements: principles and practice.* Oxford, Pergamon Press

MARCKWORDT, U., LEHR, J., (1971): Factors of transfer of Cs137 from soils to crops. In: *Int. Symp. „Die Radiologie angewendet auf den Schutz des Menschen und seiner Umwelt", Rom, 7.-10. Sept. 1971, EUR-4800, a-f-i-e, 1057 - 1066*

MAREI, A.N., et al., (1972): Effect of natural factors on cesium-137 accumulation in the bodies of residents in some geographical regions. *Health Phys. 22, 9 - 15*

MAREŠ, S., ŠILAR, J., (1978): Application of isotope techniques and well logging in investigating ground water influenced by mining. In: *Water in mining and underground works, SIAMOS, I, Granada, 931 - 947*

MARHOL, M., (1982) *: Ion exchanges in analytical chemistry.* Prague, Academia, *585 pp.*

MASSE, R., (1982a): Ruthenium and activated metals. In: *Radionuclide metabolism and toxicity.* Galle, P., Masse, R., eds., Paris, Masson, *131 - 142*

MASSE, R., (1982b): Metabolismus and toxicity of radioactive rare earths. In: *Radionuclide metabolism and toxicity.* Galle, P., Masse, R., eds., Masson, Paris, *143 - 155*

MATHIEU, G.G., et al., (1988): System for measurement of Rn at low levels in natural waters, *Health Phys. 55, 989 - 992*

MATSUSAKA, N., et al., (1988): Influence of zinc deficiency on the whole-body retention of ^{65}Zn in young and adult mice. *Jpn. J. Vet. Sci. 50, 966 - 967*

MATTHESS, G., (1973): *Die Beschaffenheit des Grundwassers.* Berlin, Gebrüder Borntraeger

MATTHIES, K., et al., (1982): Simulation des Transfers von Radionukliden in landwirt-schaftlichen Nahrungsketten. D-85764 Neuherberg, GSF - National Research Center for Environment and Health, *GSF-Report S-882*

MATTSON, L.J., (1972): Sodium-22 in the food-chain: Lichen-reindeer-man. *Health Phys. 23, 223 - 230*

MATYJEK, M., et al., (1988): Evaluation of alpha spectrometer with surface – barrier detectors for low level measurements. In: *Report on the consultants meeting on rapid instrumental and separation methods for monitoring radionuclides in food and environmental samples.* Vienna, International Atomic Energy Agency, *83 - 92*

MAY, H. MARINELLI, L. B., (1964): *The Natural Radiation Environment.* Adams J.A.S., Lowder, W.M., eds., Chicago, University of Chicago Press, *463*

MAZOR, E., et al., (1986): Tritium corrected ^{14}C and atmospheric noble gas corrected ^{4}He applied to deduce ages of mixed groundwaters: Examples from the Baden region, Switzerland. *Geochim. Cosmochim. Acta 50, 1161 - 1618*

McQUARRIE, S.A., WIEBE, L.I., EDISS, C., (1980): Observations of the performance of ESP and H# in liquid scintillation counting. In: *Liquid scintillation counting: recent applications and development, Vol. I.* Peng, C.T., Horrocks, D.L., Alpen, E.L., eds., New York, Academic Press, *291 – 300*

McAULAY, I.R., DOYLE, C., (1985): Radiocesium levels in Irish sea fish and the resulting dose to the population of the Irish Republic. *Health Phys. 48, 333 - 337*

McCARTNEY, M., et al., (1986): Global and local effects of ^{14}C discharge from the nuclear fuel cycle. *Radiocarbon 28, 2A, 634 - 643*

McCULLOCH, M.T., CHAPPELL, B.W., (1982): Nd isotopic characteristics of S- and I-type granites. *Earth Planet. Sci. Lett. 58, 51 - 64*

McINROY, J.F., et al., (1985): The U.S. transuranium registry report on the ^{241}Am content of a whole body, Part IV: Preparation and analysis of the tissues and bones. *Health Phys. 49, 587 - 621*

McINTYRE, G. A., et al., (1966) : The statistical assessment of Rb–Sr isochrons. *Journ. Geoph. Res. 71, 5459 – 5468*

MELO, D.R., et al., (1997): A biokinetic model for ^{137}Cs. *Health Phys. 73, 320 - 332*

MERGUE G.H.,(1973): Spatial distribution of ^{40}Ar/^{39}Ar ages in lunar breccia 14301. *J. Geophys. Res. 78, 3216 - 3221*

MERWIN, S.E., BALONOV, M.I., (1993): *Doses to the Soviet population and early health effects studies. The Chernobyl papers, Vol 1. Doses to the soviet population and early health effects studies.* Richland, Research Enterprise

METIVIER, H., (1982): Plutonium. In: *Radionuclide metabolism and toxicity.* Galle, P.; Masse, R., eds., Paris, Masson, *166 - 197*

MICHAEL, H.N., RALPH, E.K., (1972): Discussion of radiocarbon dates obtained from precisely dated Sequoia and Bristlecone Pine samples. In: *Proceedings of the 8th International Conference on Radiocarbon Dating, 1, Lower Hutt, New Zealand, 27 - 43*

MICHARD, A., et al., (1985): Nd isotopes in French Phanerozoic shales: external vs internal aspects of crustal evolution. *Geochim. Cosmochim. Acta 49, 601 - 610*

MICHELS,J.W., (1973): *Dating methods in archaeology.* New York, Seminar Press

MIETTINEN, J.K., (1964): Measurements of caesium-137 in Finish Lapps in 1962 - 64 by a mobile whole body counter. In: *Assessment of radioacitivity in man, Vol. 2*. Vienna, International Atomic Energy Agency, *193 - 210*

MIKHAILOV, V.N., ANDRYSHIN, I.A., et al., (1996): USSR nuclear weapon tests and peaceful nuclear explosions. 1949 trough 1990. *RFNC-VNIIEF, Sarov, 93 -*

MILINTAWISAMAI, M., et al., (1998): Application of isotope techniques to the study of groundwater pollution by arsenic in Nakorn Si Thammarat Province, Thailand. In: *Isotope techniques in the study of environmental change*. Vienna, International Atomic Energy Agency, *473 - 481*

MILLER, J.M., (1975): Incident at the Lucens reactor. In: *Operating Experiencies. Nucl. Saf. 16, 76 - 79*

MINACH, L., BRUNNER, P., (1988): Strahlenbelastung von Boden und Vegetation in Südtirol durch den Reaktorunfall in Tschernobyl. In: *IVth European Congress, XIIIth Regional Congress of IRPA, Twenty Years Experience in Radiation Protection, Salzburg, Sep. 15-19, 1986, 885 - 889*

MINENKO, L.F., et al., (1996): *Radiation doses due to Chernobyl accident in Belarus*. Minsk, EU/CIS Project ECP10

MIOTKE,F.-D., (1974): Carbon dioxide and the soil atmosphere. *Abh. Karst- u. Höhlenkunde, R. A, H. 9, München, 1 - 49*

MITCHELL J.G., (1968): The argon-40/argon-39 method for potassium-argon age determination. *Geochim. Cosmochim. Acta 32, 781 - 790*

MOKROV, Yu.G., (2002): *Reconstruction of the Thecha River contamination. Part 1. The role of the weighted particles in the process of the Techa River contamination formation in 1949-1951*. (in Russian). Библиотека Журнала "Вопросы Радиацной Безопасной" № 1, Редакцонно-Издальский Центр ВРБ, Озёрск

MOLJK, A., DREVER, R.W.P., CURRAN, S.C., (1957): The background of counters and radiocarbon dating. *Proc. Royal Soc. London 239, 433 - 445*

MONGAN, T.R., et al., (1996a): Plutonium releases from the 1957 fire at Rocky flats. *Health Phys. 71, 510 - 521*

MONGAN, T.R., RIPPLE, S.R., WINGES, K.D., (1996b): Plutonium releases from the 903 pad at Rocky Flats. *Health Phys. 71, 522 - 531*

MONGE SOARES, A.M., (1993): The ^{14}C content of marine shells: Evidence for variability in coastal upwelling off Portugal during the Holocene. In: *Isotope techniques in the study of past and current environmental changes in the hydrosphere and the atmosphere*. Vienna, International Atomic Energy Agency, *471 - 485*

MONTASER, A. ed., (1998): *Inductively coupled plasma mass spectrometry*. New York, John Wiley & Sons.

MONTEL, J.M., VESCHAMBRE, M., NICOLLET, C., (1994): Datation de la monazite à la microsonde électronique. *C.R.Acad. Sci. Paris, t. 318, série II, 1489 - 1495*

MOOK, W.G., (1980): Carbon-14 in hydrogeological studies. In: *Handbook of environmental isotope Geochemistry*. Fritz, P., Fontes, J.C., eds., Amsterdam, Elsevier Scientific Publishing Company, *49 - 74*

MOOK, W.G., (1986): Business meeting. Recommendations/resolutions adopted by the Twelfth International Radiocarbon Conference. *Radiocarbon 28, 2A, 799*

MOORBATH, S., O´NIONS, R.K., PANKHURST R.J., (1975) : The evolution of early Precambrian crustal rocks at Issua, West Greenland - geochemical and isotopic evidence. *Earth Planet. Sci. Lett., 27, 229 - 239*

MORGENSTERN, U., HEBERT, D., (1990): Carbon-extractor. In: *Isotopes in Nature, Part II*, Leipzig, Central Institute of Isotope and Radiation Research, *841 - 846*

MORIN, M., NENOT, J.C., LAFUMA, J. (1972): Metabolic and therapeutic study following administration to rats of ^{239}Pu nitrat. - A comparison with ^{238}Pu. *Health Phys. 23, 475 - 480*

MOSER, H., RAUERT, W., (1980): *Isotopenmethoden in der Hydrologie.* Berlin, Gebrüder Borntraeger

MSC, (1978): *Marshall Islands, a chronology 1944 -1981.* Micronesia Support Committee, Honolulu, Micronesia Support Committee

MÜLLER, H., (1992): Interne Strahlenexposition in der Bundesrepublik Deutschland nach dem Tschernobyl - Unfall. In: *Die Folgen von Tschernobyl für Deutschland und die ehemalige Sowjetunion, Journalisten Seminar 19.03.-19.03.1991.* Haury, H.J. , ed., D-85764 Neuherberg, GSF - National Research Center for Environment and Health, *GSF-Bericht 17/92, 13 - 26*

MÜLLER, H., PRÖHL, G., (1993): A dynamic model for assessing radiological consequences of nuclear accidents. *Health Phys. 64, 232 - 252*

MÜLLER, H., SCHEFFEL, U., (1982): Cesium. In: *Radionuclide metabolism and toxicity.* Galle, P.; Masse, R., eds., Paris, Masson, *83 - 97*

MULLER, W., MANCKTELOW, N.S., MEIER, M. (2000): Rb – Sr microchrons of synkinematic mica in mylonites : an example from DAV fault of he Eastern Alps. *Earth Planet. Sci. Lett. 180, 3-4, 385 – 397*

MÜNNICH, K.O., (1957): Messungen des C^{14}-Gehaltes von hartem Grundwasser. *Naturwiss. 44 (2), 32 - 33*

MÜNNICH, K.O., (1963): Der Kreislauf des Radiokohlenstoffs in der Natur. *Naturwiss. 50, 211 - 218*

MÜNNICH, K.O., VOGEL, J.C., (1962): Untersuchungen an pluvialen Wässern der Ostsahara. *Geol. Rdsch., 52, 611-624*

MURASE, Y., et al., (1989): Effect of air luminiscence counts on determination of ^{3}H by liquid scintillation counting. In: *Proc.15. Regional Congress of the International Radiation Protection Association (IRPA) on the Radioecology of Natural and Artifical Radionuclides.* Köln, Verlag TÜV, *509 - 513*

MUSOLINO, S.V., GREENHOUSE, N.A., HULL, A.P., (1997): An estimate by two methods of thyroid absorbed doses due to BRAVO fallout in severale northern Marshall Islands. *Health Phys. 73, 651 - 662*

MUSTONEN, R., JANTUNEN, M., (1985): Radioactivity of size fractionated fly-ash emissions from a peat- and oil fired power plant. *Health Phys. 49, 1251 - 1260*

N 1140-501, (1991): Concluding Comments of the Commission of the USSR academy of sciences on the assessment of the ecological situation in the area of activities of the production association „Mayak" of the USSR ministry of atomic energy and industry, of 12.06.90. *Radiobiologiya 72, 436 - 452*

NAIR, K.S., et al., (1997): Modeling the resuspension of radionuclides in Ukrainian region impacted by Chernobyl fallout. *Health Phys. 72, 77 - 85*

NAKAMURA, T., HIRASAWA, M., IGAKI, K., (1995): AMS radiocarbon dating of ancient oriental iron artifacts at Nagoya University. *Radiocarbon 37, 629 - 636*

NAKAOKA, A., FUKUSHIMA,M., ICHIKAWA, Y., (1985), Evaluation of radiation dose from a coal-fired power plant. *Health Phys. 48, 215 - 220*

NAS-NRC, (1957a): The effects of atomic radiation on oceanography and fisheries. Washington DC, National Academy of Sciences - National Research Council, *Publ. 551,*

NAS-NRC, (1957b): Disposal of radioactive waste on land. Washington DC, National academy of sciences - National Research Council, *Publ. 519*

NAS-NRC, (1959): Radioactive waste disposal into Atlantic and Gulf coastal waters. Washington DC, National academy of sciences - National Research Council, *Publ. 655*

NCRP 44, (1975): Krypton - 85 in the atmosphere. Accumulation, biological significance and control technology. 20014 Washington DC National Council on Radiation Protection and Measurements, , *NCRP Report No. 44*

NCRP 52, (1977): Caesium-137 from environment to man; metabolism and dose. Washington DC, National Council on Radiation Protection and Measurements, *NCRP Report No. 52*

NCRP 56, (1977): Radiation exposure from consumer products and miscellaneous sources. 20014 Washington DC,National Council on Radiation Protection and Measurements, *NCRP Report No. 56*

NCRP 58, (1978),(1985 2nd.Ed.): A handbook of radioactivity measurements procedures. Bethesda, MD, National Council for Radiation Protection and Measurements, *NCRP Report No. 58*

NCRP 60, (1978): Physical, chemical, and biological properties of radiocerium relevant to radiation protection guidelines. MD 20014 Washington DC, National Council on Radiation Protection and Measurements, *NRCP Report No. 60*

NCRP 62, (1979): Tritium in the environment. 20014 Washington DC, National Council on Radiation Protection and Measurements, *NCRP Report No. 62*

NCRP 75, (1983): Iodine-129: Evaluation of releases from nuclear power generation. MD 20814-3095 Bethesda,National Council on Radiation Protection and Measurements, *NCRP Report No. 75*

NCRP 76, (1984): Radiological assessment: predicting the transport, bioaccumulation, and uptake by man of radionuclides released to the environment. MD 20814-3095 Bethesda, National Council on Radiation Protection and Measurements, *NCRP Report No. 76*

NCRP 81, (1985): Carbon -14 in the environment. MD 20814 Bethesda, National Council on Radiation Protection and Measurements, *NCRP Report No. 81*

NCRP 90, (1988): Neptunium: radiation protection guidelines. MD 20814-3095 Bethesda, National Council on Radiation Protection and Measurements, *NRCP Report No. 90*

NCRP 95, (1987): Radiation exposure of the U.S. population from consumer products and miscellaneous sources. MD 20814 Bethesda, National Council on Radiation Protection and Measurements, *NCRP Report No. 95*

NCRP 97, (1988): Measurement of radon and radon daughters in air. MD Bethesda, National Council for Radiation Protection and Measurements, *NCRP Report No. 97*

NCRP 109, (1991): Effects of ionizing radiation on aquatic organisms. MD 20814-3095 Bethesda, National Council on Radiation Protection and Measurements, *NCRP Report No. 109*

NCRP 110, (1991). Some aspects of strontium radiobiology. MD 20814-3095 Bethesda, National Council on Radiation Protection and Measurements, *NCRP Report No. 110*

NCRP 118, (1993): Radiation protection in the mineral extraction industry. MD 20814-3095 Bethesda, National Council on Radiation Protection and Measurements, *NCRP Report No. 118*

NCRP 123, (1996): Screening models for releases of radionuclides to atmosphere, surface water, and ground. MD 20814 - 3095 Bethesda, National Council on Radiation Protection and Measurements, *NCRP Report No. 123*

NEA-OECD (1996): *Chernobyl, ten years on radiological and health impact. An appraisal by the NEA Committee on the radiation protection and public health.* Paris, Nuclear Energy Agency Organisation for Economic Co-Operation and Development, OECD

NEKOLLA, E., et al., (1995): Malignancies in patients treated with radium-224. In: *Health effects of Internally deposited radionuclides: Emphasis on radium and thorium.* Van Kaick, G., Karaoglou, A., Kellerer, A.M., eds., Singapore, World Scientific, *243 - 248*

NELSON, D.E., et al., (1986a): Radiocarbon dating blood residues on prehistoric stone tools. *Radiocarbon 28, 170 - 174*

NELSON, D.E., et al., (1986b): New radiocarbon dates on artifacts from the northern Yukon Territory: Holocene not upper Pleistocene in age. *Science 232, 749 - 751*

NENOT, J.C., (1982a): Neptunium. In: *Radionuclide metabolism and toxicity*, Galle, P., Masse, R., eds., , Paris Masson, *225 - 231*

NENOT, J.C., (1982b): Californium. In: *Radionuclide metabolism and toxicity*, Galle, P., Masse, R., eds., Paris, Masson, *232 - 238*

NERO. A.V.Jr., (1985): *What we know about indoor radon. Testimony prepared for hearings on „radon contamination: Risk assessment and mitigation research".* Subcommittee on Science Technology, U.S. House of Representatives (Oct.10.1985), Washington DC, USGPO

NEUSTUPNÝ, E. (1970): The accuracy of radiocarbon dating. In: *Radiocarbon variations and absolute chronology, Nobel symposium 12.* Olsson, I.U., ed., Stockholm, Almqvist and Wiksell, New York, Wiley Interscience Division, *23 - 34*

NEWELL, R.E., (1971): The global circulation of atmospheric pollution. *Sci. Am. 224, 32 - 42*

NICHOLS, J.P., BINFORD, F.T., (1971): Status of a noble gas removal and disposal, Oak Ridge National Laboratory Tennessee, *ORNL-TM-3515*

NICOLAYSEN, L.O., (1961): Graphic interpretation of discordant age measurements on metamorphic rocks. *Ann. N. Y. Acad. Sci. 91, 198 - 206*

NIER, A.O., (1950): A redetermination of the relative abundances of the isotopes of carbon, oxygen, argon and potassium. *Phys. Rev. 77, 789 - 793*

NIKEZIC, D., YU, K.N., (1999): Determination of deposition behaviour of Po - 218 from track density distribution on SSNTD in diffusion chamber. *Nucl. Instrum. Methods Phys. Res. A437, 531 - 537*

NIKIPELOV, B.V., MIKERIN, E.I., ROMANOV, G.N., (1990): Accident in southern Urals in 1957 and the cleanup measures implemented. In: *Proc. of a symposium on recovery operations in the event of a nuclear accident or radiological emergency.* Vienna, International Atomic Energy Agency, *Report STI/PUB/826, 373 - 403*

NIKIPELOV, B.V., NIKIFOROV,A.S., KEDROVSKY, O.L., (1992): *Practical rehabilitation of territorries contaminated as a result of implementation of nuclear material production defense pograms.* Moscow, VNIPI Prom-technologii

NILSEN, T., BØHMER, N., (1994): Sources to radioactive contamination in Murmansk and Archangelsk counties. *Bellona Report No. 1*

NILSEN, T., KUDRIK, I., NIKITIN, A., (1996): The Russian northern fleet. *Bellona Report No. 2*

NISHITA, H., ROMNEY, E.M., LARSON, K.H., (1961): Uptake of radioactive fission products by crops plants. *Agri. Fd. Chemy. 9, 101 - 106*

NISHITA, H., STEEN, A.J., LARSON, K.H., (1958): Release of Sr-90 and Cs-137 from Vina Loam upon prolonged cropping. *Soil Sci. 86, 195 - 201*

NOSHKIN, V.E., BOWEN, V:T, (1973): Concentrations and distribution of long - lived fall-out radionuclides in open ocean sediments. In: *Radioactive contamination of marine environment.* Vienna, International Atomic Energy Agency, *Report STI/PUB 313, 671*

NOUJAIM, A.A. et al.,(1976): *Liquid Scintillation Science and Technology.* New York, Academic Press.

NTIS PR-360 (1987): Health and environmental consequences of the Chernobyl nuclear power plant accident. Report to U.S. Department of Energy. Springfield, Virginia, Office of Energy Research, Office of Health and Environmental Research. National Technical Information Service, *NTIS PR-360,*

NUTI, S., (1991): Isotope techniques in geothermal studies. In: *Applications of geochemistry in geothermal reservoir development, Series of the technical guides on the use of geothermal energy.* D'Amore, F., coordinator, Rome, UNITAR/UNDP Centre on Small Energy Resources, *215 - 251*

NYAS, (1981): The Three Mile Islands nuclear accident: lessons and implications. Moss, T.H., Sills, D.L., eds., New York Academy of Sciences, *Ann. N.Y.Acad. Sci 365*

NYDAL, R., (1964): Radioaktivt Carbon-14 fra kjernefysiske eksplosjoner. *Fra Fysikkens Verden 3, Trondhjem, 1 - 4*

NYDAL, R., (2000): Radiocarbon in the ocean. *Radiocarbon 42, 81 - 98*

NYDAL, R., LÖVSETH, K., GULLIKSEN, S., (1979): A survey of radiocarbon variations in nature since the test ban treaty. In: *Radiocarbon dating.* Berger,R., Suess, H.E., eds., Berkeley, University of California Press, *313 - 323*

O BAR-YOSEF, (2000): The impact of radiocarbon dating on Old World archaeology: Past achievements and future expectations. *Radiocarbon 42, 23 - 39*

O'DELL, B.L., CAMPBELL, B.J., (1971): Metabolism and metabolic functions. In: *Metabolism of vitamins and trace elements, Comprehensive biochemistry, Vol. 21,* Amsterdam, Elsevier, *246 -248*

OBELIĆ, B., et al., (1986): Environmental ^{14}C levels around the 632 MWe Nuclear power plant Krško in Yugoslavia. *Radiocarbon 28, 2A, 644 - 648*

OECD, (1985): *Metrology and monitoring of radon, thoron and their daughter products.* Paris, Organization for Economic Cooperation and Development

OESCHGER, H., et al., (1974a): [39]Ar dating of groundwater. In: *Isotope techniques in groundwater hydrology 1974, Vol. II.* Vienna, International Atomic Energy Agency, *179 - 190*

OESCHGER. H., GUGELMANN, A., (1974b): Das geophysikalische Verhalten der Umweltisotope als Basis für Modellrechnungen in der Isotopenhydrologie. *Österr. Wasserwirtsch. 26, No. 3/4, 43 - 49*

OLSSON, I.U. (1970): The use of oxalic acid as a standard. In: *Radiocarbon variations and absolute chronology, Nobel Symposium 12.* Olsson, I.U., ed., Stockholm, Almqvist and Wiksell, New York, Wiley Interscience Division, *17*

OLSSON, I.U. (1972): The pretreatment of samples and the interpretation of the ^{14}C determinations. In: *Climatic changes in artic areas during the last ten-thousand years.* Vasari, Y., Hyvärinen, H., Hicks, S., eds., Oulu, Acta Universitatis Oulensis, *Geologica, 1, 9 - 37*

OLSSON, I.U., (1979): The importance of the pretreatment of wood and charcoal samples. In: *Radiocarbon dating.* Berger, R., Suess, H.E., eds., Berkeley, University of California Press, *135 - 146*

OLSSON, I.U., (1980): ^{14}C extractives of wood. *Radiocarbon 22, 515 - 524*

OLSSON, I.U., (1986): A study of errors in ^{14}C dates of peat and sediments. *Radiocarbon 28, 429 - 435*

OLSSON, I.U., (1991): Accuracy and precision in sediment chronology. *Hydrobiologia 214, Belgium, Kluwer Academic Publishers, 25 - 34*

OLSSON, I.U., (1992): ^{14}C Activity in different sections and chemical fractions of oak tree rings, AD 1938-1981. *Radiocarbon 34, 757 - 767*

OLSSON, I.U., POSSNERT, G., (1992): The interpretation of ^{14}C measurements on pre-Holocene samples. *Sveriges Geologiska Undersökning Ser. Ca 81, 201 - 208*

ONISHI, Y., et al., (1981): Critical Review: radionuclide transport, sediment transport, and water quality mathematical monitoring and radionuclide adsorption/desorption mechanisms. Washington, Pac Northwest Lab Richland, *NUREG/-CR-1322*

ONSTEAD, C.O., OBERHAUSEN, E., KEARY, F.V., (1960): Messungen des Kalium- und Cäsium-137 - Gehaltes der deutschen Bevölkerung. *Atompraxis 6, 337 - 341*

OSBORNE, R.V., (1966): Absorption of tritiated water vapour by people. *Health Phys. 12, 1527 -1537*

OSBORNE, R.V., (1972): Permissible levels of tritium in man and environment. *Radiat. Res. 50, 197 - 211*

OSMOND, J.K., (1980): Uranium disequilibrium in hydrologic studies. In: *Handbook of environmental isotope geochemistry*. Fritz P., Fontes, J.C., eds., Amsterdam, Elsevier Scientific Publishing Company, *259 - 282*

OTLET, R.L., WALKER, A.J., LONGLEY, H., (1983): The use of ^{14}C in natural materials to establish the average gaseous dispersion patterns of releases from nuclear installations. *Radiocarbon 25, 593 - 602*

OTTAWAY, B.S., (1986): Is radiocarbon dating obsolescent for archaeologists? *Radiocarbon 28, 732 - 738*

PA ZIQIANG (1993): *Radiological impact of coal - fired energy in China*. China National Nuclear Corporation, Beijing, Communication to the UNSCEAR Secretariat

PAČES, T., (1976): Kinetics of natural water systems. In: *Interpretation of environmental isotope and hydrochemical data in groundwater hydrology*. Vienna, International Atomic Energy Agency, *85 - 108*

PALMER, H.E., RIEKSTS, G.A., SPITZ, H.B., (1985): The U.S. transuranium registry report on the ^{241}Am content of a whole body, Part III: Gamma-ray measurements. *Health Phys. 49, 577 - 586*

PANKHURST, R.J., PIDGEON, R.T., (1976): Inherited isotope systems and the source region pre-history of early Caledonian granites in the Dalradian series of Scotland. *Earth Planet. Sci. Lett., 31, 55 - 68*

PANKHURST, R.J., SMELLIE, J.L., (1983): K-Ar geochronology of the South Shetland Island. Lesser Antarctica: apparent lateral migration of Jurassic to Quaternary island arc volcanism. *Earth Planet. Sci. Lett. 66, 214 - 222*

PAPANASTASSIOU, D.A., WASSERBURG, G.J., (1969): The determination of small time differences in the formation of planetary objects. *Earth Planet. Sci. Lett. 5, 361 - 376*

PARRISH, R.R., (1987): An improved micro-capsule for zircon dissolution in U-Pb geochronology. *Chem. Geol. (Isot. Geosci. Sect.) 66, 99 - 102*

PATCHETT, P.J., (1983): Hafnium isotope results from Mid-ocean ridges and Kerguelen. *Lithos 16, 47 - 51*

PATCHETT, P.J., et al., (1981a): Evolution of continental crust and mantle heterogeneity: Evidence from Hf isotopes. *Contrib. Mineral. Petrol. 78, 279 - 297*

PATCHETT, P.J., TATSUMOTO, M., (1980a): Lu-Hf total-rock isochron for the eucrite meteorites. *Nature 288, 571 - 574*

PATCHETT, P.J., TATSUMOTO, M., (1980b): A routine high-precision method for Lu-Hf geochemistry and chronology. *Contrib. Mineral. Petrol. 75, 263 - 267*

PATCHETT, P.J., TATSUMOTO, M., (1981b): Lu/Hf in chondrites and definition of a chondritic hafnium growth curve. *Lunar Planet. Sci. XII, 822 - 824*

PATERSON, B.A., et al., (1992a): The nature of zircon inheritance in two granite plutons. *Trans. Roy. Soc. Edinburgh Earth Sci. 83, 459 - 471*

PATERSON, B.A., ROGERS, G., STEPHENS, W.E., (1992b): Evidence for inherited Sm-Nd isotopes in granitoid rocks. *Contrib. Mineral. Petrol. 111, 378 - 390*

PATTERSON, R., (1970) : *An introduction to ion exchange.* London, Heyden

PAYNE, B.R., (1983): Introduction. In: *Guidebook on nuclear techniques in hydrology, 1983 Edition, Technical Reports Series No. 91.* Vienna, International Atomic Energy Agency, *1 - 18*

PAZDUR, A., et al., (1994): Radiocarbon chronology of Late Glacial and Holocene sedimentation and water-level changes in the area of the Gosciaz Lake basin. *Radiocarbon 36, 187 - 222*

PAZDUR, M.F., et al., (1995): Radiocarbon and thermoluminiscence studies of the karst pipe systems in Southwest England and South Wales. *Radiocarbon 37, 111 - 117*

PECSOK, R. L., SHIELDS, L. D., CAIRNS, T., McWILLIAM, I. G., (1976): *Modern methods of chemical analysis. 2nd ed.*, New York, John Willey and Sons

PELLETIER, C.A., VOILEQUE, P.G:, (1971): The behaviour of ^{137}Cs and other fallout radionuclides on a Michigan dairy farm. *Health Phys. 21, 777 - 792*

PEREVOZNIKOV, et al., (1994): Experience, problems and results of implementation of whole body counters at post-Chernobyl period. In: *Assessment of the health and environmental impact from radiation doses due to the released radionuclides, Proceedings of an International Workshop at Chiba, January 18 - 20, 1994*: National Institute of Radiological Sciences, *NIRS-M-102, 129 - 139*

PEREVOZNIKOV, O.N., et al., (1992): *Monitoring* of individual exposure of the general public. In: *Problems of radiation medicine, Vol. 4.* Kiev, Zdorov'e, *24 - 32 (in Russian)*

PEREVOZNIKOV, O.N., et al., (1993): A quantitative assessment of decreasing internal measuring efficiency: In: *Problems of radiation epidemiology of medical consequences after the accident at Chernobyl nuclear power plant.* Kiev, Dnipro, *223 - 235* (in Russian)

PERSSON, L., (1995): Ethik und radioaktiver Abfall. *Strahlenschutz Praxis 1(2), 43 - 46*

PETTINGILL, H.S., PATCHETT, P.J., (1981): Lu-Hf total rock age for the Amitsoq gneisses, West Greenland. *Earth Planet. Sci. Lett. 55, 150 - 156*

PHILLIPS, A.B., et al., (1986): Residual radioactivity in a cyclotron and its surroundings. *Health Phys. 51, 337 - 342*

PIDGEON, R.T., COMPSTON, W., (1992): A SHRIMP ion microprobe study of inherited and magmatic zircons from four Scottish Caledonian granites. *Trans. Roy. Soc. Edinburgh Earth Sci. 83, 473 - 483*

PILLAI, K.C., MATKAR, V.M., (1987): Determination of plutonium and americium in environmental samples and assessment of thorium in bone samples from normal and high background areas. *J. Radioanal. Nucl. Chem. Articles 115, 217 - 229*

PINDER, J.E., SIMMONS, J.R., LINSLEY, G.S., (1985): Normalized specific activities for Pu deposition onto foliage. *Health Phys. 49, 1280 - 1283*

PINSON, E.A., (1951): The body absorption, distribution, and excretion of tritium in man and animals. Los Alamos, New Mexico, Los Alamos Lab., *USAEC Report LA-1218*

PINSON, E.A., LANGHAM, W.H., (1957): *Physiology and toxicology of tritium in man.* J. Appl. Physiol. 10, 108 -

PLAGA, R., (1991): Silicon for ultra low level detectors and ^{32}Si. *Nucl. Instrum. Methods Phys.Res. A 309, 598 - 599*

PLATZNER, T.I., (1997): *Modern isotope ratio mass spectrometry.* New York, John Wiley and Sons.

POET, S.E., MARTELL. E.A., (1972): Plutonium-239 and americium-242 contamination in the Denver area, *Health Phys. 23, 537 - 548*

POLACH, D., (1980): *First 20 yars of radiocarbon dating, an annotated bibliography, 1948-68 (Pilot study).* Canberra, Radiocarbon Dating Research Laboratory, Australian National University.

POLACH, H., (1981): Radiocarbon concentration variations in the atmosphere and absolute chronology: how to interpret dendrochronologic evidence. Preprint of an article. In: *Australian Archaeology: A guide to field and laboratory techniques, 2nd ed.* Connah, G., ed., Canberra, Australian Institute of Aboriginal Studies, *1 - 6*

POLACH, H., (1989): ^{14}CARE. *Radiocarbon 31, 422 - 430*

POLACH, H., (1992): Four decades of progress in ^{14}C dating by liquid scintillation counting and spectrometry. In: *Radiocarbon after four decades.* New York, Springer, *198 - 213*

POLACH, H., et al., (1983a): An ideal vial and cocktail for low-level scintillation counting. In: *Advances in scintillation counting.* Mc Quarrie, S.A, Edis, C., Wiebe, L.I., eds., Edmonton, Canada, University of Alberta, *508 – 525*

POLACH, H., GOLSON, J., HEAD, J., (1981): Radiocarbon dating: a guide for archaeologists on collection and submission of samples and age reporting practices. Preprint of an article. In: *Australian Archaeology: A guide to field and laboratory techniques, 2nd ed.* Connah, G., ed., Canberra, Australian Institute of Aboriginal Studies, *7 - 26*

POLACH, H., ROBERTSON, S., (1983b): Data reduction and age calculations at the ANU radiocarbon research laboratory. ANU Radiocarbon Laboratory, Canberra *Publication LM 7, 28-*

POON, C.B., et al., (1997): Adaption of ECOSYS - 87 to Hongkong environmental conditions. *Health Phys. 72, 856 - 864*

POONG EIL JUHN, KUPITZ, J., (1998): Nuclear power beyond Chernobyl: a chaning international perspective. International Atomic Energy Agency, Vienna, *IAEA-Bulletin 381*

POPPLEWELL, D.S., (1989): Maralinga rehabilitation project study. Radiochemical analysis. Chilton, *NRPB-M199*

POPPLEWELL, D.S., HAM, G.J., Johnson, T.E., BARRY, S.F., (1985): Plutonium in autopsy tissues in Great Britain, *Health Phys. 49, 304 - 309*

POTTS, P. J., (1987): *A handbook of silicate rock analysis.* Glasgow, Blackie, *610-*

POVINEC, P., (1980): Proportional chambers for low level counting. *Nucl. Instrum. Methods 176, 111 - 117*

POVINEC, P., CHUDÝ, M., ŠIVO, A., (1986): Anthropogenic radiocarbon: Past, present, and future. *Radiocarbon 28, 2A, 668 - 672*

PRATT, R.M., (1993): Review of radium hazards and regulation of radium in industry. *Environ. Int. 19, 475 - 48*

PRICE, K.R., et al., (1985): Environmental monitoring at Hanford for 1984. Washington, Pac. Northwest Lab., Richmond, *Report PNL-5407* (UC-41&11)

PRICE, P.N., (1997): Predictions and maps of county mean indoor radon concentrations in the mid - atlantic states. *Health Phys. 72, 893 - 906*

PRICHARD, D.W., et al., (1971): Physical processes of water movement and mixing. In: *Radioactivity in the marine environment*, Washington D.C., Natl. Acad. Sci, *90 - 136*

PRINCE, C.I., KOŠLER, J., VANCE, D., GUNTHER, D. (2000): Comparison of laser ablation ICP-MS and isotope dilution REE analyses – implication for Sm – Nd garnet geochronology. *Chem. Geol. 168, 3-4, 255 - 274*

PRÖHL, G., (1990): Modelling of radionuclide transfer in food chains after deposition of strontium-90, cesium-37 and iodine-31 onto agricultural areas. D-85764 Neuherberg, GSF - National Research Center for Environment and Health, *GSF-Report 29/90*

PROVOST A., (1990): An improved diagram for isochron data. *Chem. Geol. (Isot. Geosci. Sect.) 80, 85 - 99*

QUI XOU HUA et al., (1983) : Calibration of the Chinese sucrose: *Kexue Tongbao, 9 (5), 707-713* (in Chinese)

RABON, E.W., JOHNSON, J.E., (1973): Rapid field monitoring of cesium-137 in whitetailed deer. *Health Phys. 25, 515 - 516*

RADNELL, C.J., AITKEN, M.J., OTLET, R.L., (1979): In situ ^{14}C production in wood. In: *Radiocarbon dating*. Berger,R., Suess, H.E., eds., Berkeley, University of California Press, *643 - 657*

RAES, F., et al., (1989): Radioactive measurements in Europe after the Chernobyl Accident. Part I: air. Brüssel, EEC, *EUR 12269 EN*

RAMSDELL, J.V., et al., (1996): Atmospheric dispersion and deposition of ^{131}I released from the Hanford site. *Health Phys. 71, 568 - 577*

RAUERT, W., (1980): Measurements and hydrological applications of ^{3}H and ^{14}C. In: *Interamerican symposium on isotope hydrology*. Rodríguez, C.O.N., Briceño de Monroy, C., eds., Bogotá, Colombia, *193 - 220*

REID, G.K., WOOD, R.D., (1976): *Ecology of inland waters and estuaries. 2nd ed.*, New York, Van Nostrand

REVELLE, R., SCHAEFER, M.B., (1957): General considerations concerning the ocean as a receptacle for artificially radioactive materials. Washington D.C., National Academic Science, *Publ. 551*

REVZAN, K.L., et al., (1991): Modelling radon entry into houses with basements: model description and verification. Lawrence Berkeley Laboratory, *Report. LBL - 27 742*

RICHARD, P., SHIMIZU, N., ALLEGRE, C.J., (1976): ^{143}Nd/^{146}Nd, a natural tracer: an application to oceanic basalts. *Earth Planet. Sci. Lett. 31, 269 - 278*

RIPPLE, S.R., WIDNER, T.E., MONGAN, T.R., (1996): Past radionuclide release from routine operations at Rocky Flats. *Health Phys. 71, 502 - 509*

ROBL. H.R., (1997): Entwicklung einer Methode zur retrospektiven Rekonstruktion der Schilddrüsendosis weißrussischer Kinder nach dem Reaktorunfall in Chernobyl über die Bestimmung der ^{129}I-Konzentration im Boden. *Dissertation*, Inst. Ionenphysik, D-85764 Neuherberg, Universität Innsbruck, Austria und GSF - National Research Center for Environment and Health

RODDICK, J.C., LOVERIDGE, W.D., PARRISH, R.R., (1987): Precise U/Pb dating of zircon at the sub-nanogram Pb level. *Chem. Geol. (Isot. Geosci. Sect.) 66, 111 - 121*

ROESNER, G., (1981): Measurements of actinide nuclides in water samples from the primary circuit of a nuclear power plant. *J. Radioanal. Chem. 64, 55 - 64*

ROETHER, W., (1967): Estimating the tritium input to groundwater from wine samples: groundwater and direct runoff contribution to Centra European waters. In: *Isotopes in hydrology.* Vienna, International Atomic Energy Agency, *73 - 91*

ROGERS, A.W., (1979): *Techniques of autoradiography. 3rd ed.* Amsterdam, Elsevier

ROGERS, G., et al., (1989): A high-precision U-Pb age for the Ben Vuirich granite: implications for the evolution of the Scottish Dalradian Supergroup. *J. Geol. Soc. London 146, 789 - 798*

ROGOVIN, M., FRAMPTON, G.J., (1980): A sequence of physical events. In: *Three Mile Islands, Report to the Commissioners and the Public Spec. Inquiry Group, Vol. 2 Part 2.* Washington DC, Nucl. Regul. Comm., *309 - 340*

ROMANOV, G.N., NIKIPELOV, B.V., DROZHKO, E.G., (1990): The Kyshtym accident: causes, scales and radiation characteristics. In: *Proc. of Seminar on Comparative Assessment of the Environmental Impact of Radionuclides Released during three major Nuclear Accidents: Kyshtym, Windscale, Chernobyl, EUR-13574, 3 - 22*

ROMER, R.L., WRIGHT, J.E., (1992): U-Pb dating of columbites: a geochronologic tool to date magmatism and ore deposits. *Geochim. Cosmochim. Acta 56, 2137 - 2142*

RÖMMELT, R., HIERSCHE, L., WIRTH, E., (1991): Untersuchungen über den Transfer von Caesium 137 und Strontium 90 in ausgewählten Belastungspfaden. Neuherberg, Bundesamt für Strahlenschutz, Institut für Strahlenhygiene, *ISH-155/91*

ROSENTHAL, M.W., et al., (1972): Metabolism of monomeric and polymeric plutonium in the rabbit: comparison with the mouse. *Health Phys. 23, 231 – 238*

RÖSNER, G., (1981): Measurements of actinide nuclides in water samples from the primary circiut of a nuclear power plant. *J. Radioanal. Chem. 64, 55 - 64*

ROSSI, B., STAUB, H., (1949): *Ionization chambers and counters.* New York, Mc Graw Hill

RÓZAŃSKI, K., FLORKOWSKI, T., (1979): Krypton-85 dating of groundwater. In: *Isotope hydrology 1978, II.* Vienna, International Atomic Energy Agency, *949 - 961*

RUNDO, J., (1993): History of the determination of radium in man since 1915. *Environ. Int. 19, 425 - 438*

RUSS, G.P., BAZAN, J.M., DATE, A.R., (1987): Osmium isotopic ratio measurements by inductively coupled plasma source mass spectrometry. *Anal. Chem. 59, 984 - 989*

RUSS, J., et al., (1990): Radiocarbon dating of prehistoric rock paintings by selective oxidation of organic carbon. *Nature London 348, (6303), 710 - 711*

RUTHERFORD, E., (1919a): Collision of a particles with light atoms. *Nature London 103, 415 - 418*

RUTHERFORD, E., (1919b): *Phil. Mag. 27, 538, 571, 586*

RŮŽIČKOVÁ, E., ŠILAR, J., ZEMAN, A., (1993): Flood plain of the Labe River in the Little Ice Age. In: *Application of direct and indirect data for the reconstruction of climate during the last two millenia.* Růžičková, E., Zeman, A., Mirecki, J., eds., Praha, Geological Institute of the Academy of Sciences of the Czech Republic, *63 - 70*

SALO, A., MIETTINEN, J.K., (1964): Strontium-90 and caesium-137 in arctic vegetation during 1961. *Nature 201, 1177 - 1179*

SÁNCHEZ, A.M., et al., (1992): A rapid method for determination of the isotopic composition of uranium samples by alpha spectrometry. *Nucl. Instrum. Methods Phys. Res. A 313,* 219 - 226

SAUNDERS, A.D., NORRY, M.J., TARNEY, J., (1991): Fluid influence on the trace element compositions of subduction zone magmas. In: *The behaviour and influence of fluids in subduction zones.* Tarney, J., ed., London, The Royal Society, 151 - 166

SAYRE, E.V., et al., (1981): Small gas proportional counters for the ^{14}C measurement of very small samples. In: *Methods of Low Level Counting and Spectrometry.* Vienna, International Atomic Energy Agency, 393 - 407

SCHARPENSEEL, H.W., (1979): Soil fraction dating. In: *Radiocarbon dating.* Berger, R., Suess, H.E., eds., Berkeley, University of California Press, 277 - 283

SCHERY, S.D., (1985): Measurements of airborne ^{212}Pb and ^{220}Rn at varied indoor locations within United States. *Health Phys. 49, 1061 - 1067*

SCHMIER, H., et al., (1988): Post-Chernobyl whole body counting measurements in the Federal Republic of Germany. In: *Proceedings of 7th International Congress if IRPA, Sydney 10.-17.4.1988, 1098 - 1101*

SCHNEIDER, K.H., (1991): Meßprogramm der Bundesrepublik Deutschland. Umweltradioaktivität in der republik Rußland,. Meßergebnisse des Umweltmeßwagens des Forschungszentrums Jülich GmbH vom 25 Mai bis 12 Juni 1991. Jülich, Forschungszentrum Jülich, *Report ASS-0535*

SCHÖNHOFER, F., (1991): Low-level measurements by liquid scintillation counting. In: *2nd International Conference on Methods and Applications of Radioanalytical Chemistry.* Abstracts, Kona, Hawaii, American Nuclear Society, *61*

SCHÖNHOFER, F., (1992): Measurement of ^{226}Ra in water and ^{222}Rn in water and air by liquid scintillation counting. *Health Phys. 45, 123 - 125*

SCHÖNHOFER, F., HENRICH, F., (1987): Recent progress and application of low level liquid scintillation counting. *J. Radioanal. Nucl. Chem. Articles 115, 317 - 333*

SCHÖNHOFER, F., (1989): Determination of ^{14}C in alcoholic beverages. *Radiocarbon 31,* 777 - 784

SCHUBERT, S., (1998): Radiologische Untersuchungen in Kupferschieferbergbaugebieten von Sachsen Anhalt. *Strahlenschutz Praxis 4, 40 - 44*

SCHULTZ, V., SCHULTZ, C., (1991): Bikini, Enewetak, Rongelap, Marshallese; United States nuclear weapons testing in the Marshall Islands. Livermore CA, Lawrence Livermore National Laboratory, *UCRL-ID-105719 Rev. 1*

SCHWARZ, G., RYBACH, L. (1993): Airborne radiometric survey of the environs of the Swiss nuclear installations. *Radioprotection, Spec. issue: Environmental Impact of Nuclear Installations.* Voelkle, H. Pretre, S., eds., February, 369 - 373

SEEBER, O., et al., (1998): Die Uranaufnahme erwachsener Mischköstler in Deutschland. In: *18. Arbeitstagung Mengen- und Spurenelemente.* Jena, Friedrich-Schiller Universität Jena, *924 - 932*

SEGL, M., et al., (1983): Anthropogenic ^{14}C variations. *Radiocarbon 25, 583 - 592*

SEMIOCHKINA, N., VOIGT, G.,(1998): Consumption habits of the population in Kazakhstan and at the Semipalatinsk test site. In: Initial evaluation of the radioecological situation at the Semipalatinsk test site in the republic of Kazakhstan. D-85764 Neuherberg, GSF - National Research Center for Environment and Health, *GSF-Report 10-98, 51 - 77*

SHAPIRO, J., (1981): *Radiation Protection. 2ⁿᵈ ed.*, Harvard, University Press

SHEBELL, P., HUTTER, A.R., (1995): *Environmental radiation measurements at the former Soviet Union's Semipalatinsk nuclear test site and surrounding villages. Report to the mission leaders.* Vienna, International Atomic Energy Agency

SHEBELL, P., HUTTER, A.R., (1996*): Environmental radiation measurements at the former Soviet Union's Semipalatinsk nuclear test site and surrounding villages. Environmental Measurements Lab. EML-584.* Springfield, Virginia, National Technical Information Service, US Department of Commerce

SHEPARD, F.P., (1967): *Submarine geology.* New York, Harper and Row

SHEPPARD, J.C. (1975): A radiocarbon dating primer. Wash. State Univ., Pullman, Wa., *Bulletin* 338, *1 - 77*

SHEPPARD, J.C., SYED, J.A., MEHRINGER, P.J.Jr., (1979): Radiocayrbon dating of organic components of sediments and peats. In: *Radiocarbon dating.* Berger, R., Suess, H.E., eds., Berkeley, University of California Press, *284 - 305*

SHEPPARD, S.C., EVENDEN, W.G., (1997): Variation in transfer factors for stochastic models: soil to plant transfer, *Health Phys. 72, 727 - 733*

SHIPLER, D.B., et al., (1996): Hanford environmental dose reconstruction project - An overwiew. *Health Phys. 71, 532 - 544*

SHIZUMA, K., IWATANI, K., HASAI, H., (1987): Gamma-ray scattering in the low-background shielding for Ge detector. *Radioisotopes Tokyo Japan 36, 465 - 468*

SHLEIEN, B., SLABACK, L.A., Jr., BIRKY, B.K., (1998): *Handbook of health physics and radiological health, 3ʳᵈ ed.* Baltimore, Williams and Wilkins

SHOIKHET, Y.N., et al., (1998): *The 29ᵗʰ August, 1949. Nuclear test. Radioactive impact on the Altai region population.* Barnaul, Inst. of Regional Medico-Ecological Problems

SHONO, T., (1990): *Electroorganic synthesis.* London, Academic Press

SHUKOLJUKOW, YU.A., (1970): *Fission of uranium nuclei in nature.* Moscow, Atomizdat

SIEBEL W., (1993) : Geochronology of the Leuchtenberg Granite and the associated Redwitzities. *KTB Report 93-2*, Giessen, *411 - 415*

SIEGENTHALER, U., OESCHGER, H., TONGIORGI, E., (1970): Tritium and ogygen-18 in natural water samples from Switzerland. In: *Isotope hydrology.* Vienna, International Atomic Energy Agency, *373 - 385*

ŠILAR, J., (1976): Radiocarbon ground-water dating in Czechoslovakia - first results. *Věst. Ústř. úst. geol. 51, 209 - 220*

ŠILAR, J., (1979): Radiocarbon dating of some mummy and coffin samples. In: *Multidisciplinary research on Egyptian mummies in Czechoslovakia, VIII.* E. Strouhal, *Z. Aeg. 106, 82 - 87*

ŠILAR, J., (1980): Radiocarbon activity measurements of oolitic sediments from the Persian Gulf. *Radiocarbon 22, 655 - 661*

ŠILAR, J., (1989): Radiocarbon dating of ground water in Czechoslovakia and paleoclimatic problems of its origin in central Europe. *Zeszyty Naukowe Politechniki Slaskiej, Ser. Mat.-Fiz. 61, Geochronometria 6, 133 - 141*

ŠILAR, J., (1990): Podzemní voda v hydrologickém cyklu a jako přírodní zdroj. Ground water in the hydrological cycle and as a natural resource (Engl. summary). *Vodohosp. čas. 38, 401 - 426*

ŠILAR, J., JÍLEK, P., DOBROVÁ, H., (1993): Use of isotope analysis in solving environmental problems in power engineering. *Isotopenpraxis 28, 321 - 330*

ŠILAR, J., LOŽEK, V., (1988): Dating of Holocene carbonate sediments from the Slovenská dolina valley at Valča (District of Martin). *Československý kras 39, 69 - 76*

ŠILAR, J., TYKVA, R., (1977): *Radiocarbon dating laboratory on the Charles University, Prague: Methods and results.* Povinec , P., Usatchev, S., eds. Bratislava; Slovenské pedagogické nakladatelstvo, *331 - 334*

ŠILAR, J., TYKVA, R., (1991): Charles University, Prague: Radiocarbon measurements I.. *Radiocarbon 33, 69 - 78*

ŠILAR, J., ZÁHRUBSKÝ, K., (1999): Concentración inicial de radiocarbono en aguas subterráneas del karst de Bohemia Central. In: *Contribución del estudio científico de las cavidades kársticas al conocimiento geológico.* Andreo, B., Carrasco, F., Durán, J.J., eds., Patronato de la Cueva de Nerja (Málaga), *471 - 481*

ŠILAR, J., ŠILAR, J, (1995): Using environmental isotopes for groundwater flow analysis in basinal structures. In: *Application of tracers in arid zone hydrology.* Adar, E.M., Leibundgut, C., eds., Wallingford., *IAHS Publ. 232, 141 - 150*

SILINI,G., METALLI, P., VULPIS, P., (1973): Radioactivity of tritium in mammals. Brussels, Commission European Communities, *Report EUR-5033^e*

SIMON, S.L., (1997): A brief history of people and events related to atomic weapons testing in the Marshall Islands. *Health Phys. 73, 5 - 20*

SIMON, S.L., ROBISON, W.L., (1997b): A compilation of nuclear weapons test detonation data for U.S. Pacific Ocean tests. *Health Phys. 73, 258 - 264*

SIMON, S.L., VETTER, R.J., Eds., (1997a): Consequences of nuclear testing in the Marshall Islands. *Health Phys. 73, 3 - 269*

SIMPSON, R.E., BARATTA, E.J., JELINEK, C.F., (1981): Survey of radionuclides in foods, 1961 - 1977, *Health Phys. 40, 529 - 534*

SIMPSON, R.E., BARATTA,E.J., JELINEK, C.F., (1977): Radionuclides in foods. *J. Assoc. Off. Anal. Chem. 60, 1364 - 1368*

SINE (1964): What does a C-14 date mean? *The Geochronicle, No. 2*, March, 1964, Cambridge, Mass., *4*

SINE (1981): *IAEA isotope laboratory, Sampling of water for ^{14}C analysis.* Vienna, International Atomic Energy Agency, *1 - 2*

SIRI, W., EVERS, J., (1962): Tritium exchange in biological systems. In: *Tritium in the physical and biological sciences.* Atomic Energy Agency, Vienna, International *Report STI/PUB/39, 71*

SJÖLBLOM, K., LINSLEY, G., (1998): International assessment project: summary up. Below the artic seas. Vienna, International Atomic Energy Agency, *IAEA Bulletin 40/4*

SLAT, B., KOSTIAL, K., HARRISON, G.E., (1971): Reduction in the absorption and retention of strontium in rats. *Health Phys. 21, 811 - 814*

SMITH, H., (1982): Environmental behaviour of radionuclides and transfer to man. In: *Radionuclide metabolism and toxicity.* Galle, P., Masse, R., eds., Paris, Masson, *33 - 47*

SMITH, J.T., et al., (1997): Towards a generalized model for the primary and secondary contamination of lakes by Chernobyl derived radiocesium. *Health Phys. 72, 880 - 892*

SMITH, P.E., EVENSEN N.M., YORK, D. (2000): Under the volcano : A new dimension in Ar – Ar dating of volcanic ash. *Geophys. Res. Lett. 27, 5, 585 – 588*

SMOLIAR, M.I., et al., (1996): Re-Os ages of group IIA, IIIA, IVA and IVB iron meteorites. *Science, 271, 1099 - 1102*

SNIHS, J.V., (1996): Conatmination and radiation exposure evaluation and measures in the nordic countries after the Chernobyl accident. In: *Zehn Jahre nach Tschernobyl, eine Bilanz. Seminar des Bundesamtes für Strahlenschutz und der Strahlenschutzkommission, München 6. - 7. März (1996).* Bayer, A., Kaul, A., Reiners, Chr., eds., Stuttgart, Gustav Fischer Verlag, *227 - 279*

SOBEL, H., BERGER, R., (1995): Studies on selected proteins of bone in archaeology. *Radiocarbon 37, 331 - 335*

SONETT, C.P. (1992): The present status of understanding of the long-period spectrum of radiocarbon. In: *Radiocarbon after four decades.* Taylor,R.E., Long,A., Kra, R.S., eds., New York, Springer-Verlag, *50 - 61*

SONNTAG, C., et al., (1980): Paleoclimatic evidence in apparent ^{14}C ages of Saharian groundwaters. *Radiocarbon 22, 871 - 878*

SOUDEK, P., VANĚK, T., TYKVA, R., (2000): Radiophytoremediation as a tool for water cleaning. In: *ASLO , Aquatic Sciences Meeting , Book of Abstracts.* American Society of Limnology and Oceanography, Copenhagen (Denmark), *Abstract SS25 - p05*

SPALDING, R.F., MATHEWS, T.D., (1972): Stalagmites from caves in the Bahamas: Indicators of low sea level stand. *Quaternary Res. 2, 470 - 472*

SPENCER, H., KRAMER, L., SAMACHSON, J., (1972): In: *Biomedical implications of radiostrontium exposure, Proc. Sympos. Davis Calif. Feb.22 - 24, 1971 , U.S. Atomic Energy Commission.* Goldman, M., Bustad, L.K., eds., *31 - 51*

SPENCER, H., LI, M., SAMACHSON, J., LASZLO, D., (1960): Metabolism of strontium-85 and calcium-45 in man. *Metabolism 9, 916 - 921*

SPIERS, D.D., (1968): *Radioisotopes in human body: Physical and biological aspects.* New York, Academic Press

SRDOČ, D., CHAFETZ, H., UTECH, N., (1989): Radiocarbon dating of travertine deposits, Arbucle Mountains, Oklahoma. *Radiocarbon 31, 619 - 626*

SRDOČ, D., et al., (1983): Radiocarbon dating of tufa in paleoclimatic studies. *Radiocarbon 25, 421 - 427*

SRDOČ, D., OBELIĆ, B., HORVATINČIĆ N., (1980): Radiocarbon dating of calcareous tufa: How reliable data can we expect? *Radiocarbon 22, 858 - 862*

SSK, (1987): *Auswirkungen des Reaktorunfalls in Tschernobyl auf die Bundesrepublik Deutschland. Veröffentlichungen der Strahlenschutzkommission Bd. 7.* Stuttgart, Gustav Fischer Verlag

SSK, (1989): *Leitfaden für den Fachberater Strahlenschutz der Katastrophenschutzleitung bei kerntechnischen Notfällen. Veröffentlichungen der Strahlenschutzkommission Vol.13.* Stuttgart, Gustav Fischer Verlag

SSK, (1992a): *Strahlenschutzgrundsätze für die Verwahrung, Nutzung oder Freigabe von kontaminierten Materialien, Gebäuden, Flächen oder Halden aus dem Uranerzbergbau. (Radiological protection principles concerning the safegard, use or release of contaminated materials, buildings, areas or dumps from uranium mining). Veröffentlichungen der Strahlenschutzkommission Vol.23.* Stuttgart, Gustav Fischer Verlag

SSK, (1992b): *Die Exposition durch Radon und seine Zerfallsprodukte in Wohnungen in der Bundesrepublik Deutschland und deren Bewertung. Veröffentlichungen der Strahlenschutzkommission Vol.19.* Stuttgart, Gustav Fischer Verlag

SSK, (1996a): *Zehn Jahre nach Tschernobyl, eine Bilanz. Seminar des Bundesamtes für Strahlenschutz und der Strahlenschutzkommission, München 6.-7-März (1996).* Bayer, A., Kaul, A., Reiners, Chr., eds., Stuttgart, Gustav Fischer Verlag

SSK, (1996b):10 Jahre nach Tschernobyl. Information der Strahlenschutzkommission zu den radiologischen Auswirkungen und Konsequenzen insbesondere in Deutschland. *Berichte der Strahlenschutzkommission des Bundesministeriums für Umwelt, Naturschutz und Reaktorsicherheit, Heft 4*

SSK, (2002): *Leitfaden zur Messung von Radon, Thoron und ihren Zerfallsprodukten. Veröffentlichungen der Strahlenschutzkommission Vol. 47.* Stuttgart, Gustav Fischer Verlag

STACEY, J., KRAMERS, J., (1975): Approximation of terrestrial lead isotope evolution by a two-stage model. *Eart Planet. Sci. Lett. 26, 207 - 221*

STANNARD, J.N., (1988): Radioactivity and health, a history. *DOE/RL/01830-T59,* (DE88013791), Columbus OH, Batelle Press

STAPLETON, G.B., THOMAS, R.H., (1972), Estimation of the induced radioactivity of the ground water system in the neighbourhood of a proposed 300 GeV high energy accelerator situated on a chalk site. *Health Phys. 23, 689 - 699*

STARA, J.F., et al., (1971): Comparative metabolism of radionuclides in mammals: a review. *Health Phys. 20, 113 - 137*

STATHER, J.W., (1982): Americium, curium and einsteinium. In: *Radionuclide metabolism and toxicity.* Galle, P., Masse, R., eds., Paris, Masson, *198 - 224*

STEIGER, R.H., JÄGER E., (1977): Subcomission on geochronology: convention on the use of decay constants in geo- and cosmochronology. *Earth Planet. Sci. Lett. 36, 359 - 362*

STEWARD, R.B., WIKJORD, A.G., (1980): Analytical chemistry on the trail of debris from Russian satellite Cosmos 954. In: *Chemistry in Canada.* Chemical Institut of Canada, Ottawa, *13 - 17*

STEWART, N.G., (1957): World wide deposition of long-lived fission products from nuclear test explosions, U.K. At. Energy Auth., Res.Group, *Report MP/R2354*

STÖCKLIN, G., QAIM, S.M., RÖSCH, F., (1995): The impact of radioactivity on medicine, *Radioch. Act. 70/71, 249 - 272*

STRADLING, G.N., et al., (1989): Biokinetics of plutonium-239 and americium-241 in the rat after the pulmonary deposition of contaminated dust from soil samples obtained from the former nuclear weapons test site at Maralinga: Implications for human exposure. Chilton, *NRPB-M197*

STRADLING, G.N., et al., (1992): Radiological implications of inhaled ^{239}Pu and ^{241}Am in dusts at the former nuclear test site in Maralinga. *Health Phys. 63, 641 - 650*

STRAUME, T., et al., (1996): The feasibility of using ^{129}I to reconstruct ^{131}I deposition from the Chernobyl reactor accident. *Health Phys. 71, 733 - 740*

STRAY, H., (1992) : Improved HPLC method for the separation of Rb and Sr in connection with Rb–Sr dating. *Chem. Geol.102, 129 – 135*

STROUBE, W.B., JELINEK, C.F., BARATTA, E.J., (1985): Survey of radionuclides in foods, 1978 – 1982. *Health Phys. 49, 731 - 735*

STROUHAL, E., VYHNÁNEK, L., (1976): Survey of Egyptian mummies in Czechoslovak collections. *Z. Aeg. 103, 114 - 118*

STUIVER, M., (ed.), (1986a): Calibration Issue. *Radiocarbon 28, 2B, 805 - 1030*

STUIVER, M., BECKER, B., (1993a): High-precision decadal calibration of the radiocarbon time scale, A.D.1950-6000 B.C. *Radiocarbon 35, 35 - 65*

STUIVER, M., et al., (1986b): Radiocarbon age calibration back to 13,300 years BP and the ^{14}C age matching of the German oak and US bristlecone pine chronologies. *Radiocarbon 28, 969 - 979*

STUIVER, M., POLACH, H.A., (1977): Radiocarbon 1977, discussion reporting of ^{14}C data. *Radiocarbon 19, 355 - 363*

STUIVER, M., REIMER, P.J., (1986c): A computer program for radiocarbon age calibration. *Radiocarbon 28, 1022 - 1030*

STUIVER, M., REIMER, P.J., (1993b): Extended ^{14}C data base and revised CALIB 3.0 ^{14}C age calibration program. *Radiocarbon 35, 215 - 230*

STUIVER, M., SUESS, H.E., (1966): On the relationship between radiocarbon dates and true sample ages. *Radiocarbon 8, 534 - 540*

SUESS, H.E., (1955): Radiocarbon concentration in modern wood. *Science 122, 415 - 417*

SUESS, H.E., (1965): Secular variations of the cosmic-ray produced carbon-14 in the atmosphere and their interpretions. *J. Geophys. Res. 70, 5937 - 5952*

SUESS, H.E., (1969): Tritium geophysics as an international research project. *Science 163, 1405 - 1410*

SUESS, H.E., (1986): Secular variations of cosmogenic ^{14}C on earth: their discovery and interpretation. *Radiocarbon 28*, New Haven, Connecticut

ŠULCEK, Z., POVONDRA, P., (1989): *Methods of decomposition in inorganic analysis.* CRC Press, Boca Ralton, Florida, 325-

SULLIVAN, M.F. (1980): Absorption of actinide elements from gastrointestinal tract of rats, guinea pigs, and dogs. *Health Phys. 38, 159 - 171*

SULLIVAN, M.F., et al., (1979): The influence of oxidation state on the absorption of plutonium from the gastrointestinal tract. *Radiat. Res. 80, 116 - 121*

SULLIVAN, M.F., et al., (1980): Absorption of plutonium from the gastrointestinal tract of rats and guinea pigs after ingestion of alfalfa containing Pu-238. *Health Phys. 38, 215 - 221*

SULLIVAN, M.F., et al., (1985): Further studies on the influence of chemical form and dose of absorptions of Np, Pu, Am, and Cm from the gastrointestinal tracts of adult and neonatal rodents. *Health Phys. 48, 61 - 73*

SVERDRUP, H.V., JOHNSON, M.W., FLEMING, R.H., (1963): *The Oceans: Their physics, chemistry and general biology.* Englewood Cliffs, New Jersey, Prentice-Hall

SWITSUR, R., (1986): 884 Cal.BP and all that. *Antiquity UK 60 (230), 214 - 216*

SYMONDS, J.L., (1985): *A history of British atomic tests in Australia.* Canberra, Australia, Australian Goverment Publishing Service

SZ, (1999): Indische Dorfbewohner nach Atomtest entschädigt. *Süddeutsche Zeitung 16. 03. 1999*

TADMOR, J., (1973): Deposition of ^{85}Kr and tritium released from nuclear fuel reprocessing plant, *Health Phys. 24, 37 - 42*

TAKESHITA, K., et al., (1972): Cesium-137 in placenta, urine, food and rain in Hiroshima. *Health Phys. 22, 252 - 256*

TAYLOR, R.E., (1987): *Radiocarbon dating an archaeological perspective.* Orlando, Academic Press

TAYLOR, R.E., (1992): Radiocarbon dating of bone: To collagen and beyond. In: *Radiocarbon after four decades.* Taylor, R.E., Long, A., Kra, R.S., eds., New York Springer-Verlag, *375 - 402*

TAYLOR, R.E., (2000): The contribution of radiocarbon dating to New World archaeology. *Radiocarbon 42, 1 - 21*

TAYLOR, R.E., LONG, A., KRA, R.S., eds., (1992): *Radiocarbon after four decades.* New York, Springer-Verlag

TENU, A., DAVIDESCU, F., (1998): Carbon isotope composition of atmospheric CO_2 in Romania. In: *Isotope techniques in the study of environmental change.* Vienna, International Atomic Energy Agency, *19 - 26*

TERA, F., WASSERBURG, G.J., (1972): U-Th-Pb systematics in three Apollo 14 basalt sand the problem of initial Pb in lunar rocks. *Earth Planet. Sci. Lett. 14, 281 - 304*

TERNOVSKIJ, I.A., ROMANOV, G.N., FEDOROV,E.A., (1990): Radioactive cloud trace formation dynamics after the radiation accident in the southern Urals in 1957: Migration processes. In: *Proc. of a symposium on recovery operations in the event of a nuclear accident or radiological emergency.* Vienna, International Atomic energy Agency, *STI/PUB/826, 433 - 437*

TERRY, N., BAÑUELOS, G., eds., (2000): *Phytoremediation of contamined soil and water.* Boca Raton, FL, CRC Press

TETLEY N., MCDOUGALL I., HEYDEGGER H.R., (1980): Thermal neutron interferences in the ^{40}Ar/^{39}Ar dating technique. *J. Geophys. Res. B12, 7201 - 7205*

THATCHER, L.L., PAYNE, B.R., (1965): The distribution of tritium in precipitation over continents and its significance to groundwater dating. In: *Radiocarbon and tritium dating.* Wash., Washington State University, Pullman, *604 - 629*

THIRLWALL, M. F., (1991): High – precision multi-collector isotopic analysis of low levels of Nd as oxide. *Chem. Geol. 94, 13 – 22*

THIRWALL, M.F., (1982): A triple-filament method for rapid and precise analysis of rare-earth elements by isotope dilution. *Chem. Geol. 35, 155 - 166*

THOMAS, R.H., (1970): Possible contamination of ground water system by high energy proton accelerators. *U.S. AEC Report UCRL-20131*

THOMASSET, M., (1982): Strontium metabolism and toxicity of strontium. In: *Radionuclide metabolism and toxicity.* Galle, P., Masse, R., eds., Paris, Masson, *98 - 121*

THOMPSON, C.B., McArthur, R.D., (1996): Challenges in developing estimates of exposure rate near the Nevada test site. *Health Phys. 71, 470 - 476*

THOMPSON, R.C., (1975): Animal data on plutonium toxicity. *Health Phys. 29, 511 - 519*

THOMPSON, R.C., BAIR, W.J., eds., (1972): The biological implications of the transuranium elements. Proceedings of the 11th Hanford biology symposium, Richland, 27-29 Sept. 1971. *Health Phys. 22, 533 - 960*

THOMSON, J., et al., (1995): Radiocarbon age offsets in different-sized carbonate components of deep-sea sediments. *Radiocarbon 37, 91 - 101*

TICHLER, J.K., NORDEN, K., CONGEMI, J., (1988 - 1992): Radioactive materials released from nuclear power plants. Annual Reports 1985 to 1989. *NUREG/CR-2907 and BNL-NUREG-51581, Vol. 6 - 10*

TILTON, G.R., (1960): Volume diffusion as mechanism for discordant lead ages. *J. Geophys. Res. 65, 2933 - 2945*

TOIVONEN, H., et al., (1992): A nuclear incident at a power plant in Sosnovyy Bor, Russia. *Health Phys. 63, 571 - 573*

TÖTTERMAN, L., KIVINIITY, K., SOKKANEN, R., (1972). Radioactive contamination of foetal bone. *Health Phys. 22, 193 - 195*

TRABALKA, J.R., AUERBACH, S.I. (1990): One western perspective of the 1957 soviet nuclear accident. In: Proc. of Seminar on Comparative Assessment of the Environmental Impact of Radionuclides Released during three major Nuclear Accidents: Kyshtym, Windscale, Chernobyl, *EUR-13574, 41 - 69*

TSCHURLOVITS, M., PFEIFFER, K.J., RANK,D., (1982): A comparison of different methods for determination of $^{14}CO_2$ in air. *Atomkernene. Kerntech. 40, 267 - 269*

TSOULFANIDES, N., (1983): *Measurement and Application of Radiation.* Washington, Hemisphere Publ. Comp.

TUBIANA, M., (1982): Metabolism and radiotoxicity of radionuclides, Iodine. In: *Radionuclide metabolism and toxicity.* Galle, P., Masse, R., eds., Paris Masson, *49 - 81*

TUREKIAN, K.K., (1969): The oceans, streams and atmosphere. In: *Handbook of Geochemistry.* Wedepohl, K.H., ed., Berlin, Springer

TURNER, G., (1971): Argon 40 - argon 39 dating: The optimisation of irradiation parameters. International Atomic Energy Agency, *Earth Planet. Sci. Lett. 10, 227 - 234*

TURNER, J.E., (1986): *Atoms, radiation, and radiation protection.* London, Pergamon Press

TYKVA, R., (1977). Semiconductor detectors for low-levels of alpha, beta or gamma nuclides: Counting, spectrometry and applications. In: *Low-radioactivity measurements and applications.* Povinec, P., Usatschev, S., eds., Bratislava, Slovenské pedagogické nakladatelstvo, *213 - 217*

TYKVA, R., (1980): Limits of beta counting due to sample sorption and procedures for exclusion of the counting rate instability. In: *Liquid scintillation counting: Recent applications and development, Vol. I.* Peng Chin-Tzu, Horrocks, D.L., Alpen, E.L., eds., New York, Academic Press, *225 - 233*

TYKVA, R., (1995): PC-controlled *in situ* evaluation of multilabelling with beta tracers in biological samples. *J. Radioanal. Nucl. Chem. Articles 195, 327 - 334*

TYKVA, R., (2000): Analysis of the surface technology of silicon detectors for imaging of low energy beta tracers in biological material. *Nucl. Instrum. Methods Phys. Res. A 448, 576-580*

TYKVA, R., JISL, R., (1986): Correction of the continuous scanning record of radioactivity distribution. I. Smoothing. *Nucl. Instrum. Methods Phys. Res. A 251, 160 - 165*

TYKVA, R., MACHÁČKOVÁ, I., KREKULE, J., (1992): A topographic method for studying uptake, translocation and distribution of inorganic ions using two radiotracers simultaneously. *J. Exp. Bot. 43, 1083 - 1087*

TYKVA, R., et al., (2002): Translocation studies of natural and anthropogenic radionuclides in plants. In: *7th International Conference on Nuclear Analytical Methods in the Life Sciences, Antalya, Turkey, okk of Abstracts, 159 -*

UKAEO, (1957): *Accident of Windscale No.1 pile on october 10,1957.* London, United Kingdom Atomic Energy Office, HM Stationery Off.

UKAEO, (1958): *Final Report on the Windscale accident.* London, United Kingdom Atomic Energy Office, HM Stationery Off.

UNSCEAR, (1962): United Nations Scientific Committee on the Effects of Atomic Radiation, New York, UN, *Report to the General Assembly, 17. Session, Supplement 16 (A/526)*

UNSCEAR, (1977): Sources and effects of ionization radiation. United Nations Scientific Committee on the Effects of Atomic Radiation, New York, UN, *Report to the General Assembly, with annexes*

UNSCEAR, (1982): Ionization Radiation:Sources and biological effects. United Nations Scientific Committee on the Effects of Atomic Radiation, New York, UN, *Report to the General Assembly, with annexes*

UNSCEAR, (1988): Ionization Radiation:Sources, effects and risks of ionizing radiation. United Nations Scientific Committee on the Effects of Atomic Radiation, , New York, UN, *Report to the General Assembly, with annexes*

UNSCEAR, (1993): Sources, effects and risks of ionizing radiation. United Nations Scientific Committee on the Effects of Atomic Radiation, New York, UN, *Report to the General Assembly, with annexes*

USAEC (1974): Final environmental statement related to construction and operation at Barnell nuclear fuel plant. Division of technical information, U.S. Atomic Energy Commission, Washington, *Doket No. 50-332, V-15*

USSR, (1986): *The accident at the Chernobyl'nuclear power plant and its consequences. Information compiled for the IAEA experts meeting, 25-29 August.* Vienna, International Atomic Energy Agency

VALLADAS, H., et al., (1992): Direct radiocarbon dates for prehistoric paintings at the Altamira, El Castillo and Niaux Caves. *Nature London, 357 (6373), 68 - 70*

VAN DER PLICHT, J., (1993): The Groningen radiocarbon calibration program. *Radiocarbon 35, 231 - 237*

VAN der VINK, G., et al., (1998): *False Accusations, undetected test and implifications for the CTN treaty.* Washington DC, Arms Control Association

VAN MIDDLESWORTH, L., (1954): Radioactivity in animals thyroid from various areas. *Nucleonics 12, 56*

VANCE, D., MEIER, M., OBERLI, F. (1998): The influence of high U-Th inclusions on the U-Th-Pb systematics of almandine-pyrope garnet : Results of a combined bulk dissolution, step – wise leaching and SEM study. *Geochim. Cosmochim. Acta, 62, 21 - 22, 3527 - 3540*

VERHAGEN, B.T., BUTLER, M.J., (1998): Environmental isotope studies of urban and waste disposal impact on ground water resources in South Africa. In: *Isotope techniques in the study of environmental change.* Vienna, International Atomic Energy Agency, *411 - 421*

VERHAGEN,B.T., GEYH,M.A., FRÖHLICH,K., WIRTH,K., (1991*): Isotope hydrological methods for the quantitative evaluation of ground water resources in arid and semi-arid areas - Development of a methodology. Bonn, *Research Reports of the Federal Ministry for Economic Cooperation of the Federal Republic of Germany 85*

VERPLANCKE, J., (1992): Low-level gamma spectroscopy: low, lower, lowest. *Nucl. Instrum. Methods Phys. Res. A 312, 174 - 182*

VERSCHURE R.H., MAIJER C., (1993) : A Rb - Sr isochron from a single biotite crystal. *Mineral. Mag. 57, 746 - 749*

VILLFORTH, J.C., ROBINSON, E.E., WOLD, G.J., (1969): *A review of radium incidents in the USA.* Vienna, Handl. Radiat. Accid., Proc. Symp.

VITA-FINZI, C., (1980): [14]C dating of recent crustal movements in the Persian Gulf and Iranian Makran. *Radiocarbon 22, 763 - 773*

VOELZ, G.L., GRIER, R.S., (1985): A 37 - year medical follow-up of Manhattan project Pu - workers. *Health Phys. 48, 249 -259*

VOELZ, G.L., LAWRENCE, N.P.,JOHNSON,E.R. (1997): Fifty years of plutonium exposure to the Manhattan project plutonium workers: an update. *Health Phys. 73, 611 - 619*

VOGEL, J.C., (1967): Investigation of groundwater flow with radiocarbon. In:*Isotopes in hydrology*. Vienna, International Atomic Energy Agency, *355 - 369*

VOGEL, J.C., (1970): Carbon-14 dating of groundwater. In: *Isotope hydrology*. Vienna, International Atomic Energy Agency, *225 - 239*

VOGEL, J.C., (1983): C-14 variations during the Upper Pleistocene. *Radiocarbon 25, 213 - 218*

VOGEL, J.C., EHHALT, D., (1963): The use of carbon isotopes in groundwater studies. In: *Radioisotopes in hydrology*. Vienna, International Atomic Energy Agency, *383 - 240*

VOGEL, J.C., KRONFELD, J., (1997): Calibration of radiocarbon dates for the Late Pleistocene using U/Th dates on stalagmites. *Radiocarbon 39, 27 - 32*

VOGEL, J.C., UHLITZSCH, I., (1975): Carbon-14 as an indicator of CO_2 pollution in cities. In: *Isotope ratios as pollutant source and behaviour indicators*. Vienna, International Atomic Energy Agency, *143-150*

VOIGT, G., SEMIOCHKINA, N., (1998): The radioecological situation at the Semipalatinsk test site in the republic of Kasakhstan. In: *Initial evaluation of the radioecological situation at the Semipalatinsk test site in the republic of Kazakhstan*. D-85764 Neuherberg, GSF - National Research Center for Environment and Health, *GSF-Report 10-98,1 - 50*

VOLKENING, J., WALCZYK, T., HEUMANN, K.G., (1991): Osmium isotope ratio determinations by negative thermal ionization mass spectrometry. *Int. J. Mass Spectrom. Proc. 105, 147 - 159*

VÖLKLE, H.R., (2000): Tokaimura I: Bewertung der Ereignisse aus der Sicht des Strahlenschutzes. *Strahlenschutz Praxis 6, 58 – 64*

VUGRINOVICH, R. G. , (1981) : A distribution – free alternative to least – squares regression and its application to Rb/Sr isochron calculations. *Math. Calc. Geol. 13, 443 – 454*

WALCYK, T., HEBEDA, E.H., HEUMANN, K.G., (1991): Osmium isotope ratio measurements by negative thermal ionization mass spectrometry (NTI-MS). *Fres. J. Anal. Chem. 341, 537 - 541*

WALFORD, G.V., COOPER, J.A., KEYER, R.N., (1976) : Evaluation of standardized Ge (Li) gamma ray detectors for low-level environmental measurements. *IEEE Trans. Nucl. Sci. NS 23 (1),734 - 737*

WALKER, A.J., OTLET, R.I., LONGLEY, H., (1986): Applications of the use of hawthorn berries in monitoring ^{14}C emissions from a UK nuclear establishment over an extended period. *Radiocarbon 28, 2A, 673 - 680*

WALKER, R.J., MORGAN, J.W., NALDRETT, A.J., LI, C., (1991): Re-Os isotopic systematics of Ni-Cu sulfide ores, Sudbury Igneous Complex, Ontario: evidence for a major crustal component. *Earth Planet. Sci. Lett. 105, 416 - 429*

WALTERS, W.H., RICHMOND, M.C., GILMORE, B.G., (1996): Reconstruction of radioactive contamination in the Columbia River. *Health Phys. 71, 556 - 567*

WASSENAAR, L.I., ARAVENA, R., FRITZ, P., (1989): The geochemistry and evolution of natural organic solutes in groundwater. *Radiocarbon 31, 865 - 875*

WASSENAAR, L.I., ARAVENA, R., FRITZ, P., (1992): Radiocarbon contents of dissolved organic and inorganic carbon in shallow groundwater systems. In: *Isotope techniques in water resources development 1991*. Vienna, International Atomic Energy Agency, *143 - 166*

WEBB, J.H., (1949): The gogging of photographic film by radioactive contaminants in cardboard packaging materials. *Phys. Rev. 76, 375 - 380*

WEBB, S.B., SHAWKI, A.I., WHICKER, F.W., (1997): a three - dimensional spatial model of plutonium in soil near Rocky Flats, Colorado. *Health Phys. 73, 340 - 349*

WEISS, W., ROETHER, W., (1975): Der Tritiumabfluss des Rheins 1961-1973. *Dt. Gewässerkdl. Mitt. 19, 1-5*

WENDORF, F., (1992): The impact of radiocarbon dating on North African Archaeology. In: *Radiocarbon after four decades.* Taylor,R.E., Long, A., Kra, R.S., eds., New York, Springer-Verlag, *310 - 323*

WENDT, I., (1984): A three-dimensional U-Pb discordia plane to evaluate samples with common lead of unknown isotopic composition. *Isot. Geosci. 2, 1 - 12*

WENDT, I., CARL, C., (1991): The statistical distribution of the mean square weighted deviation. *Chem. Geol. 86, 275 - 285*

WENDT, J.I., WENDT, I., TUTTAS, D., (1993): Determination of U-Pb ages of zircons by direct measurement of the $^{210}Pb/^{206}Pb$ ratio. *Chem. Geol. 106, 467 - 474*

WENG, P.S., BECKNER, W.M., (1973): Cesium-137 turnover rates in human subjects of different ages. *Health Phys. 25, 603 - 605*

WETHERILL, G.W., (1956): Discordant uranium-lead ages. *Trans. Am. Geophys. Un. 37, 320 - 326*

WHICKER, F.W., et al., (1996): Ingestion of Nevada test site fallout: internal dose estimates. *Health Phys. 71, 477 - 486*

WHICKER, F.W., NELSON, W.C., GALLEGOS, A.F., (1972): Fallout ^{137}Cs and ^{90}Sr in trout from mountain lakes in Colorado. *Health Phys. 23, 519 - 527*

WHITEHEAD, D.C., (1984): The distribution and transformation of iodine in the environment. *Environ. Int. 10, 321 - 339*

WHO, (1996): *Health consequences of the Chernobyl accident. Results of the IPHECA pilot projects and related nationals programmes.* Scientific Report International Programm on the Health Effects of the Chernobyl Accident. Souchkevitch, G.N., Tsyb, A.F., eds., Geneva, World Health Organization

WICK, R.R., CHMELEVSKY, D., GÖSSNER, W., (1995): Current status of the follow - up of radium-224 treated ankylosing spondylitis patients. In: *Health effects of Internally deposited Radionuclides: Emphasis on Radium and Thorium.* Van Kaick, G., Karaoglou, A., Kellerer, A.M., eds., Singapore, World Scientific, *165 - 169*

WICK, R.R., GÖSSNER, W., (1993): History and current uses of ^{224}Ra in ankylosing spondylitis and other diseases. *Environ. Int. 19, 467 - 473*

WIECHEN, A., (1972): Ursachen des hohen Cs-137 - Gehaltes der Milch von Moorböden. *Milchwissenschaft 27, 82 - 84*

WIGLEY, T.M.L., (1976): Effect of mineral precipitation on isotopic composition of ^{14}C dating of groundwater. *Nature 263, No. 5574, 219 - 221*

WILKENING, M.H., (1956): Variation of natural radioactivity in the atmosphere with altitude. *Trans. AGU 37, 117 - 180*

WILKINSON, D.H., (1950): *Ionization chambers and counters.* London, Cambridge, University Press

WILSON, M., (1989): *Igneous Petrogenesis.* London, Unwin Hyman

WINKELMANN, I. et al. (1995). Überwachung der Umweltradioaktivität vom Hubschrauber aus. In : 9. *Fachgespräch zur Überwachung der Umweltradioaktivität.* Bundesamt für Strahlenschutz, Oberschleisheim, Germany, *94 - 97*

WINKELMANN, I., et al., (1994): Radioaktivitätsmessungen im Südural im Jahre 1992. In: *Erste deutsche Aktivitäten zur Validierung der radiologischen Lage im Südural.* Burkhard, W., ed., Bundesamt für Strahlenschutz, Inst.für Strahlenhygiene, Neuherberg, *BfS-ISH-166/94, 2 - 10*

WINKLER, R., RÖSNER, G., (1989): Proportional counter for low-level [241]Pu measurement. *Nucl. Instrum. Methods. Phys. Res. A 274, 359 - 361*

WINN, W.G., BOWMAN, W.G., BONI, A.L., (1988): Ultra-clean underground counting facility for low-level environmental samples. *Sci. Total Envir. 69, 107 - 144*

WIRDZEK, S., KAZIMIR, D., (1985): Radioactivity of frequently used building materials in Slovakia. In: *Low-level counting and spectrometry.* Povinec, P., ed., Bratislava, Veda, *275 - 280*

WIRTH, E., (1980*): Literaturstudie zur Problematik des Transfers langlebiger Radionuklide aus dem Boden in die Pflanze.* Neuherberg, Bundesamt für Strahlenschutz, *STH-Berichte 1/1980*

WITTELS, M.C. (1966): Stored energy in graphite and other reactor materials. *Nucl. Saf. 8, 134*

WOGMAN, N.A., LAUL, J.C., (1982):Natural contamination in radionuclide detection systems. In: *Natural radiation environment.* Vohra, W.G., et al., eds., New York, John Wiley and Sons, *384 - 390*

WOHLFARTH, B., BJÖRCK, S., POSSNERT, G., (1995): The Swedish time scale: a potential calibration tool for the radiocarbon time scale during the Late Weichselian. *Radiocarbon 37, 347 - 359*

WOLF, M., (1993): Datierung des Bordeaux „1787 Lafitte Th.J.,, durch Kohlenstoff-14- und Tritiumanalysen. D-85764 Neuherberg, GSF - National Research Center for Environment and Health, *Jahresbericht 1992, Institut für Hydrologie, 17 - 23*

WOLF, M., SINGER, C., (1991): Synthesis of propane via hydrogenation of propadiene, propine or propene for low-level measurement of tritium.(In German). D-85764 Neuherberg, GSF - National Research Center for Environment and Health, *Annual Report 1990 of the GSF Institute of Hydrology No. GSF-HY-1/91, 150 - 153*

WÖLFLI, W., et al., (1984): Proceedings of the 3rd International Symposium on Accelerator Mass Spectrometry. *Nucl. Instr. Methods, B52 (2,3), 211-630*

WOLTERBEEK, H.T., VAN DER MEER, A.J.G.M., (1996): A sensitive and quantitative biosensing method for the determination of γ - ray emitting radionuclides in surface water. *J. Environ. Radioact. 33, 237 - 254*

WOODARD, H.Q., (1970): The biological effects of tritium. New York, Health Safety Lab., U.S. Atomic Energy Commission, *Report HASL-229*

WOODHEAD, D.S., (1973): The radiation dose received by plaice (pleuronectes platessa) from the waste discharged into the North - East Irish Sea from the fuel reprocessing plant at Windscale. *Health Phys. 25, 115 - 121*

WRIGHT N., LAYER P.W., YORK D., (1991): New insights into thermal history from single grain [40]Ar/[39]Ar analysis of biotite. *Earth Planet. Sci. Lett. 104, 70 - 79*

XU, X.G., REECE, W.D., (1996): Sex-specific tissue weighting factors for effective dose equivalent calculations. *Health Phys. 106, 81 - 86*

YABLOKOV, A.V., ed., (1994): *Plutonium in Russia.* Moscow, Centre for Ecological Politics

YACHMENOV, V.A., ISAGEVA, L.W., (1996): Environmental monitoring in the vicinity of the Mayak atomic facility. *Health Phys. 71, 61 - 70*

YAMAMOTO, M., TSUKATANI, T., KATAYAMA, Y., (1996): Residual radioacitivity in the soil of the Semipalatinsk nuclear test site in the former USSR. *Health Phys. 71, 142 - 148*

YATES, T., (1986): Studies of non-marine mollusks for the selection of shell samples for radiocarbon dating. *Radiocarbon 28, 457 - 463*

YIN, Q.Z., et al., (1993): ^{187}Os - ^{186}Os and ^{187}Os - ^{188}Os method of dating: introduction. *Geochim. Cosmochim. Acta 57, 4119 - 4128*

YORK, D. (1969) : Least - squares fitting of a straight line with correlated errors. *Earth Planet. Sci. Lett. 5, 320 - 324*

YU-FU, YU., SALBU, B., BJORNSTAD, H.E., (1991): Recent advances in the determination of low level plutonium in environmental and biological materials. *J. Radioanal. Nucl. Chem. Articles 148, 163 - 174*

YURTSEVER, Y., (1983): Models for tracer data analysis. In: *Guidebook on nuclear techniques in hydrology, 1983 Edition.* Vienna, International Atomic Energy Agency, *Tech. Report Series 91, 381 - 402*

ZDESENKO, YU.G., et al., (1985): Preliminary results of neutrinoless double beta-decay of ^{76} Ge, (in Russian). *Izv. Akad. Nauk SSSR, Ser. fiz. 49, 862 – 867*

ZECHNER, J., MÜCK, K., (1996): Die Strahlenexposition der österreichischen Bevölkerung nach dem Reaktorunfall und die Maßhahmen zu deren Reaktion. In: *Zehn Jahre nach Tschernobyl, eine Bilanz. Seminar des Bundesamtes für Strahlenschutz und der Strahlenschutzkommission, München 6.-7-März (1996).* Bayer, A., Kaul, A., Reiners, Chr., eds., Stuttgart, Gustav Fischer Verlag, *321 - 341*

ZINCHENKO, G.S., et al., (1997): Mathematical modeling of radioactive fallouts from the Semipalatinsk nuclear test. *Russian Meteorology and Hydrology 8, 1 -9*

ZITO, R., et al., (1980): Possible subsurface production of carbon 14. *Geophys. Res. Lett. 17, 235 - 238*

ZOJER, H. (1992): Identification of palaeowaters by means of environmental isotope correlation. In: *Isotope techniques in water resources development 1991.* Vienna, International Atomic Energy Agency, *625 - 627*

ZOUARI, K., CHKIR, N., CAUSSE, C., (1998): Pleistocene humid episodes in southern Tunisian chotts. In: *Isotope techniques in the study of environmental change.* Vienna, International Atomic Energy Agency, *543 – 554*

Abbreviations

A.D.	*Anno Domini*
AGR	*graphite moderated and gas cooled reactor*
AMS	*accelerator mass spectrometry*
ANU	*secondary Sucrose standard of the Australian National University for 95 % modern radiocarbon activity*
B.C.	*before christ*
B.P.	*years before present (1950)*
BABI	*basaltic achondrite best inicial ($[^{87}Sr]/^{86}Sr]$) = 0.698990*
BF	*bioaccumulation factor*
BWR	*boiled light-water moderated and light-water cooled reactor*
CC	*composition of average crustal reservoir*
CF	*concentration factor*
Ch-Suc	*secondary Sucrose standard of China for 95 % modern radiocarbon activity*
CHUR	*chondrite uniform reservoir*
cph	*counting rate (counts per hour)*
cpm	*counting rate (counts per minute)*
cps	*counting rate (counts per second)*
DIC	*dissolved inorganic carbon*
DM	*depleted mantle*
DOC	*dissolved organic carbon*
dpm	*disintegrations per minute*
DVB	*divinylbenzene*
EC	*electron capture*
EEDC	*equilibrium-equivalent decay-product concentration*
FBR	*fast breeder*
FEP	*fluorinated ethylene propylene*
FWHM	*peak at its half maximum*
GCR	*graphite moderated, gas cooled reactor*
GM	*Geiger-Müller counter*
GSMS	*gas-source mass spectrometry*
HIMU	*subducted oceanic crust*
HPGe	*high purity germanium single crystals*
HPLC	*high performance liquid chromatography*
HWR	*heavy water moderated and cooled reactor*

ICP	*inductively coupled plasma analysis*
ICP-MS	*inductively coupled plasma mass spectrometry*
IUGS	*International Union of Geological Sciences*
K_d - value	*distribution coefficient*
LSC	*liquid scintillation counting*
LWGR	*light water cooled and graphite moderated reactor*
LWR	*light water cooled and normally water moderated reactor*
MORB	*mid ocean ridge basalts*
MSWD	*mean squared weighted deviation*
NBS	*National Bureau of Standards*
NOx	*New NBS Oxalic acid standard*
N-TIMS	*negative-ion thermal ionisation mass spectrometry*
OIB	*ocean island basalt*
OR-ratio	*observed concentration ratio*
Ox	*NBS Oxalic acid standard*
PAEC	*potential alpha-energy concentration*
PC	*personal computer*
PDB	*^{13}C standard Peedee Belemnite limestone from South Carolina*
PET	*positron emission tomography*
PGE	*platinum group element*
pmc	*% of the mordern carbon standard acticity*
POPOP	*1,4-di-[2-(5-phenyloxazol)]-benzene*
ppm	*part per million*
PPO	*2,5-diphenyloxazol*
p-T-t	*pressure-temperature-time*
PWR	*pressurized water moderated and cooled reactor*
RBMK	*high power reactor of boiling water type, graphite moderated*
REE	*rare earth elements*
RTG	*radioisotope thermal generator*
SEM	*secondary electron multiplier*
SHRIMP	*sensitive high mass-resolution ion microprobe*
SI	*système internationale d'unités*
SIMS	*secondary ion mass spectrometry*
SMOW	*the world standard of mean ocean water*
SNAP	*satellite nuclear auxiliary power supply*
T.R.	*tritium ratio*
T.U.	*tritium unit corresponds to the specific activity of 0.118 Bq·kg-1*
TFE	*teflon - tetrafluoroethylene*

TIMS	*mass spectrometer with thermal ionisation source*
TNT	*trinitrotoluene*
UR	*uniform reservoir*
vol.%	*volume percent*
WL	*working level*
WR	*whole rock sample*
wt. %	*weight percent*
XRF	*X-ray fluorescence*

Index

A

absolute time scale 184
absorption see resorption, transfer
  caesium 60, 61
  iodine 54, 55
  mammals 54, 60, 63, 64, 67
  plants 42-44, 54, 63, 67
  plutonium 67, 69
  rocks 114, 141
  strontium 63, 64
accelerator
  contamination 19, 142
  radionuclide production 2, 19
accelerator mass spectrometry (AMS)
  169, 173, 185-188, 195, 211, 322, 332, 333
accident 49, 53, 59, 66
  [137]Cs source Goiânia 144
  [192]Ir source Morocco 144
  [60]Co source Mexico 144
  aircraft with nuclear weapons 131, 132
  Kyshtym 87, 88
  Lake-Karachay 88
  nuclear powered ships .. 134, 136, 138
  nuclear powered submarines   136, 137
  Oak Ridge 85
  reactor 26, 110
    Chernobyl 61, 117-123, 125-130
    Fermi reactor 113
    Lucens 114
    SL-1 Idaho Falls 112, 113
    Sosnovy Bor 131
    Three Mile Island 115-117
    Windscale 111, 112
  Rocky Flats 84
  satellites with nuclear power   132, 133
  Techa 85, 87
  Tomsk 7 89
actinoides, human 48
activation products 18, 22-25, 53, 71, 93, 106, 107, 121, 122
activity 2, 3, 5, 6
  building materials 277
  concentration 7
  constructional materials 277
  detector 274, 289, 290, 291
    background 274
    counting rate 274
    detection efficiency 274, 275
    measurement geometry 275

    response 275
    selection 289
  evaluation 275
  human body 309
    spectrum 309
  measurements 273, 282, 283, 288, 290, 291, 295
  mother's milk 309
  specific 7, 287
  unit (Bq) 6
  unit (Ci) 6
  unit (dps, dpm, dph) 6
activity detector
  applied types 291
  ionization 291
adiabatic lapse rate **29-31**
admixture
  allochthonous carbon 174
  calcium carbonate 189
  foreign material 160
  modern carbon 163
  non active carbon 163, 164
  tritium 181
adsorption 38, 44
aerosol 28, 32, 54, 58, 59, 68, 106, 114
age
  absolute 163
  apparent 204, 258
  relative 182
air pollution monitoring 206
aircraft surveys [137]Cs 310
algae 169
aluminium
  [26]Al 333
  emission 76
  human 48
  industry 76, 77
americium
  behaviour 50
  concentration factor 41
  deposition 66
  nuclear weapon tests 66, 102
  smoke detector 144
  emission reactor 106
  formation 23
  nuclear weapon tests 97
anticompton 302
ANU sucrose standard 160, 161
aquifers
  admixing of modern water 200

Aquitaine 192
Chad Basin 192
coastal 203
confined 190, 199, 206, 208
Czech Republic 190, 192, 204
exponential model 200
geochemical processes 196
groundwater recharge Czech Rep. 205
hydrodynamic processes 196
identifying 167
infiltration model 200
injection model 200
Lorraine 192
model completely mixed reservoir 200
Paris Basin 192
piston flow model 200
tertiary 204
aragonite 166, 189
Ar-Ar dating 216, 219-222
example 223, 224
limitation 222, 223
orthogneisses 224
archaeology 147, 164, 182, 184-185, 188
archival documents 209
argon
^{36}Ar 221
atmosphere 219
neutron activation 220
^{37}Ar half-life 216
^{38}Ar 216
spiking 218
^{39}Ar 221
cosmic radiation 213
dating 213, 214
half-life 213, 216, 219
neutron activation 219, 220
reservoirs 213
thermal waters 214
^{40}Ar 212, 216, 217, 218, 219, 221
neutron activation 220
^{41}Ar reactor 103
Ar-Ar analytical technique 219
laser probe technique 221-223
natural isotopes 216
step heating technique 221, 222
ash 208
radioactivity 72, 74
volcanic 223
atmosphere 25, 27, 28, 168
^{137}Cs 59, 95
^{14}C 15, 17, 18, 52
^{22}Na 15, 17, 18

^{39}Ar 213
^{3}H 15, 17, 18, 51
^{7}Be 15, 17, 18
^{85}Kr 57, 83, 214
carbon 52
carbon dioxide concentration 13
debris 32
emissions 28, 72, 74, 76, 77, 81-84, 92,
93, 95, 97, 106, 108, 109, 116, 133
greenhouse effect 13
intermixture 29, 30, 32, 34
iodine 5.............. 3, 54, 83, 84, 93, 106
natural nuclear processes 15
natural produced radionuclides 15, 17, 18,
71
nuclear weapon tests 25, 28, 62, 64, 66,
92, 93, 95-99, 102
plutonium 66, 84, 102, 133
pressure 29, 30
radon 57, 58, 72, 77
reactor 106, 116
reprocessing 108, 109
strontium 62, 64, 95
temperature 29, 30, 31
transport 32, 34
Australian National University Sucrose
Standard 160, 161
autoradiography 302

B

background 158, 162, 274, 275, 278, 280,
281, 285, 300, 305
^{232}Th 276
^{238}U 276
^{40}K 276
radon 279
sources 276
spectra 281, 282
barium human 48
basalt 231, 247, 262, 266, 267, 269, 271
becquerel (Bq) 6
behaviour
^{106}Ru 50
^{129}I 54
^{131}I 48
^{137}Cs 40, 49, 59, 60, 61
^{14}C 52
^{241}Am 50
^{3}H 51
^{54}Mn 50
^{55}Fe 50

^{60}Co 49, 50
^{65}Zn 50
californium 50
inert gases 103
neptunium 50
plutonium 65, 68, 69
radioactive waste 140
radiocerium 50
radionuclides 25, 50
   air 28-30, 32
   aquatic systems 34, 140
   biosphere 26, 65
   ecological systems 25, 26
   living systems 45, 46, 48, 61, 62, 64,
      68, 69
   model 49
   soil 40, 44, 60
   water 34
strontium 40, 49, 61, 62, 64
beryllium 48, 331
  ^{10}Be 333
  ^{7}Be 15, 17, 18
biblical chronology 187
bicarbonate 168, 170, 172, 193, 197
  extracting from water 167
  groundwater 154, 167, 197
  isotopic exchange 166
  water 159, 166
bioaccumulation factor 41
bioluminescence 298
bioremediation 310
biosensors 310
biosphere 13, 23-27, 52, 61, 71, 72, 75, 80,
  82, 100, 132, 140, 141
body irradiation 8, 9
Bohemian Karst 198
Bohemian Massif 202, 205, 224, 233, 248
Boltzmann constant 326
bomb-produced
  ^{32}Si 212
  tritium 196, 210
bone 155, 164, 169, 170, 172, 173
**bristlecone pine radiocarbon age
  calibration** 176
building materials activity 277, 278
BWR 82, 103-107

C

caesium
  ^{134}Cs
    Chernobyl accident 59
    emission 106, 131
    reactor 26, 106
  ^{137}Cs
    aquatic systems 41, 49
    behaviour 49, 60, 61
    biological half-life 60, 61
    bone 61
    Chernobyl accident 125-130
    concentration factor 41
    cosmos 954 133
    criminal dealing 145
    deposition 66, 123, 125
    emission 60, 106, 108, 109, 112, 113,
      131
    fission product 25
    food-chain 59-61
    foodstuffs 60, 61, 86, 102, 108, 109,
      125-127
    Goiânia 144
    half-life 59
    intake 101
    Kyshtym 87
    Lake Karachay 88
    living systems 59
    measurement 307
    metabolism 60, 61
    milk 101, 102, 125-127
    nuclear weapon tests 39, 59, 66, 74,
      95-97, 100, 102
    plants 61
    pregnancy 60, 61, 128
    reactor 21, 26, 88, 89, 105, 106
    reprocessing 108
    resuspension 49
    retention 128, 129
    SL-1 Idaho accident 113
    soil ...40, 60, 95, 96, 123, 125-127
    stratosphere 59
    Techa 85, 86
    troposphere 59
    waste 141
    *whole body content 61, 95, 100-102,
      128-130*
    Windscale accident 112
  air 59
  decorporation 60, 61
  fallout 59
  human 60, 61
  ocean 60
  sediment 60
  source 143
  turnover 60, 61

waste 87
calcite 154, 163, 166, 189
  isotopic composition 166
calcium 62, 63
  ^{40}Ca 220
  ^{41}Ca 333
  ^{42}Ca 220
  ^{47}Ca particle accelerator 143
  carbonate
    dating 159
    sediments 151
  human 48
calcium carbonate ...188, 189, 203, 204
  cement 171
  concentration of ^{13}C 153
  dating 160
  precipitated from groundwater 168
  samples 166, 171
  secondary
    cement 171
    layers 171
  soil 161
  stalagmite 176
calcium phosphate bone 172, 173
calibrated
  age 176, 178
  years 176
calibration
  ^{14}C and ionium stalagmite 176
  curve 174, 177, 178, 298
  curves high-precision 185
  detection system 318-322
  programs 176
  radiocarbon age 174, 176
  tables 176
  time range 176
californium behaviour 50
Calvin-Benson cycle 155
Cap de la Hague 108, 109
carbon
  ^{13}C 153, 196, 203
    corrections 161
    groundwater 197, 208, 209
    tufa 154
  ^{14}C, see radiocarbon   17, 18, 52, 71, 72
    activity 205
    activity of plant material 198
    air 206
    apparent age 159
    atmosphere 150, 151, 157, 159, 160, 175
    behaviour 26, 52, 103

biological half-life 52
closed system 159
combustion 206
concentration factor 41
cosmogenic origin . 15, 17, 18, 150, 151
decay 158
dioxide 13, 52, 104, 151
dormant reservoir 151
dynamic reservoir 151
environmental tracer 174
fallout 26
global inventory 18
goundwater 209
groundwater 151, 197, 208
half-life 17, 51, 52, 158, 178
human 48, 52, 100
initial concentration . 152, 159-161, 163
Libby half-life 158
measurement 298
natural cycle 151, 152, 158
nuclear weapons 25, 167, 175, 196, 210
ocean 18, 35, 41, 52, 150, 151
open system 157, 159
production rate .. 17, 150, 151, 159
reactor 23, 24, 104, 105, 206
reference standard 160
reprocessing 108, 109
reservoir 157, 158
residence time 206
rock 150, 151
sediment 18
soil 151, 193
standard modern activity   160, 162, 163, 175, 198
synthetic and natural products . 210
synthetic ethanol 210
technology 209
waste 142
waste air 103
wine 210, 211
allochthonous 160
atmosphere 152
dioxide 150, 154, 206
  atmosphere 189, 193
  emanations 185
  environmental pollution 206
  magmatic origin 154, 160
  postvolcanic origin 163
  soil 193, 197

vents 168
   volcanic origin 160, 163
extracting from water 167
inorganic 159, 161
ocean 152
organic 159
reservoir 151, 152, 157, 158
carbonate 153, 166, 168, 170, 171
   soil 193
   water 197
Carpathian System 202
carrier-free 7
cathodoluminiscence 255, 257
cerium
   ^{140}Ce 238
   ^{144}Ce
      behaviour 50
      emission 83, 87, 131
      fallout 26
      fission product 25
      Kyshtym accident 87
   primordial 14
change climatic 190, 191
charcoal 164, 166, 170, 171, 186
charge of electron 291
Chelyabinsk 65 Mayak 85
chemiluminescence 298
Chinese Charred Sucrose (Ch-
   Suc)standard 160, 161
chlorine ^{36}Cl 196, 283, 333
Christian calendar 184
chromatography high-performance liquid
   324
chromium ^{51}Cr particle accelerator 143
chronology 182
   absolute 183
   biblical 187
   glacial 169
   groundwater recharge 192
   Minoan 185
   relative 182
   varve 183
chronometry 183, 184
chronostratigraphy 182
clay mineral 39, 44, 63, 77, 142
clay stone 205, 206
climatic
   changes 190, 191
      Aquitaine Basin 192
      Chad Basin 192
      Lorraine Basin 192
      Paris Basin 192

      Styrian Basin 190, 191
      Syr Darya artesian basin . 190, 191
      Syrian Desert 190, 191
   events 201
   fluctuations 194
coal 71, 72, 74, 151
   radioactivity 72
coastal aquifers 1 92
cobalt 41
   ^{57}Co 143
   ^{58}Co
      emission 106
      half-life 24
      reactor 24, 107
      river 88, 89
   ^{60}Co
      accident 144
      aquatic systems 49
      behaviour 50
      fallout 26
      half-life 24
      measurement 287, 307
      particle accelerator 143
      reactor 24, 106, 107
      sediment 136
      soil 95
      source 144
      spectrum 301
   human 48
   nuclear weapon tests 97
   reactor 107
   source 143
   stable 44
cogenetic samples 149, 263, 268, 322
collagen 172, 173
combustion 72, 74
common lead method 259
concentration factor 41, 62, 67
concepts of chronology 182
concordia
   curve 251, 252, 255-258
   method 250
confidence level 163, 315, 316
conservative tracer 181
constructional materials activity 277, 278
contamination . 160, 163-165, 168, 169,
   171, 173, 174, 181, 209
   ^{208}Tl 276
   ^{226}Ra 276
   ^{32}Si 300
   ^{40}K 276, 278
   algae 169

allochthonous calcium carbonate . 160, 171
allochthonous carbon 160
atmospheric 215
calcium carbonate 169
dust 164
extracted gas samples 213
foreign carbon 170
groundwater 207
groundwater samples 203
microorganisms 164
moulds 164, 169
preservatives 169
primary 160, 168
radon 276
secondary 160, 163, 164, 168, 169
conventional time scale 184
copper
human 48
industry 76, 77
stable 41
corals 166, 176, 189
correction factor 161
initial ^{14}C activity 197
correlation εNd-εSr 246
cosmic radiation .15, 17, 18, 51, 53, 56, 195, 212-214, 275, 276, 305
cosmochemistry 147
counting
curve 291, 292
device 282
efficiency 293, 300, 302
plateau 293
counting rate 284
cps,cpm, cph 274
counts per minute (cpm) 159
Crassulacean acid metabolism cycle 155
criminal dealing radioactive material 145
Cross arrangement 329
curie (Ci) 6
cycle
Calvin-Benson 155
Crassulacean acid 155
Hatch-Slack 155
photosynthetic 155

D

Daly detector 330, 332
dating
absolute 163, 183, 184
chronometric 183

events 166, 167, 182
radiometric 183
relative 182, 184
de Vries effect 174
dead interval 314, 315
decay 26, 56, 83, 87, 106, 133, 139-141, 143
constant 3, 4, 158, 311
law 311
product 5, 6, 106
rate 274
series 5, 6, 10, 15, 16, 21, 57, 72, 74
actinium 15, 16
thorium 15, 16
uranium/radium 15, 16, 80
statistical nature 315
type 2, 14
deep-sea 26, 35, 60, 67, 99, 131, 140
delta ^{13}C value . 153-156, 160, 161, 192, 197, 203, 205
delta ^{18}O value ... 156, 190-192, 203-206
delta ^{2}H value 190-192, 203, 204
dendrochronological
age 174, 176
calibration 159, 160, 174, 184
corrected age 158, 185
dendrochronology 183, 206
depleted mantle
age 243, 244, 269
deposition 72, 121
^{137}Cs 123, 125
^{3}H 51
^{90}Sr 87, 88, 123
Chernobyl 121, 123, 125
dry 54
fallout 100
iodine 121
Kyshtym accident 87, 88
nuclear weapon tests 62, 66, 100
ocean 34, 142
organ 26
plants 45, 54, 59, 63, 95
plutonium 66, 67, 100, 123
salt 142
sediment 67
soil 38, 63, 66, 85, 87, 88, 95, 100, 121, 123, 125
Techa 85
wash-out 32
waste 142
wet 54
desert 26, 38, 90, 99

detection efficiency 274, 285, 296
detector
   detection efficiency 274, 275
   gas filled 290, 291
   germanium 278, 279, 281, 290, 291, 296, 297, 299-301, 306, 308, 310
   ionization 290, 291
   measurement geometry 275
   NaI (Tl) 296, 297, 306, 308, 310
   noise 284
   proportional 211
   response 275
   scintillation 290, 295-297
   semiconductor 298-300
   semoconductor 290, 296
   silicon 290, 297, 299-301
   track 302
   well-type 302
diffusion 28, 34, 35, 38, 58, 8-83, 222, 255
discordant age 252
discordia 252, 255-257
dissolved carbon
   inorganic (DIC) 154, 198, 203
   organic (DOC) 198
distribution 28
   ^{14}C 71
   ^{226}Ra 35
   ^{230}Th 35
   ^{85}Kr 57
   area 34
   atmosphere 57, 58
   biosphere 71, 72
   body 47, 48, 67-69
   coefficient 39
   cosmogenic radionuclides 18
   events 311, 314
      binomial 312
      Gaussian 313
      normal 313
   fallout 66
   lake 37
   living systems 45, 47
   particle sizes 32
   plutonium 66-69, 84
   radionuclides 34, 38-40, 45, 47, 48, 57, 58, 68, 72, 84
   radon 58
   soil 38-40, 45, 66
   temperature 31
   water 40
dose 7, 305
   absorbed 8, 9

   committed effective 8, 10
   committed equivalent 8, 10
   effective 8, 9, 306
   effective equivalent 306
   equivalent 8-10
   measurement 306
   unit (Gy) 8
   unit (Sv) 8
   weighting factor 8-10
driftwood 165, 166
dripstone 171, 189

E

ecological system 25-27, 40, 45
ECOSYS 49
Elovy Islands 224
emission 57, 72, 74
   ^{106}Ru 108, 109
   ^{131}I 93
   ^{137}Cs 59, 108, 109
   ^{14}C 206
   ^{241}Am 65
   ^{3}H 108, 109
   ^{85}Kr 108, 109
   ^{90}Sr 108, 109
   air 27, 49, 77, 81
   carbon dioxide 13
   Chernobyl 59
   fuel element production 82
   iodine 106
   Lucens 114
   nuclear weapons ... 52, 83, 93, 97, 102
   plutonium 84, 102
   radionuclides 49, 52, 57-59, 62, 65, 76, 77, 80-84, 92, 93, 97, 102, 103, 106, 107, 112, 114
   radon 57, 58, 80
   reactor .52, 59, 62, 103, 106, 107, 114, 206
      air 104, 105
   reprocessing 59, 62, 103, 107-109, 112, 206
   soil 57, 58
   water 27, 49
environment
   ^{129}I 53
   ^{137}Cs 59, 60
   ^{14}C 51, 52
   ^{3}H 51, 52
   ^{54}Mn 50
   ^{60}Co 50

^{65}Zn 50
^{85}Kr 57
^{90}Sr 62
behaviour radionuclides 25, 26, 46, 50
concentration factor 41
contamination 13
fission products 25
natural radionuclides 72, 74, 75, 79-82
nuclear weapons 13, 82, 83, 91
plutonium 65-67
pollution from accidents 110-114, 117,
        118, 121, 122
primordial radionuclides 14
radionuclides 26, 27, 46, 50, 82, 83, 91,
        101, 103, 107-118, 121, 122, 134, 136,
        138-140, 142, 143, 145
radon 75, 79, 81
uranium 80-82
environmental
changes 194
isotope
        hydrology 201
        hydrosphere 193
tracer 181
equilibrium 6
factor 10
radiocarbon
        dynamic 159
        dynamic 157
radon 10
secular 6
transient 6
erosion calcium carbonate 189
error
relative 317
statistical 311, 316
errorchron 149, 270
estuary 36
evaluation radiocarbon dating 184
evaporation 53, 139, 140
evaporation method 254
evolution
Hf model 269
paleoclimatic 190
Pb model 260, 261, 262

F

fallout 27
air 95, 121, 128
nuclear weapon tests 32, 34, 59, 74, 82,
        91, 92, 95, 97, 100

reactor 32, 59, 112, 121, 128
stratosphere 32, 34
troposphere 3 2
Faraday cup 330, 331
FBR 103, 104, 106, 107
fertilizer 72, 74, 75
figure of merit 274, 283, 285
filament 254, 258, 264, 269, 326-328
fission 23, 82, 131, 143
^{232}Th 61
^{233}U 20, 21
^{235}U 18, 20, 51, 53
^{239,241}Pu 20, 61
^{239}Pu 21
energy 25
fast neutrons 61
products 18, 20-23, 25, 53, 56, 59, 61, 71,
        87, 89, 104, 106-108, 111, 113, 114,
        121, 122, 131
reaction 18, 131
spontaneous 18, 56
ternary 51
thermal 53
Flandrian transgression 192
Caen region 192
fluorine
^{18}F 20
food-chain 26, 28, 38, 40-42, 45, 46, 48-51,
        53, 54, 57, 59-61, 63, 64, 67, 91-93, 101,
        103, 112, 121
foodstuffs 164
fossil groundwater 206
fractionation
carbon isotopes 153, 154, 161, 198
hydrogen isotopes 190, 191
oxygen isotopes 190, 191
fresh-water tufa deposits 189
friction layer 28, 35
fuel conversion 81

G

gallium ^{67}Ga 20
gas 72
amplification 292, 294
combustion 72, 74
Gaussian distribution 312, 313
Geiger-Müller (GM) counter .292-294,
        306
geochronology 182
errors 149
geochronometry 183, 324

geological events 183
geology 182, 183, 188
glaciation events 190
gneiss 224, 231, 235, 236, 270, 271
grass 170
gray (Gy) 8
groundwater 164, 167, 194, 201, 202
^{14}C 208, 209
^{32}Si 212, 213
^{39}Ar 213
age 163, 164, 193, 196, 202-204
flow systems 199, 205, 207
identifying modern and fossil 207
origin 194
residence time 193, 196, 198, 199, 208, 215
transition time 199
tritium 200, 208, 209
groundwater dating 152, 159, 193, 202, 203, 205
^{129}I 196
^{13}C correction 161
^{14}C 203, 204, 205, 206
^{234}U/^{238}U ratio 196
^{234}U-^{238}U 195
^{32}Si 195, 196, 212
^{36}Cl 195, 196
^{39}Ar 195, 196, 213
^{3}H-^{3}He 195
^{81}Kr 195, 196
^{85}Kr 195, 214
BaCO$_3$ samples 172
bomb-produced ^{14}C 196
bomb-produced ^{3}H 196
correction factor 161
initial ^{14}C activity 197
dispersion model 201
dissolved organic carbon 198
exponential model 200
final ^{14}C activity 196
geochemical factors 197
geochemical models 197
geochemical process 160
hydrodynamic model 199
infiltration model 200
initial ^{14}C activity 163, 197, 198
injection model 200
ion exchanger samples 172
magmatic CO$_2$ 168
mean residence time 200
model completely mixed reservoir 200
piston-flow model 199

radiocarbon 195, 198
radiocarbon age 200, 202
sample 165
sampling 167
stable isotope methods 196
tritium 181, 195

H

hafnium
ε-parameter 270, 271
^{176}Hf 267, 268
^{177}Hf 267, 268
initial isotopic ratio 268, 269
isotopic ratio 268
model ages 270
natural isotopes 267
half-life 3, 4
Hardley´s cell circulation 34
Hatch-Slack cycle 155
helicopter surveys 310
hemisphere 101
nuclear weapon tests 100, 211
radionuclides ... 25, 32, 34, 59, 62, 100
transport 32, 62
Hiroshima 24, 90
Holocene 151, 188, 194, 201
temperature changes 188
humic acids 168, 169, 171
HWR 23, 82, 103, 104, 106, 107
hydrogen 50
^{2}H 203, 204
human 48
hydrogeology 147, 201, 202
hydrological cycle 193, 194, 201
hydrology 190, 196, 200, 201, 213
hydrosphere 26

I

industrial effect 174
ingestion 47, 51
inhalation 50, 51
inherited component 255
initial isotopic ratios 149
inorganic carbon 172, 186
bone 173
internal gas detector 293-295
iodine 53
^{123}I 20
^{129}I 27, 53, 333
air 53

behaviour 54
cosmogenic origin 17
emission 108, 109
food-chain 54
generated 53
half-life 53
living systems 54
migration 54
natural 53
nuclear weapon tests 53
reactor 26, 53
reprocessing 53, 108, 109
residence time 54
soil 53, 54
transfer 54
water 53
[131]I 24, 27, 53
air 83, 93
Chernobyl 121
concentration factor 41
cosmos 954 133
deficiency 55
emission 83, 93, 106, 112, 113
fission 25
food-chain 54, 93, 112
living systems 54
measurement 307
milk 56, 83
nuclear weapon tests 53, 56, 66, 93, 100
reactor 26, 53, 104, 106
reprocessing 83
SL-1 Idaho accident 113
Sosnovy Bor 131
thyroid gland 54, 55, 56, 112
Windscale accident 112
absorption 54, 55
aerosol 54
atmosphere 53, 54
behaviour 54
body 54
Chernobyl 121, 122
deficiency 54, 55
deposition 54
emission 106, 121, 122, 131, 143
fission 53, 106
food-chain 48, 50, 53
ingestion 54
inhalation 54
living systems 53, 54
medical application 143
metabolism 54, 55

ocean 53
plants 54
reactor 106
reprocessing 108
residence time 54
soil 54
thyroid gland 54, 55
ion
charged 328
source 326
trajectory 329
exchange techniques 324, 325
ionization
first potential 326
second potential 326
thermal 328
ionization chamber 291, 306
ionosphere 29, 30
iridium ^{192}Ir accident Moroco 144
iron 41, 50
[55]Fe
behaviour 50
foodstuffs 50
nuclear weapon tests 50, 66
particle acclerator 143
soil 50
meteorites dating 265, 266
irrigation 38-40, 72
isochron 149
calculation 149
diagram 227, 228
Lu-Hf 268, 271
method 226, 233, 250
Pb-Pb 260
Pt-Os 264
Re-Os 266
slope 149
Sm-Nd 239, 248
whole-rocks 228, 229, 238, 248
isotope
dilution analysis 227, 240, 253, 254, 264, 322, 334, 335
hydrology 193, 197

K

K-Ar dating 216-218, 222
example 223, 224
limitation 222, 223
K-Ca dating 216
K$_d$-value 39
Kola Penisula 224

Kos Island 223
Krasnoyarsk 26 89
krypton 56
  ^{84}Kr 56
  ^{85}Kr 78, 213
    accident Three Mile Island 117
    atmosphere 214, 215
    behaviour 26, 57
    cosmic radiation 214
    dating 214, 215
    distribution 57
    emission 108, 109, 117
    fallout 26
    food-chain 57
    half-life 32, 56, 214
    nuclear reactor 56
    nuclear weapon tests 56
    production 56
    reactor 104, 105
    reprocessing 56, 108, 109
    reservoir 215
    rock 214
    water 215
  atmosphere 57
  behaviour 57, 59
  measurement 287
  ocean 56, 57
  reprocessing 108
  soil 56

L

lake 26, 27, 34, 36, 37, 40, 62, 88, 89, 95,
  133, 139, 140
Langmuir-Saha equation 326
law
  faunal assemblages 182
  superposition 182
lead
  ^{204}Pb 249, 250
  ^{206}Pb 249, 250
  ^{207}Pb 249, 250
  ^{208}Pb 249, 250
  ^{210}Pb 72, 77
    emission 72, 74
  ^{212}Pb 78
  human 48
  initial isotopic ratio 250, 251, 260
  isotopic ratio 250, 251, 258
  loss 252, 255-258, 262
  model age 250, 259
  natural isotopes 249

rock 250
lead-lead age 251, 252
least-square regression 149
lithosphere 25, 26
Lu-Hf dating 267-270
  analytical technique 268, 270
  isochron 268, 271
lutetium
  ^{176}Lu decay 267, 268
  natural isotopes 267
LWGR 103, 106, 107
LWR 23, 24, 103

M

manganese 41, 44
  ^{54}Mn
    behaviour 49, 50
    fallout 26
    particle accelerator 143
Manhattan project 79
mantle array 246
marine carbonate 166
Marinelli beaker 302
marine-water intrusion 192, 203
mass spectrometry . 215, 219, 240, 254,
  268, 295, 322, 326-328, 330
  gas-source 322
  inductively coupled plasma ..264, 322
  ion thermal ionization 322, 324-326, 329,
    330, 333-335
  negative-ion thermal ionization 322
  secondary ion 322
  thermal ionization 264
maximum detectable radiocarbon age
  .... 162, 163
mean life 4
mean squared weighted deviation .149
measurement geometry 275, 302
melting point 327
mercury
  ^{203}Hg 143
  human 48
  shielding 280
mesosphere 29, 30
metabolism 44, 48, 50-52, 63, 67, 71
migration .28, 35, 38, 45, 46, 48, 54, 62,
  139, 140
minimum detectable radiocarbon age
  .... 162, 163
mining industries radioactivity 75
Miocene age 203

mobilization 44
model 34, 57, 60, 61, 69
   coastal waters 36
   ecological systems 45, 49
   ECOSYS 49
   estuaries 36
   foodstuffs 49
   ocean 35
   Radfoot 49
   resuspension 49
   river 36, 49
   soil migration 62
   strato-/troposphere 32, 34
modern carbon 206
modern standard sample 157, 158, 174, 198, 205
modern water 200, 208
molybdenite 265, 266
monazite 268
   sand 76, 77
   thorium 253
   uranium 253
monitor
   radiation 305, 307
   radioactivity 303
   systems 306, 307
multichannel analyzer 282
mummy 166, 185

M

Nagasaki 24, 65, 90
natural ethanol 210, 211
NBS oxalic acid standard 160, 161, 174
Nd model age 243-245
neodymium
   ε-parameter 242, 246, 247, 267, 271
   ^{143}Nd 237
   ^{143}Nd/^{144}Nd initial ratio ....
   .... 239, 241, 242
   ^{143}Nd/^{144}Nd ratio 244, 246-248
   ^{144}Nd 14, 237, 238
   ^{150}Nd spiking 240
   CHUR 243
   depleted mantle 243, 244
   isotopic evolution 241, 243-245
   natural isotopes 237
   Nd-Os dating 267
   Nd-Sr isotopic systematic 249
neptunium ^{239}Np
   behaviour 50
   emission 83, 131

production 23, 83
neutron reaction 20, 21
Nier arrangement 329
niobium
   ^{95}Nb 85, 87, 131
   Tomsk accident 89
noble gases, reactor 104, 105
noise 284, 302
normalization ^{14}C by ^{13}C 153, 154
normalized sample 161
nuclear bomb effect 174, 175
nuclear power plant surroundings 310
nuclear transformations 2, 3, 19
nuclear weapons 24, 25, 107, 112, 131, 132
   Atlantic Ocean 99
   emission 52
   fallout 82, 83, 100, 102
   Pacific Ocean 99
   production 81, 82, 83
      Albuquerque 83
      Chelyabinsk 40/65 83
      Hanford 83
      Krasnoyarsk 26 .... 83, 89
      Mayak/Ozyorsk 87, 88
      Oak Ridge 83, 85
      Rocky Flats 84
      Tomsk 7 .... 83, 88
   test series 82, 83, 195
   test sites 90
      Alamogordo 90
      Christmas Islands 92, 98
      Emu & Maralinga 97
      India 99
      Johnston Islands 92
      Lop Nur 98
      Malden Islands 98
      Marshall Islands 91, 92
      Mururoa & Fangataufa 98
      Nevada 92, 93, 98
      Novaya Zemlya 96, 97
      Pakistan 99
      Reganne 98
      Semipalatinsk 94, 95
      South Africa 99
      Trimouille Islands (Monte Bello) 97
nuclide 2

O

ocean .26, 27, 32, 34, 37, 40, 41, 52, 133
   ^{14}C cycle 194

caesium 60
carbon 52
contamination 141
depth 34
diffusion processes 34, 35
floor 222
food-chain 40, 41
iodine 53
krypton 56, 57
mixed region 34
nuclear weapon 99, 131
phytoplankton 52
plutonium 67
pycnocline 34
radon 58
strontium 62
surface 34
thermocline 34
transport 34, 35
turbulent currents 35
vessel 135, 136
volume 34
oil 71, 72, 74, 151
Oklo 18, 71
ooids 189
oolithic
  cement 189
  limestone 171
osmium
  ^{186}Os 263-265
  ^{187}Os 263, 264
    half-life 263
  ^{187}Os/^{186}Os initial ratio 264, 266
  ^{188}Os 263, 265
  ^{190}Os spiking 264
  natural isotopes 263
  γ-value 264, 267

P

paints luminous 78, 144
paleoclimatic
  episodes Africa 192
  events 188
  evolution 190
paleoclimatology 190, 194
paleotemperature 205
paper 170
  problems of ^{14}C dating 210
parchment 210
particle accelerator 19
PDB standard 153, 154, 156

peat 72, 74, 168, 170, 190, 205
pedogenesis 192
Peedee Belemnite limestone ... 153, 156
phosphorescence 298
phosphorus
  ^{32}P 212
    particle accelerator 143
  human 48
photomultiplier 294, 295, 297
photosynthetic cycle 155
pictographs ^{14}C dating 186
piston-flow model 199, 200
pitchblende 1
plankton
  phytoplankton 40, 41, 52, 67
  zooplankton 40, 41
plants
  ^{137}Cs 61
  autotrophically 26
  contamination 42
  decomposition 41
  food-chain 42, 45-47, 54
  plutonium 67, 132
  strontium 63
  transfer 38, 41, 42-44
  translocation 42
plasmalemma 42, 43
platinum ^{190}Pt half-life 264
Pleistocene 201, 204, 206
Pleniglacial 188
plume 31, 121
plutonium 13
  ^{238}Pu 65, 96, 102, 133, 143
  ^{239}Pu 20, 21, 23, 53, 56, 61, 65, 68, 83,
    87-89, 95-97, 102, 121-123, 131, 132,
    138, 145, 306, 307
    concentration factor 41
    waste 141
  ^{240}Pu 65, 87, 95, 96, 102, 121-123
  ^{241}Pu 20, 23, 65, 102, 121, 122
  ^{242}Pu 65
  absorption 67
  accident 84, 85, 89, 131-133
  activation product 25
  atmosphere 84
  batterie 133, 143
  behaviour 65-69, 132
  biological half-life 69
  bomb 24, 65, 90, 94, 97, 131
  bone 68, 69
  criminal dealing 145
  deposition 66

emission 49, 83-85
excretion 68, 69
fallout 26, 92, 102
foodstuffs 67, 102
gastrointestinal tract 68, 69
half-life 68
incorporation 50
inhalation 68
kidney 68, 69
liver 68, 69
lung 68
lymph nodes 68
metabolism 67
nuclear fuel 23
nuclear weapons 65, 66, 67, 83, 84, 100, 102, 107
nuclear wepons 66
ocean 67
plants 67, 132
production 56, 87-89, 97, 111, 112
reactor 65, 105, 121, 122
reprocessing 107, 108
sediment 67, 97
soil 66, 84, 95-97, 100
spleen 68
transfer 67
waste 138
Poisson distribution 312
polar ice 213
polonium human 48
potassium 60
    ^{39}K 214, 216, 219
    ^{40}K 14, 72, 74, 216, 220
        background 276
        body 308
        decay 217, 218
        decay constant 216-218
        emission *72, 74*
        phosphate rock 75
        rock 218
    ^{41}K 216
    biosphere 71
    body 102
    K-Ar analytical technique 218
    K-Ca analytical technique 216
    muscle 47
    natural isotopes 216
    ocean 41
    plants 44
    transport 44
prehistoric paintings 186
preservatives 169

primordial radionuclides 14, 242, 260, 276, 278
proportional detector 211, 214, 215, 292, 293, 294
Pt-Os
    isochron 264
    ratio 264
PWR 21-23, 82, 103-107, 134
pycnocline 34, 35

Q

Quaternary geology 164, 182, 190, 201
quenching 295, 297, 298

R

Radfoot-model 49
radiation hazard 8, 9, 74
radioactivity 2, 3, 6
    artificial 1
    discovery 1, 77
    fish 47
    foodstuffs 49
    in liquid releases 103, 107
    in waste air 103
    mining industries 75
    nuclear weapon tests 95
    phosphate industries 74
    reactor 110, 114, 115
    rocks 14
    sediment 136
    waste 134
    water 47, 108, 109
radiocarbon
    analyses 182
    apparent age 175
    bomb-produced 167
    dynamic equilibrium 157
    environmental tracers 196
    food industry 210
    groundwater flow system 205
    groundwater resources 201
    limits dating 162
    separating of groundwater components
        208
radiocarbon age 157, 158, 163, 165, 166, 171, 174, 175, 178, 191, 198, 200, 201, 203-205
    apparent age, marine shells 189
    bacterial activity 169
    calculation 161

calibration 159, 176
changes of ^{14}C activity 174
converting into calendar years 159
groundwater 202, 212
   Cretaceous basin of Třeboň 190, 192
   Styrian Basin 190, 191
   Syr Darya artesian basin . *190, 191*
isotopic fractionation 155
precision 162
reporting 178
rounding off 178, 179
radiocarbon dating .157, 158, 182, 185,
  188, 190, 193, 194, 201, 214, 333
affecting factors 151, 174
amount of sample 169, 170
basic conditions 163
bibliography 182
blood 173
bone 155, 169, 171-173
calcium carbonate 188
cellulose 210
charcoal 155, 166, 171, 186
Christian calendar 184
concept 157
conches 159
conventional time scale 184
counting error 162
dating errors 198
delta ^{13}C correction 155
detectable age
   maximum, minimum 162
errors 155, 184
evaluation 184
final ^{14}C activity 174, 189
forgeries 209
grass 163, 167
groundwater 194, 197
information value 167
inherent error 160
inorganic carbon 188, 190
iron 187
marine shells 166, 189
materials 150
measurement 288, 295
modern standard 157
molluscs 159
mortar 160, 171
mummy 185
organic carbon 190
paper 166, 210
parchment 166, 209, 210
pictographs 186

pottery 187
proteins 173
reference sample 167, 168, 186
representativeness of samples 165
reservoir effect 152, 159, 167, 189
sample treatment
   chemical pretreatment 171
   collection 164
   preparation 164, 166, 170, 171
   storing 164, 169
   transport 169
   wrapping 169
sampling record 164, 165
shells 159
shroud of Turin 187
soil 192
speleothems 171, 189, 190
statistical deviations 184
textiles 166, 210
tooth enamel 173
tufa 190
vellum 209
wood 151, 154, 165, 166, 171, 210
writing materials 166, 209
radiochemical procedures 287, 296, 318
radiochronology 147
general concept 148
radiochronometry 172
radiometric dating 183
groundwater 181, 199
radionuclide
aquatic systems 34, 36, 37
artificial 1, 2, 19, 20, 23, 25, 26, 82
atmosphere .. 15, 17, 18, 28, 30, 32, 34
behaviour 25, 26
biokinetics 7
cosmogenic 15, 17, 18
decay 3-5
existence 2
food-chain 26
in off-air 103
naturally generated 15
primordial 14
   hier 14
produced in
   accelerator 19, 20
   nuclear reactor 21-24
   nuclear weapon test 13, 25, 26
secondarily occuring 15
soil 15
spectrometry 273
transfer 27, 28

radium 6, 77
  ^{224}Ra 77, 78, 303
  ^{226}Ra 15, 35, 72, 74, 75, 77, 78, 80, 82,
    143, 144, 303, 308
    emission 72, 74
    fertilizer 75
    phosphate rock 75
  ^{228}Ra 72
    emission 72, 74
  decay series 16, 80
  dial painters 78, 308
  human 48
  incidents 78
  medicine 78, 143
  rock 58, 75
  smoke alarm 144
  waste 80
radon 56-58, 276, 289, 302
  ^{219}Rn 16, 57, 58
  ^{220}Rn 15, 16, 303
    emission 72
  ^{222}Rn 15, 16, 57, 58, 75, 77, 79, 80-82,
    303
    emission 72
  air 304
  atmosphere 57, 58, 72, 77
  background (indoor) 276, 279
  coal mining 72
  combustion 74
  decay series 15, 16
  drinking water 58, 59
  exposure 10, 11
  fertilizers 75
  houses 57, 58, 59
  inhalation 59
  measurement 303, 304
  medicine 78
  melting facility 77, 81
  ocean 58
  progenies 15, 16, 58, 59
  radiation protection 303
  rock 58
  soil 57, 58
  uranium mining 79, 81
  varies mines 77
  waste dumps 77, 80
  water 304
rain 32, 38, 44, 51, 57, 59, 71, 121
rain-out 32
rare earths human 48
rate of sedimentation 171, 188
Rb-Sr dating 224- 230, 270

age 237
analytical techniques 227
example 233-237
isochron diagram 226
whole-rocks isochron 228, 229
reaction
  activation 23, 83
  fission 18, 20, 21, 23, 71, 131
  neutrons 20, 21, 23, 83
  thermonuclear 83
reactor
  accident 110
    battery 133
    Chernobyl 27, 49, 59, 117-123, 125-
      130, 309
    Fermi reactor 113
    icebreaker Lenin 134
    Lucens 114
    SL-1 Idaho Falls 113
    Sosnovy Bor 131
    Three Mile Island 115-117
    Windscale 111, 112
  activation products 23, 24
  activity inventory 21, 22
  fission products 21
  monitoring 307
  normal emission 103-107
  ships 134-136
  type 65, 81, 82, 94
recrystallization 171, 189, 229, 255
  calcium carbonate
    fossil 166
    shell 166
redeposition
  calcium carbonate 189
redistribution 28, 88
reference source 274, 318-322
  calcium carbonate sample 153
  modern radiocarbon 160, 161, 174
  world standard of ^{13}C 153
Re-Os dating 263, 265, 266
  analytical techniques 264
  isochron 263, 266
  isotopic system 263
  ratio 264, 265
reprocessing 107, 108
reservoir ...159, 160, 166, 167, 174, 194
  chondrite 264
  chondrite uniform (CHUR) 242-244, 246,
    247, 267, 270
  crustal 245
  dormant 151, 152

dynamic 151, 152
Earth's 259, 261
uniform (UR) 232, 233, 246, 247
resorption, see absorption, adsorption,
transfer 45, 46
gastro-intestinal tract 47, 48, 67
men 67
rodents 67
skin 67
resuspension 38, 49, 54
rhenium
¹⁸⁵Re 263, 265
spiking 264
¹⁸⁷Re 263, 265
decay 263
natural isotopes 263
river 26, 27, 34, 36, 40, 49, 62, 143
Chemilshikov 88, 89
Columbia 49, 83
Irtysh 94
Ob 88
Pripyat 117
Savannah 54, 83, 107, 131
Techa 85, 86, 87
Tom 88, 89
Yenisey 89
rock dating 223-225, 227, 231, 232, 234-
236, 238, 241, 244, 246, 248, 250, 264,
270, 271
rubidium
⁸⁵Rb 224
⁸⁷Rb 14, 224, 227, 232
decay 225, 226
half-life 224
natural isotopes 224
Rb-Sr ratio 225-229, 231, 232, 234-236,
246, 247
ruthenium
¹⁰³Ru emission 131
¹⁰⁶Ru
behaviour 50
emission 83, 85, 87, 112, 131
fallout 26
nuclear weapon tests 66
reactor 26, 27
reprocessing 108, 109
accident Tomsk 89
emission 89, 121, 122
human 48
reprocessing 108

S

salivary gland, iodine uptake 54
samarium
¹⁴³Sm 237
¹⁴⁴Sm 237
¹⁴⁷Sm 237
decay 239
half-life 237, 238
¹⁴⁷Sm/¹⁴⁴Nd ratio 239, 240, 242, 245, 248
¹⁴⁸Sm 237, 238
half-life 237, 238
¹⁴⁹Sm spiking 240
natural isotopes 237
Sm decay scheme 238
Sm-Nd dating
analytical techniques 240
ratio 245
sample
calcium carbonate 171
collection 164
contamination 285, 286
marine origin 152
record 164, 165
representativeness 165
stability 285
storing 164
treatment 285, 286
types 164
sandstone 205, 206
scintillation liquid 296-298
scintillation counter 210, 211, 285, 295,
297, 298
sea level stands
¹⁴C dating stalagmites 190
seawater Sr composition 230
secondary calcite 166
secondary electron multiplier 254,
258,325, 330,-332
sediment 14, 27, 40, 41, 52, 171
¹⁴C 52
Azores 136
caesium 60, 86, 97
carbonate 189
calcium 161
Quaternary 163, 164
Chemilshikov 88, 89
Chernaya Bay 97
cobalt 136
dating 229, 230
fresh-water 152
glaciofluvial 163

glaciolacustrine 183
Kara Sea 141
lake 156, 171, 188, 190
   fine-grained detrital 189
   marine 171, 188
   oolithic 171, 189
   organic 168, 171
   pelagic 188
   phosphate 75
   plutonium 67, 97, 132
   Quaternary 194, 207, 208
   strontium 62, 86
   Techa River 85
   travertine 171
   Yenisey 89
selectivity constant 324
selfabsorption ..275, 287, 292, 295, 297
Sellafield 108, 109, 206
semiconductor 298, 299
sensitive high mass-resolution 254, 257
sequoia radiocarbon age calibration 176
shell 164, 166, 170
   ^{13}C correction 161
   mineral composition 166
Shetland Islands 223
shielding
   activity 279
   anticoincidence 281, 282
   anticompton 300
   iron 280
   lead 281, 282
   mechanical 279
   mercury 280
   various linings 281, 282
shroud of Turin 187
sievert (Sv) 8
signal-to-noise ratio 284
silicon
   ^{32}Si
      bomb-produced 212
      cosmic origin 212
      cosmogenic origin 17
      dating 212, 213
      groundwater samples 212
      half-life 212
      in soil water 212
      initial concentration 212, 213
      sandy aquifers 212
   detector 299, 300
single grain dating 254
Sm-Nd dating 237-239, 248
   closure temperature 241

isochron 239
   whole-rock dating 241
smog particle sampling 206
smoke detector 78, 144
sodium ^{22}**Na particle acclerator** 143
soil 31, 32, 34, 38, 40, 42, 72, 155
   ^{129}I 53, 54
   ^{137}Cs 40, 60, 61
   ^{40}K 14
   ^{55}Fe 50
   ^{85}Kr 56, 57
   ^{90}Sr 40, 62
   adsorption 44
   amount of sample 170
   atmosphere 197
   behaviour radionuclide 37-39, 44
   decay series 15
   deposition 38
   diffusion 38
   food-chain 47, 49, 61
   horizon 37, 38
   K_d-coefficient 39
   natural nuclear processes 15
   natural radionuclide 38
   pH 44
   radon 57, 58
   strontium 18
   thorium 14
   transfer factor 45
   transfer to plants 42-45, 47, 49, 54, 60, 63,
      310
   uranium 14
   water 44
Solomon Islands 224
spectrometric device 282
spectrometry 287
   energy resolution (FWHM) ..284, 313
   radiation 283, 307
   radionuclide 273, 292, 296, 297, 300, 305
   scintillation 285, 293, 297, 298
   semiconductor 282, 301
spike 323, 334, 335
stable isotope in groundwater 205
stalagmite 176
standard deviation 315
statistical errors 311, 316
stochastic effects 7, 8, 9
stratosphere ... 15, 18, 29, 30, 32, 34, 59
strontium 61, 62, 63
   ε-parameter 233, 246, 247
   ^{84}Sr 225
   spiking 334, 335

^{86}Sr 225, 227, 232, 328, 334, 335
  initial 225, 226
^{87}Sr 225, 227, 232, 334, 335
  initial 225, 226
^{87}Sr/^{86}Sr
  initial ratio 226, 227, 229, 231, 233-235
  ratio 228, 229, 235
  ratio carbonate 230
  upper mantle 231, 232
^{87}Sr/^{86}Sr ratio 236, 237, 246, 247
^{88}Sr 225, 328, 334, 335
^{89}Sr half-life 61
^{90}Sr 21
  air 62, 83, 95
  aquatic systems 49, 62
  behaviour 40, 42, 49, 62
  bone 101, 102
  Chernobyl accident 123
  concentration factor 41
  cosmos 954 133
  deposition 66, 88, 96, 112, 123
  emission 83, 108, 109, 112, 113
  fallout 26, 62, 95
  fission 25, 61
  food-chain 26, 42, 62, 63, 86, 87, 102
  half-life 61
  intake 101
  Kyshtym accident 87, 88
  Lake Karachay 88
  measurement 287
  milk 101, 102
  nuclear weapon tests 62, 95, 96, 100, 102
  reactor 26, 27, 62
  reprocessing 62, 108, 109
  SL-1 Idaho accident 113
  soil 18, 40, 62, 96
  Techa 85, 86
  thermal generator 143
absorption 63, 64
artic food-chain 64
behaviour 64
bone 64
emission reactor 106, 121, 122
excretion 64
human 48
internal isochron method 233
intestinal tract 64
isotopes
  evolution 231, 237
  homogenisation 228, 233

  stable 44, 61, 63
milk 64
natural isotopes 225
nuclear weapon tests 64, 97
plants 63
ratio
  chondrite meteorite 231
  observed 62, 63
seawater 230
soil 63
UR parameter 233
Windscale accident 1 12
Suess effect 174
sulphur ^{35}S particle accelerator 143
surface-barrier 299
synthetic ethanol 210

T

target 24, 142
technology
  ^{14}C 209
  tritium 209
tectonic evolution 223
tellurium 24, 93
  reactor 26
temperature
  atmosphere 29-31, 58
  blocking 222, 224, 228
  change
    Holocene 188
    Upper Pleistocene 188
  ocean 34, 58, 99
  plume 31
  power plants 72
  radioactive waste 87, 139, 140
  reactor 111, 113, 115, 119, 121
Tera-Wasserburg projection 258
Th/Pb ratio 250, 259, 260
thallium
  ^{201}Tl 20
  human 48
thermocline 34, 62, 67
thorium 19, 72, 77, 143
  ^{227}Th 16
  ^{228}Th 16, 72, 82, 249
  ^{230}Th 16, 80-82, 249, 335
    ocean 35
  ^{231}Th 16
  ^{232}Th 14-16, 72, 74, 77, 81, 82, 249, 335
    background 276
    emission 72, 74

fission 61
half-life 249
phosphate rock 75
sand 76, 77
²³⁴Th 16, 81, 249
decay constant 250, 251
decay series 15, 16, 249
glas 78
natural isotopes 249
rock 58
spallation 3
Welsbach mantle 78
thoron 302
thyroid gland 54, 55
blocking 55, 121
Chernobyl accident 121
fetus 55
hormones 54, 55
iodination 54
iodine uptake 47, 48, 54, 55
fallout 54, 55
inhalation 100
milk 56, 83, 112
Windscale accident 112
thyroxine 55
titanite 268
thorium 253
uranium 253
Tokaimura ....
.... 108, 109
transfer see distribution, ingestion, inhalation, resorption
air 28, 51, 52
concentration factor 41, 42
food-chain 26, 28, 40-43, 45-49, 51
plants 38, 40-47, 52
radionuclide 28, 31, 32, 34-36, 38-49, 51, 52, 59, 67, 68, 87-90, 131, 132, 135, 140, 142-145
sediment 41, 52
soil 38, 39, 42, 43, 44, 45, 47, 52
transfer factor 43-45
water 40, 41, 47, 51, 52
transport see behaviour, food-chain, transfer
tree
rings 174, 183, 187, 211
trunks 165
tritium 50, 203
accelerator 143
activity 205
analyses 182

atmosphere 179, 181
behaviour 51
air 51
water 51
biological half-life 51
bomb efect 180
concentration factor 41
cosmogenic origin 17
cosmogenic orign 179
dating 181, 182, 193, 194
food industry 181
environmental tracer 181, 196
food industry 210
food-chain 26, 51
global inventory 18
groundwater 208, 209
groundwater flow system 205
groundwater resources 201
half-life 17, 51, 179
hemisphere 180
hydrosphere 179
input function 181
Lucens accident 114
luminous paints 78, 144
measurement 288, 298
metabolism 51
monitors 308
natural 15, 17, 18, 51
nuclear weapon test 26, 51, 83, 180, 181, 196
ocean 18
precipitation by rain 179
production rate 17, 18, 179
rain 210
ratio (TR) 179
reactor 23, 51, 103-105, 107, 114
reprocessing 83, 108, 109
residence time 206
sampling of water 181
sealed tube 78
sediment 18
seeds 144
separating of groundwater components
.... 208
spring injection 180, 208
technology 209
thermonuclear explosions 180
troposphere 180
unit (T.U.) 6, 179, 298
waste 141
water 107, 208
wine 210, 211

tropopause 29, 32, 34
troposphere 15, 17, 18, 29, 30, 32, 34, 52, 58, 59
true ages 163

U

U/Pb ratio 250-252, 259, 260
U-Pb dating 250-252, 255-257, 259, 270
  examples 262
upper mantle
  ^{87}Sr/^{86}Sr ratio 231
  ^{87}Sr/^{86}Sr ratio 231, 232
  age 232
Upper Pleistocene 188
  temperature changes 188
uranium 79
  ^{233}U 20, 21
  ^{234}U 15, 77, 79, 81, 82, 249
    half-life 249
  ^{235}U 14, 15, 16, 18, 20, 21, 51, 53, 56, 61, 79, 81-83, 98, 107, 121, 122, 145, 249, 250
    half-life 249
  ^{236}U 20
  ^{238}U 14, 15, 16, 18, 23, 65, 72, 74, 76, 77, 79, 81, 82, 83, 84, 107, 249, 250, 335
    background 276
    emission 72, 74
    fertilizer 75
    half-life 249
    phosphate rock 75
  ^{239}U 23
  biosphere 72
  bomb 24
  Chernobyl accident 117, 121, 122
  colour 77, 79
  criminal dealing 145
  decay 251
  decay constant 250, 251
  decay series 249, 303
  depleted 84
  enrichment 81-83, 88
  Fermi reactor accident 113
  fission 18, 53, 56, 71
  intake 79
  kidney 79
  Lucens accident 114
  mill 80, 81, 310
  mine 58, 79-81, 200
  natural isotopes 14-16, 18, 56, 249
  natural reactor 18

  nuclear weapon 90, 98
  nuclear weapon test 97
  production 81
  reactor 111, 133, 134
  refinery 80, 81
  reprocessing 88, 107, 108
  SL-1 reactor accident 112
  soil 58
  Tomsk accident 89
  waste 82
  Windscale accident 111
U-Th-Pb dating 249, 250, 252, 253, 255-257
  analytical techniques 253

V

varve chronology 183
vellum 210

W

wash-out 32, 42, 57, 85, 86
waste 78, 131, 138-140, 144
  air 114, 116
    filter 84
  disposal .. 135, 136, 138-142, 207, 209
  enrichment facility 82, 88
  level
    high 139, 140
    low 139, 140
    medium 139
  liquid 85, 87-89, 104, 105, 107-109, 134, 135
  plutonium extraction 87-89
  reprocessing 89, 108, 109
  solid 134, 135
  uranium
    mill 80
    mine 58, 77
water
  final isotopic composition 193
Weichselian 188
whole body counter 303, 308, 309
whole-rocks
  isochron 228, 229
wiggles calibration curve 177
wine
  ^{14}C 211
wine dating
  ^{14}C 210, 211
  ^{3}H 210, 211

wood 165, 166, 171, 210, 211
working function 326, 327
Würm 188

X

xenon
   ^{133}Xe
     reactor 104, 105
   ^{135}Xe
     reactor 104
xenotime 268
   hafnium 268
   thorium 253
   uranium 253
X-ray fluorescence 227

Y

years A.D. 160
years B.P. 158

Z

zinc 41
   ^{65}Zn 50, 83, 88, 89, 106, 143
   human 48
   industry 76, 77
   plants 44
zircon 252, 254-258, 262, 268, 270, 271
   evaporation technique 258
   thorium 253
   uranium 253
zirconium . 108, 113-115, 117, 119, 121
   ^{95}Zr 85, 87, 131
   accident Tomsk 89
   sand 76, 77

εNd-εSr correlation 246
ε-parameter
   hafnium 270, 271
   neodymium ... 242, 246, 247, 267, 271
   strontium 233, 246, 247
   osmium 264, 267
μ, κ, ω parameters 260-262